KU-166-347

Arms Procurement Decision Making

Volume II: Chile, Greece, Malaysia, Poland, South Africa and Taiwan

sipri

Stockholm International Peace Research Institute

SIPRI is an independent international institute for research into problems of peace and conflict, especially those of arms control and disarmament. It was established in 1966 to commemorate Sweden's 150 years of unbroken peace.

The Institute is financed mainly by the Swedish Parliament. The staff and the Governing Board are international. The Institute also has an Advisory Committee as an international consultative body.

The Governing Board is not responsible for the views expressed in the publications of the Institute.

sipri

Stockholm International Peace Research Institute
Signalistgatan 9, S-169 70 Solna, Sweden
Cable: SIPRI
Telephone: 46 8/655 97 00
Telefax: 46 8/655 97 33
E-mail: sipri@sipri.se
Internet URL: http://www.sipri.se

Arms Procurement Decision Making

Volume II: Chile, Greece, Malaysia, Poland, South Africa and Taiwan

Edited by
Ravinder Pal Singh

sipri

OXFORD UNIVERSITY PRESS
2000

OXFORD
UNIVERSITY PRESS

Great Clarendon Street, Oxford OX2 6DP
Oxford University Press is a department of the University of Oxford.
It furthers the University's objective of excellence in research, scholarship,
and education by publishing worldwide in

Oxford New York
Athens Auckland Bangkok Bogotá Buenos Aires Calcutta
Cape Town Chennai Dar es Salaam Delhi Florence Hong Kong Istanbul
Karachi Kuala Lumpur Madrid Melbourne Mexico City Mumbai
Nairobi Paris São Paulo Singapore Taipei Tokyo Toronto Warsaw
and associated companies in Berlin Ibadan

Oxford is a registered trade mark of Oxford University Press
in the UK and certain other countries

Published in the United States
by Oxford University Press Inc., New York

First published 2000

British Library Cataloguing in Publication Data
Data available

Library of Congress Cataloguing-in-Publication Data
Data available

ISBN 0-19-829580-4

Typeset and originated by Stockholm International Peace Research Institute
Printed in Great Britain on acid-free paper by
Biddles Ltd., Guildford and King's Lynn

Contents

Preface

This volume is the result of the study conducted as part of the SIPRI Arms Procurement Decision Making Project, initiated in 1993 under the leadership of Ravinder Pal Singh. The initial results of the research, covering China, India, Israel, Japan, South Korea and Thailand, were published in the first volume in 1998. In the second and concluding phase of the study, covering six more countries, the scope of the project was broadened somewhat to identify barriers to and opportunities for harmonizing countries' arms procurement decision-making processes with the requirements of public accountability.

This SIPRI project is unique in its exploration of arms procurement decision making at different levels in countries where such issues have not been adequately researched. By drawing on contributions from national experts, it elucidates both national characteristics and a complex range of factors that influence decisions. The military typically has a preference for ambiguity and avoidance of public debate. The authors of the chapters in this volume, while taking a comprehensive view of security decision making, have therefore had a difficult task in analysing the influences of various actors in military security decision-making processes.

The goals of the project were: (*a*) to identify ways of facilitating institutional reforms which will harmonize with the requirements of good governance and the needs of national security, through democratic oversight of security decision making; and, as in the first volume, (*b*) to advance the debate on the development of norms for public accountability in arms procurement decision making. It is believed that stronger norms of public accountability in the process of decision making will help to promote restraint in arms procurement. Moreover, if the public interest and the military's perceptions of national security needs can be harmonized, this can be expected to foster the confidence not only of countries' own society but also of other countries. It is also assumed that recommendations made by national experts will be more acceptable and durable in the process of introducing arms procurement restraints than conventional arms control initiatives, which are seen as driven by the West.

This major SIPRI study will contribute to the debate on building transparency in defence budgets and arms procurement decision making. By examining the influence of democratic control and oversight of the military, it should also provide useful information about methods by which arbitrary actions by governments can be avoided.

I would like to thank both Ravinder Pal Singh and Eve Johansson, SIPRI editor, for their innovative approach to research and editorial work. Finally, the Arms Procurement Decision Making Project would not have been possible without generous support from the Ford Foundation, to whom I would like to express SIPRI's most sincere thanks.

Adam Daniel Rotfeld
Director of SIPRI
September 2000

Acknowledgements

This volume appears thanks to the efforts and contributions of numerous experts from diverse backgrounds and disciplines.

Adam Daniel Rotfeld, Director of SIPRI, was more than helpful in facilitating the research and encouraging the project. Grateful acknowledgements are due to numerous other colleagues at SIPRI: Björn Hagelin, for reading the book and offering useful comments; Ian Anthony and Elisabeth Sköns, for facilitating contacts with researchers in some of the countries that participated in the study; and Zdzislaw Lachowski, Evamaria Loose-Weintraub, Pieter D. Wezeman and Siemon T. Wezeman, for reviewing the country studies and providing comments and suggestions. Invaluable research support and help in managing the whole project was provided by research assistants Eva Hagström and Oscar Schlyter. I would also like to thank Nenne Bodell and the library staff for building up the country database, Ingvor Wallin for facilitating communications, so essential for a project of this nature, and Cynthia Loo for indefatigable help. That six very different original drafts have been transformed into the present volume is thanks to SIPRI editors Rebecka Charan and Eve Johansson.

This research would not have been possible without the participation of experts in countries where public discussion of issues such as those covered in this study is traditionally viewed with scepticism. The belief of these researchers in public accountability in a sensitive area of national security decision making placed them, on occasions, in unenviable positions in respect to the security establishments in their countries. The contributors are listed in annexe C. I owe them profound thanks for their exceptional dedication to research, to which the success of this work is due. I would in particular like to thank the institutions and leading experts who cooperated and facilitated the research in their countries, notably Dr Francisco Rojas Aravena, Director of the Facultad Latinoamericana de Ciencias Sociales (FLACSO), Chile; Dr Theodoros Stathis, Member of Parliament of Greece; Professor Haji Ahmad Zakaria, Universiti Kebangsaan Malaysia; Puan Siti Azizah Abod, Malaysian Ministry of Defence; Ambassador Janusz Reiter, Centre for International Relations, Poland; Dr Andrzej Karkoszka, former Deputy Minister of Defence, Poland, who gave most generously of his time; Paul Graham, Institute for Democracy in South Africa; Abba Y. Omar of Armscor, South Africa; and Dr Hung-mao Tien, Institute for National Policy Research, Taiwan. I would also be failing in my debts if I did not offer acknowledgement to the anonymous reviewers of the country studies.

Finally, I would like to thank the Ford Foundation for its generous financial support, which was instrumental in the setting up of this project.

Ravinder Pal Singh
SIPRI Project Leader
May 2000

Acronyms

AAC	Armament Acquisition Council (South Africa)
AAM	Air-to-air missile (Taiwan)
AASB	Armament Acquisition Steering Board (South Africa)
AFC	Armed Forces Council (Malaysia)
AIDC	Aviation Industry Development Center (Taiwan)
AIFV	Armoured infantry fighting vehicle
ANC	African National Congress (South Africa)
APC	Armoured personnel carrier
APD	Armaments Programmes Directorate (Greece)
Armscor	Armaments Corporation of South Africa
ASEAN	Association of South-East Asian Nations
ASMAR	Astilleros y Maestranzas del Armada (Chile)
ASW	Anti-submarine warfare
BBN	Biuro Bezpieczenstwa Narodowego (Bureau of National Security, Poland)
BTWC	Biological and Toxin Weapons Convention (1972)
CGS	Chief of the General Staff (Taiwan)
CODELCO	Corporación del Cobre (Copper Corporation, Chile)
CODF	Chief of the Defence Forces (Malaysia)
CONSUDENA	Consejo Superior de la Defensa Nacional (National Defense Superior Council, Chile)
CONSUSENA	Consejo Superior de Seguridad Nacional (National Security Superior Council, Chile)
COSENA	Consejo de Seguridad Nacional (National Security Council, Chile)
CPD	Concertación de los Partidos por la Democracia (Chile)
CPI	Consumer price index
CSBC	China Shipbuilding Corporation (Taiwan)
CSIR	Council for Scientific and Industrial Research (South Africa)
CSIST	Chung Shan Institute of Science and Technology (Taiwan)
CWC	Chemical Weapons Convention (1993)
DCAC	Directorate for Conventional Arms Control (South Africa)
DID	Defence Industry Directorate (Greece)
DOD	Department of Defence (South Africa)
DSTC	Defence Science and Technology Centre (Malaysia)
EBO	Hellenic Arms Industry
EAB	Hellenic Aerospace Industry
EEZ	Exclusive economic zone
ELBO	Hellenic Vehicle Industry
EMPAE	Medium-term Programme of Development and Modernization (Greece)
EMU	Economic and Monetary Union
ENAER	Empresa Nacional del Aeronautica (Chile)

EU	European Union
EYP	Ethniki Ypiresia Pliroforion (National Intelligence Service, Greece)
FAMAE	Fabricas y Maestranzas del Ejercito (Chile)
FLACSO	Facultad Latinoamericana de Ciencias Sociales
FMS	Foreign Military Sales (US)
FY	Fiscal year
GDA	General Directorate of Armaments (Greece)
GDP	Gross domestic product
GNP	Gross national product
GSH	General Staff Headquarters (Taiwan)
GSR	General Staff Requirements (Malaysia)
HAFGS	Hellenic Air Force General Staff
HAGS	Hellenic Army General Staff
HNDGS	Hellenic National Defence General Staff
HNGS	Hellenic Navy General Staff
IDF	Indigenous Defense Fighter (Taiwan)
JCAFC	Joint Chiefs of the Armed Forces Committee (Malaysia)
JMCC	Joint Military Co-ordinating Council (South Africa)
JSCD	Joint Standing Committee on Defence (South Africa)
JUSMAG-G	Joint US Military Assistance Group in Greece
KMT	Kuomintang
KOK	Komitet Obrony Kraju (Country Defence Committee, Poland)
KSORM	Komitet Spraw Obronnych Rady Ministrow (Defence Committee of the Council of Ministers, Poland)
KYSEA	Kivernitiko Simboulio Exoterikon kai Aminas (Government Council on Foreign Affairs and Defence, Greece)
LCC	Life-cycle costing
MAC	Mainland Affairs Council (Taiwan)
MAF	Malaysian Armed Forces
MBS	Modified Budgeting System (Malaysia)
MIGHT	Malaysian Industry Government Group of High Technology
MK	Umkhonto we Sizwe (South Africa)
MND	Ministry of National Defense (Taiwan)
MOD	Ministry of Defense (Chile, South Africa, Malaysia)
MOD	Ministry of National Defence (Greece)
MoND	Ministry of National Defense (Poland)
MTCR	Missile Technology Control Regime (1987)
MTDC	Malaysian Technology Development Corporation
NAO	National Accounting Office (Taiwan)
NATO	North Atlantic Treaty Organization
NCACC	National Conventional Arms Control Committee (South Africa)
NDC	National Development Council (Malaysia)
NGO	Non-governmental organization
NIK	Najwyzsza Izbo Kontrol (Highest Chamber of Control, Poland)
NPT	Non-Proliferation Treaty (1968)
NSB	National Security Bureau (Taiwan)
NSC	National Security Council (Malaysia, Taiwan)
OSCE	Organization for Security and Co-operation in Europe

PAIZ	Panstwowa Agencja Inwestycji Zagranicznych (Polish Agency for Foreign Investment)
PLA	People's Liberation Army
PPBS	Programme Performance Budget System (Malaysia)
PPBS	Planning, Programming and Budgeting System (Taiwan)
PRC	People's Republic of China
PYRKAL	Greek Powder and Cartridge Company
QA	Quality assurance
R&D	Research and development
RBN	Rada Bezpieczenstwa Narodowego (National Security Council, Poland)
RMAF	Royal Malaysian Air Force
SADC	Southern African Development Community
SADF	South African Defence Force
SAM	Surface-to-air missile
SANDF	South African National Defence Force
SEKPY	Hellenic Manufacturers of Defence Materials Association
SIRIM	Standards and Industrial Research Insitute Malaysia
SLD	Sojusz Lewicy Demokratycznej (Democratic Left Alliance, Poland)
SSM	Surface-to-surface missile
TA	Technology assessment
TECRO	Taiwanese representative office in Washington, technical section
TMD	Theater Missile Defense
TRA	Taiwan Relations Act (USA)
UMNO	United Malay National Organization
UNROCA	United Nations Register of Conventional Arms
WEU	Western European Union
WTO	Warsaw Treaty Organization

1. Introduction

Ravinder Pal Singh

I. Background

The SIPRI Arms Procurement Decision Making Project was designed to investigate the possibilities of and potential for introducing restraints on national arms procurement and to examine the assumption that greater accountability of decision makers to the public would encourage such restraint by introducing checks on the security sector—the military, the military research and development (R&D) establishment and officials in the executive branch—which in many countries has a considerable degree of autonomy. With this objective, the processes that lead up to arms procurement decisions were examined. The central question of the project was whether or not individual countries' arms procurement decision-making processes could be rendered more accountable to the elected representatives of the public and the broader interests of the society. The security sector helps the government in defining its defence policy and identifying the context for its arms procurement decisions. In the classic model of the separation of powers, the legislature can provide essential checks and balances. Two related questions were examined: (*a*) the degree of professionalism and institutionalization of democratic oversight; and (*b*) whether greater accountability will help to achieve a better balance between the military's perception of its arms procurement needs and the broader priorities of the society.

There are many aspects to security, including economic security, and many ways of assuring it besides military capability. Where public debate on the alternatives to military capability for national security is not encouraged, the military tends to take the lead in identifying national security problems and implementing the solutions that its capacities permit. The challenge therefore is to find out how to consolidate peace by institutionalizing security policy and decision-making processes that interpret security from the perspectives of the broader interests of society.

In the first phase of the project studies were conducted on the arms procurement decision-making processes in China, India, Israel, Japan, South Korea and Thailand, leading to publication of volume I in 1998.[1] The criteria used in selecting the countries participating in the two phases of the study included: (*a*) their significance in their respective regions, based on economic potential, size and population; (*b*) their significance as recipients of major conventional arms in the 1990s; (*c*) the inadequacy of published research on their arms

[1] Singh, R. P., SIPRI, *Arms Procurement Decision Making, Vol. I: China, India, Israel, Japan, South Korea and Thailand* (Oxford University Press: Oxford, 1998).

procurement decision-making processes; and (*d*) the transitions they are making in their political systems, moving from relatively centralized decision making to structures more appropriate to a democratic political environment. This transition process is reflected in changes that have been made or are being made in their security decision-making and arms procurement processes.[2]

The second phase of the project was initiated in 1996 to broaden the sample of countries. Some of the countries participating in this second phase, such as Chile, Poland, South Africa and Taiwan, have been making the transition from non-representative political systems to more democratic systems in the past decade. An interesting contrast where the pace of democratic change is concerned is provided by Greece, which moved away from control by a military junta after 1974. For somewhat different reasons, Malaysia, which moved towards a highly centralized political administration after the race riots of 1969, enacting draconian laws to control dissent and public criticism, is now experiencing calls for reform among the educated middle classes. Studies of security decision making in countries which are in the process of democratization are of particular interest for the light they shed on issues of public accountability and public availability of information generally, because security matters are always sensitive.

II. Strengthening the role of society in monitoring defence policy

The discussion that follows on the ways in which the role of society in monitoring and reviewing security-sensitive decision making can be strengthened proceeds by testing countries' decision-making processes against the yardstick of a traditional system of separation of powers under which the executive and the military are accountable to the legislature and the legislature is accountable to the electorate.

Public accountability is described by the experts participating in the project in terms of its legal, political and financial aspects. It faces two barriers: (*a*) the predominance of the executive power over the legislature in many countries; and (*b*) the lack of or weaknesses in a statutory right to information, which is necessary if the military, officials and others involved in arms procurement are to be held accountable.[3] The argument commonly offered by the military that confidentiality helps the decision-making process function smoothly is true only up to a point, and only where the technical elements of decision making are concerned. Where oversight in the public interest is concerned, secrecy is an impediment to the accountability of the security sector. Because the need for secrecy remains unquestioned the genuine case for it where it is appropriate has never been properly made.

[2] Many other countries could have been included in this study using these criteria. Among the main candidates were Argentina, Brazil, Egypt, Indonesia, Iran, Iraq, Pakistan, the Democratic People's Republic of Korea (North Korea), Romania, Saudi Arabia, Turkey and Ukraine.

[3] Chih-cheng Lo, 'Secrecy versus accountability: arms procurement decision making in Taiwan', SIPRI Arms Procurement Decision Making Project, Working Paper no. 116 (1998), p. 8.

Table 1.1. Oversight of military functions and components of military power

Military functions	Components of military power				
	Manpower	Arms procurement	Weapon deployment	Facilities and bases	Financial resources
Operational policies	L, E	L, E	E	L, E	L, E
Operational plans	E	E	E	E	E
Assessment of needs	L, E	L, E	E	L, E	L, E
Assessment of capabilities	L, E	L, E	E	L, E	L, E
Processes and methods for decision making	L, E	L, E	L, E	L, E	L, E

Notes: L = legislative oversight; E = executive oversight.

The assumption that transparency and accountability are inherently inconsistent with secrecy has not been tested. Complete transparency is not a condition of accountability, nor is secrecy in military matters the same thing as non-accountability. Accountability in combination with secrecy is both possible and practicable.[4] Table 1.1 suggests which components and functions of military power are less sensitive and which more sensitive, and indicates the role that oversight either by the executive or by the legislature through its defence committee could have. Where questions of the employment and deployment of weapons are concerned, the military has professional reasons for maintaining secrecy.

The question is who outside the military can monitor the security sector and scrutinize security-related decisions with a degree of professionalism that can equal that of the military and without conflict of interest. How can the performance of arms procurement decision making be measured and monitored? The procurement agency has a vested interest, since its decisions reflect on its professionalism, and the user (the military) has a vested interest in acquiring a new weapon. In assessing military needs and capabilities, and in particular the processes and methods for determining those needs, there seems to be a genuine requirement for validation by agencies external to the military organization, such as the parliament, other statutory authorities such as audit bodies, or vigilance commissions. The question whether and how such an examination by public representative bodies or audit organizations can be undertaken needs to be decided on the basis of professional reasons and not by the corporate interests of the military. The military, meanwhile, tends to react against research such as this. It even put up barriers against this study of a small window of its decision-making processes because its whole culture resists any study of its decision-making methods.

[4] 'Non-sensitive matters such as the processes and management of the defence sector may be open to public scrutiny, but . . . there is clearly a case for technical detail, e.g. specifications or operational parmeters to be classified.' Griffiths, B., 'Arms procurement decision making', SIPRI Arms Procurement Decision Making Project, Working Paper no. 105 (1997), p. 20.

Democratic oversight would involve review and scrutiny of defence policies and arms procurement decisions and decision-making processes in the executive, the military and other statutory authorities. It has two elements, a public representative element and a consultative one. The former involves security-related decisions in the broader context of societal priorities; the latter brings the benefit of the professional capacities that exist in a society and the perspectives of different specializations to the scrutiny of decisions.

This study assumes, moreover, that checks on arms procurement decision making by a professional process of democratic oversight that is institutionalized in national parliaments will help promote voluntary restraints on arms procurement and will be more enduring and acceptable in this regard than regional or international arms control initiatives, which are perceived as driven by the West. A positive attitude on the part of national parliaments would encourage respect for the public's right to be provided with information in a professional manner rather than as political expediency dictates. It would promote transparency and accountability and encourage acceptance of the elected representatives and understanding of their role and motives among the military leadership.

III. The policy-making and review cycle

A typology describing the stages of the policy-making cycle and selected elements of decision making is discussed below. The principle of the separation of powers means that decision making and implementation remain in the hands of the executive. However, in all other stages of the decision-making cycle there can be a role for the parliamentary bodies and statutory authorities to monitor, review and scrutinize the executive's policies, its decisions and the methods it employs to arrive at those decisions. In that sense, the role of the parliamentary defence committee on arms procurement issues will be that of a watchdog.

The challenge is designing a method by which the constitutional role of the legislature can be exercised in a purposeful and professional manner. If a rigorous method is not formalized then policy making and decision making will amount to no more than political rhetoric, and the use of bureaucratic or military discretion not to implement policies or to change decisions that have been taken can become a norm. Identification of the various stages and functions in the policy-making cycle, such as those described below, should simplify the policy and decision process, provide a framework within which policy in the broader sense of the term can be implemented and monitored, and allow alternative perspectives to be considered and thereby help to harmonize policy or decision making with the society's broader and diverse perspectives.

Five stages can be identified in a comprehensive security policy-making and review cycle. Arms procurement policies and decisions can be examined in the framework of such a structure. The cycle is a theoretical description of the functions and responsibilities that should be undertaken by the executive,

Table 1.2. The policy-making and review cycle

1. Research on and assessment of problems and policy options

(*a*) determining the entire range of external security problems facing a country; determining the need to define a policy to address those problems; and devising methods to identify priorities among the problems so defined;
(*b*) identifying methods, frameworks and processes for policy making, policy implementation, monitoring, review and scrutiny, and adjusting policy;
(*c*) building up information and data on policy options; and
(*d*) building up information and data on alternative methods of policy implementation.

2. Examining policy alternatives

(*a*) forecasts of alternative scenarios and assessment of the methods of implementing alternative policies;
(*b*) advanced research to examine the impact of alternative policies on each of the alternative scenarios; and
(*c*) analysis of the strengths and weaknesses of each policy and the opportunities they offer in advancing national security and society.

3. Decision making and implementation

(*a*) deciding on policy and defining responsibilities, resources and time frames for implementation;
(*b*) selecting methods for policy monitoring and review and for carrying through a change or adjustments in policy; and
(*c*) defining decisions that would need to be taken in order to implement the policy, and setting objectives.

4. Policy evaluation and review

(*a*) periodical scrutiny of the objectives and results; monitoring of effectiveness in terms of costs and benefits; and evaluation of the outcome to assess the effectiveness of implementation;
(*b*) review of policy implementation, methods, resources and priorities, and assessment of the impact of policy on problems; and
(*c*) meta-evaluation—examining the evaluation process itself, to validate the objectives of policy, methods, assumptions, and supporting data and processes.

5. Policy reassessment, adjustment or termination

(*a*) decision on continuation of policy; corrections by the executive;
(*b*) decision on policy modification—major corrections and adjustments; and
(*c*) decision on termination of policy. A decision to stop the policy means initiating a new policy, which involves going back to stage 2.

the legislature and the statutory audit authorities in keeping with the principles of public accountability, checks and balances, and separation of powers. It presupposes that both the executive and the legislative branches accept: (*a*) that policy making is an organizational process which gives roles and responsibility to all the actors who have a legitimate part in the affairs of a democratic state; (*b*) that sufficient resources are made available to operationalize the policy and implement its objectives; and *(c)* that public accountability norms are applied to guide the policies and help avoid failures in delivering the objectives.

A monitoring and review process in the sensitive area of defence policy will also provide a useful precedent in policy monitoring and review in other sectors

Table 1.3. The four major themes of this study

1. Military and politico-security issues
(*a*) effects of security threats and operational doctrines on force planning;
(*b*) influence of foreign and security policies on arms procurement;
(*c*) the relationships between national security, military security and military capability objectives; and
(*d*) the determinants of recipient dependence on a single source or a predominant arms supplier and the effects on political autonomy and foreign and domestic policy.

2. Budget, financial planning and audit issues
(*a*) arms procurement budget planning, methods for costing, pricing and tendering, and offset policies;
(*b*) balancing arms procurement with national socio-economic imperatives; and
(*c*) methodologies for military audit in terms of the performance, operability and serviceability of the selected system.

3. Techno-industrial issues
(*a*) equipment modernization and the building of a national defence industry;
(*b*) technology assessment (TA);
(*c*) trends in weapon system development from a R&D perspective; and
(*d*) range and level of participation of the private sector in the national defence industrial base.

4. Organizational behaviour and public-interest issues
(*a*) domestic considerations and élite motivations in the choice of equipment or sources;
(*b*) the institutionalization of decision-making processes, principles of good governance, accountability and legislative oversight;
(*c*) arms procurement and organizational behaviour of the structures at the top levels; and
(*d*) sociology of national decision-making behaviour, including predominant attitudes or cultural codes that shape decisions.

of policy making. This is particularly useful in countries making the transition from relatively closed decision-making structures to political systems that embody a system of checks and balances.

IV. The scope, method and conduct of the study

The scope of this study is limited to the decision-making processes relating to arms procurement by the state, through domestic production and imports, focusing on the procurement of major conventional weapon systems.[5]

The research was conducted in tiers: topics were selected and research questionnaires were prepared; primary papers were written by experts from the individual countries and discussed at workshops; researchers drew on these papers to write country studies; these studies were reviewed; and the final chapters were scrutinized by the volume editor.

[5] Major conventional weapon systems are defined as: aircraft; armoured vehicles; artillery; guidance and radar systems; missiles; and warships. For further detail, see Wezeman, P. D. and Wezeman, S. T., 'Transfers of major conventional weapons', *SIPRI Yearbook 1998: Armaments, Disarmament and International Security* (Oxford University Press: Oxford, 1998), pp. 369–70.

In order to understand the rationale of different interest groups and constituencies that play or should play a role in arms procurement decision making, the country studies examined four main themes (see table 1.3). These four areas of interest were then broken down into 15 topics using an interdisciplinary approach. There were no changes to the topics identified in the first phase of the project. The topics were then presented as sets of questions to be addressed.[6]

Experts on the topics specified in the research questionnaire were identified with the help of local research institutes, researchers and national experts in different disciplines and specializations both within and outside government. These experts were invited to contribute working papers which they presented at workshops in their countries. They included political leaders; legislators; serving or retired officials in the military and ministries of defence, finance and foreign affairs; functionaries in the military R&D and production organizations from government or industry; government auditors; representatives of national procurement agencies; representatives of the media; and experts in constitutional and international law.[7] Different perspectives were essential if the complexity and dissenting viewpoints that characterize decision making were to be shown.

By its nature, the subject required a broad, in-depth analysis of many political, military, economic, technical, industrial, organizational and cultural variables. In order to gain an even broader understanding, papers were commissioned from economists and sociologists. National experts from the countries involved were assumed to be best able to identify strengths, weaknesses and opportunities, and to add value to the debate on the problems in their regions.

The contributors were asked to address specific questions in the different subject areas and other aspects that they considered important. They based their work mainly on open sources but were also encouraged to draw from their own experience. Interviews were also conducted in order to benefit from the personal insight and experience of individuals in these countries. Some were not able to discuss certain aspects either because the skills and capacities to address such issues publicly have not been fully developed in their countries or because there was insufficient information or expertise. While the resulting country studies are uneven as regards detail, the lack of detail in some areas also constitutes a finding: namely, that the standard of research on security issues that is available to the public and its elected representatives, and consequently the quality of the public debate, are uneven.

The authors of these papers were asked to analyse the role and functions of the different agencies and organizations involved in arms procurement decision making. Although the final structures of the papers varied somewhat, depending on the individual authors' judgement, all authors were asked in the first section of their paper to provide a general description of the national arms procurement policy-making processes and procedure as seen from the point of view of the

[6] For a presentation of the questions guiding the preparation of the workshop papers, see annexe A in this volume.

[7] For a list of the contributors to this study, see annexe C in this volume.

author or the organization he or she represented. While highlighting declared government policies and statements, the authors were asked to describe the specific role and function of their organization in the arms procurement process, the role of other organizations, their relative influences and the relevance of other external factors or actors. The second section of each paper was an account of the contributor's own perspective and a prescription for an 'ideal type' of decision-making structure and process for his or her country. Any political or national characteristics which have a particular bearing on the way arms procurement decisions are made were identified. In the third section contributors analysed the differences between the actual process and the ideal type. They were invited to elaborate on the reasons for the differences, review the barriers and recommend measures for building public accountability in arms procurement decision-making processes. The research was complemented by input from a wider group of experts during the workshops and in interviews.

The resulting 63 papers, which were presented at workshops in the respective countries, are deposited in the SIPRI Library. Supplemented by published material and government reports, they are the primary source of data for the chapters in this volume.[8]

A researcher selected for each country was then invited to write a country study on the basis of the workshop papers, other secondary research materials and his or her own commissioned research. The country researcher evaluated the general descriptions of the decision-making structures given in the first section of each working paper and analysed the different interests involved. Chapters 2–7 were developed in this way. To facilitate access to the appropriate levels of the government agencies and to specialists, a country adviser in each country was asked to review the country study. The country adviser was a senior person who also helped in identifying specialists, facilitated the organization of the country workshop and coordinated the in-country reviews.

The review process for each country study included the soliciting of two or three reviews in the country, internal reviews by the Project Leader and SIPRI researchers, and external reviews by one or two experts who independently and anonymously provided comments.

The entire pool of experts consulted forms a substantial network for providing professional resources to national publics, legislatures and opinion makers. The network is also of great potential use to the international research community specializing in one or more of the themes under study in this project. One added benefit of the project that was not envisaged at the early stages is the horizontal networking which developed between participants from different countries.

The cut-off date for new information was 31 December 1999 except in the case of South Africa, where it was January 1999.

[8] Abstracts of the workshop papers commissioned for this book are included in annexe B in this volume.

2. Chile

Francisco Rojas Aravena[*]

I. Introduction

Latin America is a relatively stable region in terms of the relationships between the states and levels of interstate conflict. The level of sophistication of weapon systems in the region is low. Arms procurement decision-making processes are not very transparent, as they have traditionally been limited to a very closed group of persons. The introduction of democratic systems has made an important contribution in this area, but much remains to be done in most of the larger Latin American countries with regard to building public accountability in arms procurement and encouraging mutual trust and security through the development of a more transparent arms procurement information system.

Since the return to democratic rule in 1990, Chile has been going through a series of economic, social, political and military reforms. The armed forces are being modernized and civil–military relations transformed. However, the military and the individual branches of the armed services still enjoy considerable autonomy. The system of arms procurement decision making is closed and compact, decision-making power being concentrated in the military and the executive branch.

An explicit and transparent defence policy is a priority of the present government. One of the main problems in building a systematic framework for accountability in the arms procurement process is the lack of expertise among civilians to monitor and scrutinize the military's recommendations. Although a number of studies have been carried out, there is no critical systematic knowledge available to provide policy guidelines and facilitate monitoring of military security policies and processes.

This chapter examines Chile's arms procurement processes and priorities in the upper-level state institutions. Section II examines the framework of defence policy and the bodies involved, and section III the present arms procurement system. Section IV considers Chile's unique system for funding arms procurement, the defence budget process and socio-economic aspects, section V the Chilean defence industry and related research and development (R&D), and section VI the organizational and public-interest aspects. Section VII presents conclusions and two recommendations.

* This study is based on a joint Facultad Latinoamericana de Ciencias Sociales (FLACSO)–SIPRI workshop held in Santiago in June 1997. Ten working papers were prepared by Chilean researchers in connection with the workshop. They are not published but are deposited in the SIPRI Library. Abstracts appear in annexe B in this volume.

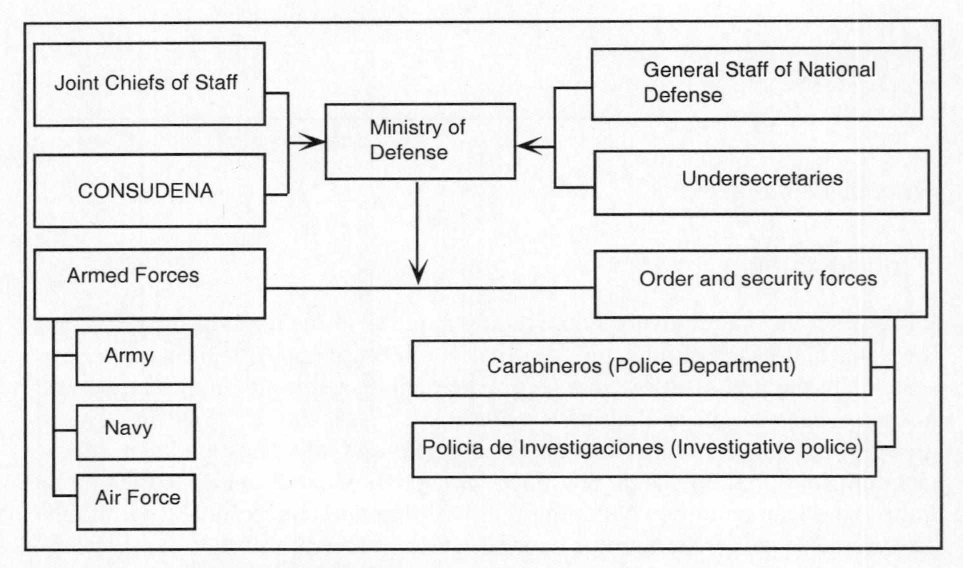

Figure 2.1. The Chilean Ministry of Defense

Note: CONSUDENA = Consejo Superior de la Defensa Nacional (National Defense Superior Council).

Source: Chilean Ministry of Defense, *Book of the National Defense of Chile* (Ministry of Defence: [Santiago], Sep. 1998), p. 96 (in English).

II. The background to arms procurement decision making

The framework for decision making

Although the Ministry of Defense (MOD) is one of the oldest ministries in Chile, it is the only one not governed by a formal organic law.[1] All branches of the armed forces that are responsible for external defence and security are attached to the MOD, as are the agencies responsible for domestic law and order—the police force, both plain-clothes (the investigative force) and military (the Carabineros). Other organizations, including academic institutions and the Civil Aviation Board, also form part of the MOD (see figure 2.1).

According to the constitution the president, through the MOD, is formally responsible for control and administration of the defence institutions and establishment. The armed forces are therefore subject to the authority of the Minister of Defense for administrative and budget purposes. However, they enjoy a high degree of functional autonomy, in decisions on arms acquisition as well as in other matters.[2] Since the return to democratic rule in March 1990, the Minister

[1] In Chile here are 4 categories of legislation: (*a*) the constitution, (*b*) organic laws which describe the organization and the constitutional setting of a given ministry or agency, (*c*) qualified quorum laws and (*d*) simple laws.

[2] 'Ley no. 18.948 Orgánica constitucional de las fuerzas armadas' [Organic Law on the Armed Forces], *Diario Oficial* [Official gazette], 27 Feb. 1990. See, e.g., Robledo, M., 'Domestic considerations and

of Defense has been a civilian. The ministry has three undersecretaries, each responsible for one of the branches of service. There are also two undersecretaries responsible for the police force. All these appointed positions are held by civilians. The Minister of Defense has a cabinet of advisers, both civilian and military, for different functions. Most of the staff of the MOD come under the undersecretaries' offices.

Defining defence policy

Since the mid-1990s, Chile has endeavoured to make its defence policy explicit and transparent. This has been the main priority of President Eduardo Frei's administration and the improvement of transparency remains a long-term objective.[3] The political platform of the second administration of the present ruling coalition, the Concertación de los Partidos por la Democracia (CPD), states that this policy should be explicit, coordinated and implemented consistently across the board in all state actions.[4]

Efforts have therefore been made to establish a national defence community to discuss, analyse and make proposals on the main aspects of national strategic development.[5] One of the main tasks undertaken was the publication in August 1997 of the *Libro de la Defensa Nacional de Chile*,[6] which stated the country's defence policy for and was the first attempt to bring a degree of coherence and transparency to it. Historically, such policies have been excluded from public debate and discussion because of the 'reserved' nature of this subject. The book is comparable to a White Paper on defence. Prepared under the direction of a former Minister of Defense and with contributions from academics,[7] state agencies and members of the Congress, it provides a common conceptual framework which both civilians and the military can use as a reference tool.[8]

The process of defining Chile's defence policy has run parallel to and been linked with another process—that of modernization of the armed forces and replacement of the weapon systems of the different branches of service. The links between the two processes have not, however, been strong enough and have essentially been limited to the upper levels of the political leadership.

actors involved in the decision-making process of arms acquisition in Chile, 1990–97', SIPRI Arms Procurement Decision Making Project, Working Paper no. 69 (1997).

[3] Frei Ruiz-Tagle, E. (President), 'Exordium', in Chilean Ministry of Defense, *Libro de la Defensa Nacional de Chile* [Chilean national defence book] (Ministerio de Defensa Nacional: [Santiago], 1997), p. 13.

[4] Concertación de los Partidos por la Democracia, 'Un gobierno para los nuevos tiempos: bases programaticas del segundo gobierno de la Concertación, 1994–2000' [A government for the new times: the programmes of the second government of the Concertación], Santiago, 1993.

[5] The concept of a 'defence community' was developed by former Minister of Defense Edmundo Pérez Yoma and is defined as the convergence of civil and military actors in a shared field of interest and cooperation with regard to defence issues. It gained importance through the elaboration of the *Libro de la Defensa Nacional*.

[6] *Libro de la Defensa Nacional de Chile* (note 3).

[7] Regular participants in the workshop that produced the *Libro de la Defensa Nacional* included 11 academic centres, 8 of which were civilian, 2 military and 1 civilian run by the state.

[8] Pérez Yoma, E., 'Presentación del Libro de la Defensa Nacional de Chile el 8 de enero de 1998 en Iquique', *Fuerzas Armadas y Sociedad*, vol. 12, no. 4 (Oct.–Dec. 1997).

Institutional mechanisms to foster transparency are therefore still in the early stages of a slow process of development. Progress has been made in enhancing interaction between the different branches of the armed forces: modern warfare and new weapon systems require enhanced interoperability. However, the autonomy which the armed forces enjoy in arms procurement has made it more difficult for the MOD to gain a broader perspective on the capability-building process.

Several decision-making channels were institutionalized again during Edmundo Pérez Yoma's period as Minister of Defense (1994–98), or, more precisely, a regular functioning of the existing institutional framework, such as the Board of Commanders-in-Chief, the Consejo Superior de Seguridad Nacional (National Security Superior Council, CONSUSENA) and the Consejo Superior de la Defensa Nacional (National Defense Superior Council, CONSUDENA), is being attempted. These bodies are discussed further below.

Security perceptions, concepts and doctrines

From the late 1880s until 1911, Chile sought regional superiority. This was followed by a more cautious and less aggressive policy in the region, which continued until the end of World War II and the start of the cold war.[9] From 1947, defence policy began to be strongly influenced by Chile's role in the hemispheric security system. Like the other countries in the region, during the cold war Chile began to depend greatly on US military assistance. In the late 1970s democracy and human rights became increasingly important concerns and influenced the military assistance relationship between the USA and Latin America, leading to the US arms embargo of 1976 on several military regimes in Latin America.

For national defence assessments, a 'global political and strategic approach' was employed after 1990 in evaluating the foreign and domestic context and strategic risks, including diplomacy, military security, and political, social and economic aspects. The actors involved in the preparation of this analysis were the armed forces' staffs and the defence general staff. With democratization, the decision-making process began to involve civilians on the staff of the MOD, the ministers participating in CONSUSENA and the President, who has final responsibility.

Five factors affect threat perceptions in Chile: (*a*) history and tradition; (*b*) the subregional security environment; (*c*) a tradition of professionalism in the Chilean armed forces; (*d*) the influence of the historical powers, the USA and the UK, in the cycle of tension in the region; and (*e*) inertia in the bureaucratic procedures, which preserves the traditional threat perceptions. The national defence objectives are: 'to preserve the independence and sovereignty of the country; to maintain Chile's territorial integrity; to contribute to the preservation of institutionality and the Rule of Law; to safeguard, strengthen,

[9] Navarro, M., 'The influence of foreign and security policies on arms procurement decision making in Chile', SIPRI Arms Procurement Decision Making Project, Working Paper no. 66 (1997), p. 1.

and renew [the country's] historical and cultural identity; to create the critical external security conditions . . . ; to contribute, in a well-balanced and harmonious way, to the development of *National Power*; . . . . [and] to contribute to the preservation and promotion of international peace and security, in accordance with national interests'.[10]

The Chilean political system, which has moved from authoritarianism to democracy, seeks to define national security on the basis of a strong non-offensive deterrence policy, respect for international law and the inviolable nature of treaties, a high degree of political, diplomatic and military credibility, and professionalism in the civilian and military bureaucracies. The current debate on civil–military relations centres on the scope and constitutional role of the armed forces. In a country like Chile, with great problems caused by the extreme geography, deterrence and strategic balance play a crucial role.[11] The legal framework clearly defines who can participate in decision making and how decisions are to be implemented. Chile has a long record of support, backing and respect for international law. Its security with regard to defining its national boundaries is based entirely on legally established treaties.

Security threats to Chile are perceived from a coalition of actors from neighbouring countries. With advances in technology, threats from the Pacific Ocean involving extra-continental powers are also being considered. The Latin American countries share common threats from poverty and underdevelopment, and economic issues are therefore also becoming important in defence and foreign policy. However, given Chile's size, resources, level of development and political inclinations, there is no tendency in the country towards the use of force. Indicators of perceived security threats to Chile show that, in contrast to other countries in the region, it is subject to a relatively limited number of threats and is thus able to implement a preventive policy.[12] Throughout the 20th century it has implemented a defensive–dissuasive policy expressed in an approach that seeks to protect the regional status quo.

Early-warning systems are essential to avoid the escalation of tensions when no major threats are perceived. For this reason, preventive security is essential in bilateral, regional and hemispheric relations.[13] In this regard Latin America

[10] Chilean Ministry of Defense, *Book of the National Defense of Chile* (Ministry of Defense: [Santiago], Sep. 1998), p. 29 (in English). See also Rojas Aravena, F. and Fuentes, C., 'Civil–military relations in Chile's geopolitical transition', ed. D. R. Mares, *Civil–Military Relations: Building Democracy and Regional Security in Latin America, Southern Asia and Central Europe* (Westview Press: Boulder, Colo., 1998), pp. 165–87.

[11] Chile's geography, population concentrations and lack of strategic depth in the east–west axis hamper operations along a north–south axis. Rojas Aravena, F., 'Chile y el gasto militar: un criterio historico y juridico de asignacion' [Chile and military expenditure: historical and judicial criteria of allocation], ed. F. Rojas Aravena, *Gasto Militar en America Latina: Procesos de Decisiones y Actores Claves* [Military expenditure in Latin America: decision processes and the major actors] (FLACSO and CINDE: Santiago, 1994), pp. 239–78.

[12] Varas, A. and Fuentes, C., *Defensa Nacional, Chile 1990–1994: Modernización y Desarrollo* [National defence, Chile, 1920–1994: modernization and development] (FLACSO: Santiago, 1994).

[13] Paz y Seguridad en las Américas, *Políticas de Seguridad Hemisféricas Cooperativas: Recomendaciones de Políticas* [Politics of cooperative hemispheric security] (FLACSO and Wilson Center: Santiago), no. 1 (Mar. 1995).

CONSUSENA
Consejo Superior de Seguridad Nacional (National Security Superior Council)
- President
- Minister of Internal Affairs
- Minister of Defense
- Minister of Foreign Affairs
- Minister of Finance
- Minister of Economic Affairs
- Commander-in-Chief, Army
- Commander-in-Chief, Navy
- Commander-in-Chief, Air Force
- Director of Frontier and Boundaries
- Chief of National Defense Staff

CONSUDENA
Consejo Superior de la Defensa Nacional (National Defense Superior Council)
- Minister of Defense
- Minister of Finance
- Minister of Foreign Affairs
- Commander-in-Chief, Army
- Commander-in-Chief, Navy
- Commander-in-Chief, Air Force
- Under-Secretary of War
- Under-Secretary, Navy
- Under-Secretary, Air Force
- Chief of Army High Command
- Chief of Navy High Command
- Chief of Air Force High Command
- Chief of National Defense Staff

COSENA
Consejo de Seguridad Nacional (National Security Council)
- President
- President of the Senate
- President of the Supreme Court
- Commander-in-Chief, Army
- Commander-in-Chief, Navy
- Commander-in-Chief, Air Force
- Director General of Police
- Comptroller General

Consultative members
- Minister of Internal Affairs
- Minister of Foreign Affairs
- Minister of Economic Affairs
- Minister of Finance

CAPE
Comite Asesor Politica Exterior (Foreign Policy Advisory Committee)
- Minister of Defense
- Under-Secretary of Foreign Affairs
- Chief of the National Defense Staff
- Commander-in-Chief, Army
- Commander-in-Chief, Navy
- Commander-in-Chief, Air Force
- Director of Frontier and Boundaries
- Director of Planning
- Director of Legal Affairs
- Team of experts in international affairs and representatives of different political currents

Figure 2.2. The defence advisory agencies in Chile

has embarked on a serious commitment as a result of two regional conferences on confidence-building measures, held in Santiago in 1995 and in El Salvador in 1998. At a bilateral level, in November 1995 Argentina and Chile established a formal and permanent mechanism of dialogue in defence issues, the Comité Permanente de Seguridad Chileno-Argentina, which is attended by officials from the Ministry of Foreign Affairs and the MOD as well as military and civilian advisers.

Coordination of foreign and defence policies

In the elaboration of policy, a conceptual and operative framework is being developed to harmonize the main actors involved in the defence community. The drafting of the *Libro de la Defensa Nacional* also created the opportunity for different actors to meet personally, regardless of their backgrounds.

There has traditionally been no formal coordination between foreign and defence policy, although the CPD has indicated the need for it and it became an explicit goal in the political platform for the period 1994–2000.[14] When the new democratically elected government took office in 1990, it faced circumstances inherited from the previous regime which prevented it from bringing in new staff and forced it to keep the bureaucratic organization which had, in many cases, been designed and decreed by the military government. Significantly, the Chilean armed forces managed to maintain a measure of institutional independence from the political powers. Coordination is now done at the stage of the dialogue between the different ministers or within institutional structures dedicated to this purpose. A series of laws were enacted to create a legal framework to provide the basis for relationships between different agencies.[15] The agencies created are CONSUSENA, CONSUDENA, the Consejo de Seguridad Nacional (National Security Council, COSENA) and the Foreign Policy Advisory Committee (CAPE), which is linked to the Ministry of Foreign Affairs and in which different actors participate (see figure 2.2). CONSUSENA defines the long-term projects, assesses national security needs and is responsible for ensuring that necessary resources are provided. Its main role is in the management of crises more than strategic planning. CONSUDENA's main role is approving all procurement applications. COSENA is one of the legacies of the authoritarian regime, a purely political institution based on the constitution.

In practice these agencies have operated as self-contained divisions and not as an integrated network designed for coordinated national decision making. The analyses carried out by the state bureaucracy are characterized by insularity and consequently lack of routine communication and integration.[16]

Coordination occurs in times of crisis and through a merger of posts and hierarchies during international crises. However, these decisions are restricted to the highest levels of the decision-making system. During the military regime the fact that the officials in charge of foreign policy were military officers led to a merger of policies but did not involve any exchange between the two sectors.[17] In the current dispensation lack of coordination is reinforced by the mistrust that still exists between the civilians who are in power and the military who ruled for 17 years. The different compositions of the ministries of defence and foreign affairs produce different perceptions and priorities in terms of goals, interests and approaches.

[14] Concertación de Partidos por la Democracia (note 4).

[15] Fuentes, C., 'Política exterior y de defensa: propuesta para su coordinación' [Foreign and defence policies: proposal for their coordination], Working paper, FLACSO, Santiago, 1995.

[16] Meneses, E., 'Percepciones de amenazas militares y agenda para la politica de defensa' [Perceptions of military threats and the agenda for the defence policy], eds R. Cruz Johnson and A. Varas, *Percepciones de Amenaza y Politicas de Defensa en America Latina* [Perceptions of military threats and defence policies in Latin America] (FLACSO and Centro de Estudios Estrategicos de la Armada (CEEA): Santiago, 1993).

[17] Fuentes (note 15); and Durán, R., 'Política de defensa y política exterior: notas para una presentación temática' [Defense policy and foreign policy: notes for a thematic presentation], *Fuerzas Armadas y Sociedad*, vol. 6, no. 1 (Jan.–Mar. 1991).

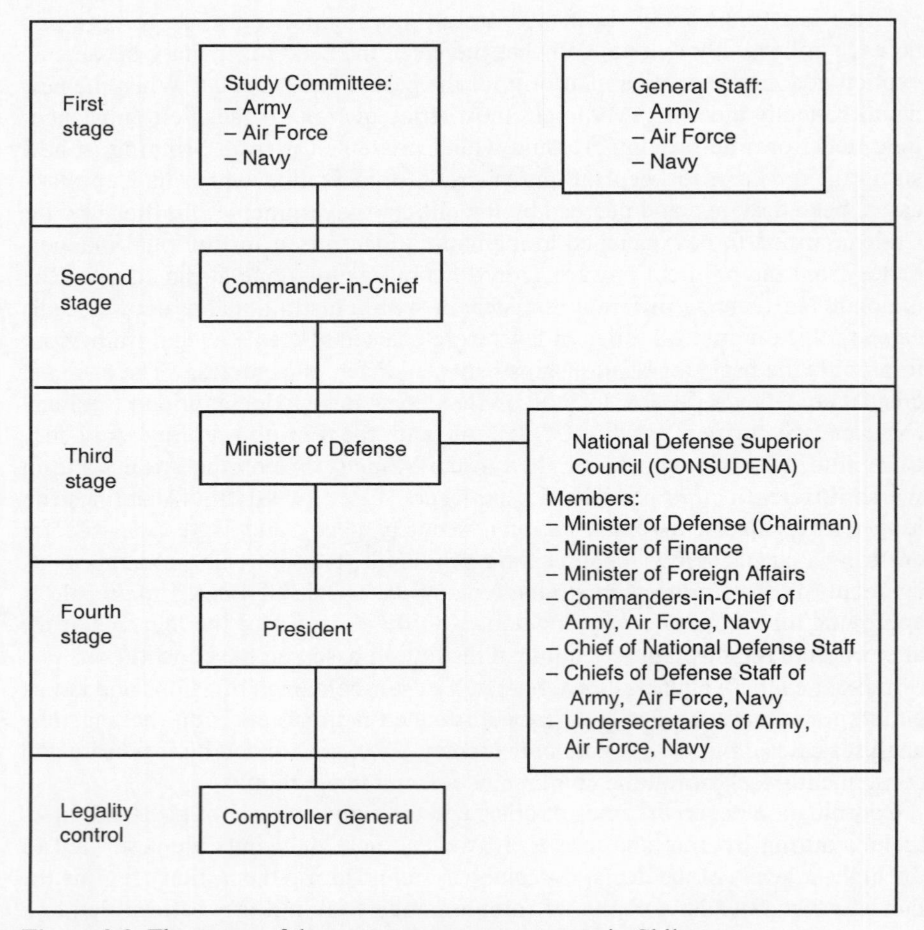

Figure 2.3. The stages of the arms procurement process in Chile

No permanent channels have been established at different working levels to routinely link defence and foreign policy processes. There are no systematic assessments of decisions adopted in a particular area and their effects on areas of responsibility of different ministries, such as might form part of a working method used to generate comprehensive input for the government.

In order to enhance cooperation the Frei Administration assigned a Ministry of Foreign Affairs official to the National Defense Staff and a high official of the National Defense Staff to the Ministry of Foreign Affairs: both work as links facilitating coordination.

This situation changed slightly when CONSUSENA was revived by the CPD at the start of its second term in office in 1994. It began to meet regularly and thus to operate as a permanent consultative entity. It also established an inter-ministerial working group that studies international security issues. Regarding arms procurement, it continues to play a formal role.

III. The present arms procurement system

The system has remained in force since 1990 and is not expected to change in the near future. Figure 2.3 shows the stages of the arms procurement process.

Arms procurement decision making in Chile takes place primarily in the individual armed services. Essential studies on technical development, strategic assessments and financial analyses are carried out by the service that intends to make a purchase. A Study Committee, whose composition depends entirely on the Commander-in-Chief, so that there is no standard composition, is appointed by the High Command of the military service that needs to procure weapon systems. The committee examines the options available on the market and interacts directly with the Commander-in-Chief of the armed service in question. The final decision is formally adopted by the respective Commander-in-Chief as part of his prerogative and no other military authority can intervene.

Civilian authorities formally begin to participate only after the process within the armed services is completed. However, in recent years they have been informed of the steps being taken in the decision-making sequences in the armed forces before their final decision is made.

CONSUDENA is the second actor in the process of purchasing weapons for the armed forces. It acts as an advisory body to the Minister of Defense and its duties include: (*a*) analysing and determining the needs of the armed forces and making proposals for arms procurement; (*b*) authorizing arms procurement from specially allocated funds and the necessary investments to cover these needs; and (*c*) monitoring compliance with the procurement and investment plans. It has played a mainly formal role in essential matters but an important function in administrative matters and can be defined as 'the missing agency' in the crucial stages of the procurement decision-making process.[18]

CONSUDENA's procedure is formal and the analysis focuses on the role of the individual armed service and its needs and not on the armed forces as a whole. Funds are allocated directly to each service of the armed forces and the sensible procedure is therefore for each service to prioritize the projects, analyse the financial options and present to CONSUDENA its own project in which no other institution has been involved. In this framework CONSUDENA is expected not to question proposals submitted by the different armed services.

The MOD must then approve the decisions.

Formally, the President has the right to veto those decisions, but a veto might generate a high level of tension, which should be avoided in the still maturing Chilean democracy. In effect, the final decision still lies with the relevant branch of service. Access to information has become easier than it was under the previous regime and a more fruitful dialogue has ensued between senior civilian and military decision makers that has made it easier to reach consensus. During the long period of military rule (1973–90) the military's decisions were never vetoed, since the offices of President and Commander-in-Chief were held

[18] Pattillo, G., 'The decision-making process in acquisition of arms systems: an approach', SIPRI Arms Procurement Decision Making Project, Working Paper no. 67 (1997), p. 11.

by one person, General Augusto Pinochet. After the return to democratic rule, the main focus was on restoring an institutional dialogue. However, analysis of the decision-making chain shows that the crucial aspects are still controlled by the military service that carries out the study and provides the relevant evaluation.

The most prominent non-participant throughout the process is the legislative branch. The law allocating funds to the armed forces for arms procurement—the 'Copper Law' of 1958, revised in 1985—does not allow the Congress to involve itself in the study and approval of arms purchases. Intervention by the Congress in arms procurement matters is legally precluded: it is allowed only limited participation in matters related to the regular administrative budget of each branch of the armed forces. The Copper Law is examined further in section IV below.

It can therefore be said that the arms procurement decision-making process is concentrated in the hands of a few players.

This is the established method, but no step of the procedure can be analysed in detail in the absence of publicly available information and detailed studies on the matter. Public analyses are limited to press reports and recurring patterns are difficult to identify, as the procurement procedures follow no regular cycles.

The order for two French–Spanish Scorpene submarines, for a total cost of approximately $400 million, illustrates procurement decision making in the navy. After almost seven years of analysis, it was decided in 1997 to buy this new model, which was still in the prototype stage. The decision was criticized since the model, tailor-made for the Chilean Navy, had not been tested and would commit 35 per cent of the navy's budget for long-term commitments for the next 25 years while adding only 15 per cent to the fleet's overall firepower.[19] The bids were studied by the navy and, although the Minister of Defense had a part in the process, his participation was limited. The final decision was taken by the navy's newly appointed Commander-in-Chief.

The current administration aims to enhance coordination between the three armed services so that weapon systems are better integrated within the overall defence framework. This will require effective functioning of CONSUDENA in examining and determining the needs of the armed forces, making proposals for arms procurement, authorizing purchases, monitoring compliance with procurement plans and controlling purchases from special funds.

Transparency of arms procurement decisions

Once procurement of an item is approved, the only institutions authorized to have knowledge of it are the secretary of CONSUDENA, the undersecretaries of the respective military institutions who make the purchases, the Controlaria General de la Republica (Office of the Comptroller General), who processes the supreme decrees (but does not know of or have to approve expenditure of funds provided under the Copper Law), the Treasury, which receives the funds for

[19] Meneses, E., 'El fin de una era' [The end of an era], *El Mercurio de Santiago*, 18 Jan. 1998, p. D8.

procurement from the Central Bank and makes payments to each individual institution, and the Central Bank. These last three institutions receive authorized copies of the confidential supreme decrees.[20]

The purchase of the Scorpene submarines mentioned above has highlighted the present methods of the military institutions, which are surrounded by secrecy. Secrecy in decision making aims to preserve and safeguard the military's interests. It is argued that the military is the only institution with the necessary professional knowledge to decide on such matters. However, committees established by the Congress or the executive to investigate particular decisions have reported that: (*a*) in certain situations, other actors would have made the same decision; (*b*) major secrets are actually very few; (*c*) material is frequently classified as 'reserved' in order to hide mistakes and poor decisions or to protect the independence of certain groups; (*d*) military bureaucracies tend to confine knowledge to a closed group which frequently develops traits of 'group thinking'; and (*e*) knowledgeable external observers using open sources have found contradictions in the operational rationale used by these agencies and deficiencies which they were later forced to acknowledge.[21]

Consequently the present government intends to create a civilian bureaucracy to handle defence matters, including arms procurement. The purpose is to avoid errors of judgement, which involves industrial, strategic and political considerations. The existing process, which is more public than the process under the military regime, has led to a national debate involving Chilean experts, members of the Congress and military personnel. Although this does not change the main actors involved in decision making, the debate has given greater political legitimacy to the process and produced more information in the media.

The influence of foreign governments and weapon suppliers

Chile's arms procurement policies were redefined in the context of the cold war. The air force and navy revived their tradition of cooperation with the UK, which had been disrupted since 1933. From 1947 and until the US embargo of 1976, procurement policies were strongly influenced by the relationship with the USA. The advent of the military government in 1973 cut off all links with the Soviet Union and its allies, and the US embargo was joined by the West European countries. This determined arms procurement policy during the 1980s and even influences present-day considerations. The Chilean Government promoted domestic industry in priority sectors such as munitions, equipment, maintenance and repairs. Crises in 1974–79 with neighbouring countries along Chile's northern borders (Bolivia and Peru) and with Argentina in 1978 underlined the military consequences of suppliers' failure because of the arms embargoes to comply with arms contracts.

In 1979 the British Government lifted its embargo and a year later the USA, headed by newly elected President Ronald Reagan, began to limit the scope of

[20] Robledo (note 2), p. 13.
[21] Meneses (note 19).

its embargo on Chile. Agreements with several foreign companies to upgrade existing military equipment allowed state-owned companies to survive.

As Chile is not self-sufficient in weapons and is unlikely to become so, it relies on maintaining a diversity of suppliers, which may not be the most cost-effective solution but avoids creating a dependence which could lead to strategic vulnerability. By the end of the cold war a bureaucratic procedure and practice of buying from different providers were well established. Moreover, the defence procurement system was strongly protected by the complex legislation enacted and amended by the military government (the constitution, the Organic Law on the Armed Forces and so on).

The three branches of service have had different policies on the purchase of US military equipment. The army virtually rejected outright the possibility of renewing the relationship with the USA during the 1990s. The navy agreed to consider US offers on a case-by-case basis, but procurement so far has been in the most part from the UK or other European countries because of the complexity of the US systems.

Changes in the processes for selection of weapon systems

Chile is currently introducing a new process for the selection of arms to be procured. The policy adopted by the MOD under the second administration of the CPD requires all significant weapon purchases to be subject to a detailed study of the options available. Since there is currently no single superior entity responsible for coordination, the new policy also aims to centralize decisions on arms procurement through meetings between the officials in the executive branch responsible for authorizing arms procurement decisions and the military staff who formulate the proposals. Examples are the navy's decision to buy the two Scorpene submarines mentioned above, the air force's proposals for replacements for its obsolete A-37 combat aircraft,[22] the army's procurement in 1999 of 200 Leopard IV tanks from the Netherlands,[23] and the joint venture between the state-owned Fabricas y Maestranzas del Ejercito (FAMAE) and the British Royal Ordnance to manufacture the Rayo multiple rocket launcher.[24]

Choosing and testing the arms selected

No established method exists for choosing arms and testing those selected. This is performed by the study committee responsible for examining available alternatives in each branch of service. Similarly, there is no central technology assessment team responsible for analysing the choices and evaluating them in general. The degree of institutionalization of aspects of evaluation of weapon technology is low and there is no permanent committee in charge of evaluation.

[22] Replacements for the A-37 combat aircraft are presently being considered from the USA, France and Sweden—F-16 from General Dynamics, the F/A-18 manufactured by McDonnell Douglas, the JAS 39 Gripen manufactured by Saab of Sweden and the French Mirage 2000-5 manufactured by Dassault.

[23] *Jane's Armour and Artillery 1999/2000* (Jane's Information Group: Coulsdon, 1999), p. 36.

[24] *Jane's Armour and Artillery 1999/2000* (Jane's Information Group: Coulsdon, 1999), p. 774.

An ad hoc committee is created, the members of which are chosen according to the type of armaments to be purchased and represent the branch of the armed forces that is to acquire the equipment.

The fact that the volume of arms purchased by the individual branches of the armed forces is small could be the reason for the failure to establish a special agency to evaluate and test the arms to be purchased. Because each branch is independent in procurement matters, a central organization responsible for this task is not really warranted. The absence of an independent authority or a central authority within the MOD makes the exercise of accountability difficult.

IV. The defence budget and socio-economic aspects

The Chilean armed forces rely on three direct sources of funding:[25] (*a*) government funding under the annual national budget law; (*b*) the funds allocated for purchase of armaments under the Copper Law, a reserved law providing financing from a tax on copper revenues through the Corporación del Cobre (CODELCO)[26]; and (*c*) other sources, which include the lease or sale of land and other property owned by the armed forces to the private sector, with certain restrictions. The first of these represents over 80 per cent of the funds available.

Funds for personnel and operations are covered in the national budget. The national budget law is subject to approval by the Congress and the defence budget benefits from one exclusive provision not granted to any other government agency: there is a minimum 'floor' that stipulates that the government's contribution must be at least equal to the funds received by defence agencies in the 1989 budget, adjusted for inflation according to the consumer price index (CPI).[27] Policies are therefore determined to some extent by the views prevailing at that time, which were based on the military's apprehensions regarding civilian decisions about military expenditure.[28]

The non-military forces mentioned above are included in the MOD budget.

The defence budget process

The *Libro de la Defensa Nacional*[29] explains how the state funds spent on defence are calculated.

Allocations to the three armed services in the 1990s followed the same distribution patterns employed since the late 1970s and are proportional to the 'historical share' which has characterized defence expenditure in Latin America.[30] In Chilean pesos, approximately 40 per cent of funds are allocated to

[25] Pattillo (note 18), p. 2.
[26] Law no. 13.196 of 29 Nov. 1958, revised by Reserved Law no. 18.445, 7 Oct. 1985. CODELCO is a state-owned company created when large-scale mining was nationalized in 1971. It owns most of the operating copper mines in Chile. See below in this section.
[27] Law no. 18.948 of 1990 (note 2).
[28] Rojas Aravena (note 11), p. 248.
[29] *Libro de la Defensa Nacional* (note 3), section VI, chapter 3, p. 200.
[30] Pattillo (note 18), p. 3; and Rojas Aravena (note 11), p. 6.

the army, 35 per cent to the navy, 17 per cent to the air force and 8 per cent to decentralized institutions (bodies which come under the armed forces but are responsible for civil activities, such as the Instituto Hydrográfico and the Dirección de Aviación Civil). In dollars the percentages are slightly different: 39 per cent to the navy, 27 per cent to the air force, 19 per cent to the army and 15 per cent to decentralized institutions.[31]

The defence part of the national budget is prepared in two distinct stages. It is drafted by the executive and then submitted to the Congress for approval. Budget preparation begins with a study by a special financial unit dedicated to this task in each defence agency. The proposal is sent for analysis to the relevant undersecretary (for the army, the navy or the air force) and then to the Minister of Defense, who submits the request to the Minister of Finance, who is responsible for the entire national budget. The MOD request is presented as a package, but this does not mean that it is coordinated or integrated.

The defence budget covers basically four types of expense: (*a*) salaries and allowances for military personnel; (*b*) goods and consumer services—recurring expenses such as maintenance, fuel and minor munitions; (*c*) transfers including expenses such as health care for military personnel and their dependants; and (*d*) investment which involves the purchase of real estate and other property subject to inventory. Arms procurement is not included.[32] Arms procurement expenditure is thus not subject to congressional debate. The budget heads do not reflect programmes or projects to be carried out by the different agencies of the defence sector. It is therefore difficult to understand the rationale for these expenses, and because of the nature of the Chilean political system, the constitution and the national budget law, the legislature has little control over this matter.

The role of the Congress is limited to approving, cutting or rejecting items. It cannot make its own estimates and is not authorized to increase budget items. The 'historical allocation criterion' remains unchanged. Moreover, 'permanent laws'—such as the Copper Law—cannot be modified by initiatives introduced by the Congress: according to the constitution, laws establishing a permanent revenue can only be modified by presidential initiative. Nor does the constitution give the Congress authority to monitor such expenses. The Office of the Comptroller General is responsible for examining budgetary expenses as part of the auditing of all government accounts.

The executive initiative

Given the strongly presidential nature of the Chilean political system, the executive plays a very important role in preparing the laws. Article 62 of the constitution establishes that the President has the sole right of initiative to introduce

[31] In the national budget for different ministries there is a part set in the national currency and another part set in US dollars because certain activities require expenditures in a foreign currency.

[32] Gaspar, G., 'Military expenditures and parliamentary control: the Chilean case', SIPRI Arms Procurement Decision Making Project, Working Paper no. 64 (1997), p. 2.

any bills related to the financial or budgetary administration of the state, including any amendment to the national budget law. He also has the exclusive initiative in two other matters which have a direct bearing on defence—the sale of state-owned property and any determination regarding air, land and sea forces required—and only he can initiate any amendments to permanent laws which allocate funds to the state, such as the Copper Law, and to those that create, amend or suppress public services or gainful employment.

The Minister of Finance and the Director General of the National Budget Office are responsible for the budget operation. Different sectors compete for funds. Each year, in March, preparatory work for the next year's budget begins. The initial research work starts in the Budget Directorate, an agency under the Ministry of Industry. The National Budget Office prepares the overall budget framework, which includes defence, in compliance with the provisions of the Organic Law on the Armed Forces and other relevant regulations which tend to reaffirm the 'historical allocation criterion'.[33] Staffing decisions are clear evidence of the application of this criterion.

The Ministry of Finance sends its budget framework to the different ministries in mid-June. The Minister of Defense sends the framework to the armed forces and the budget is coordinated by the Undersecretary of the Army. Each branch of the armed forces examines the budget proposal and states its needs: this must be done within one month. Interaction between civilian and military authorities is carried out at this stage through the three relevant undersecretaries. The outcome of these meetings and the accompanying technical analyses are determining factors in the subsequent negotiations conducted by the Minister of Finance with the Minister of Defense.

Debate might arise at this point because of different interpretations of the budget needs. At the first level of decision making, in the upper echelons of the MOD, the minister and the undersecretaries decide on the suitability of, timeliness of and amount of information required to support requests for budget increases. Once the process in the MOD is complete, coordination with the National Budget Office and the Ministry of Finance begins to include new items not included in the original proposal. This operation, which lasts around two months, forces the MOD to rank according to priority any additional requests received. This is where the most important decisions as to budget allocations are made. Finally, the President and the Minister of Finance deliver a consolidated budget to the Congress for approval.

Congressional approval

Article 64 of the constitution regulates the processing of the national budget law by the Congress: 'The National Budget Bill shall be submitted by the President of the Republic to the Congress at least three months prior to the date on which it must be in force. Should the Congress fail to approve the bill in a period of 60

[33] Rojas Aravena (note 11), p. 241.

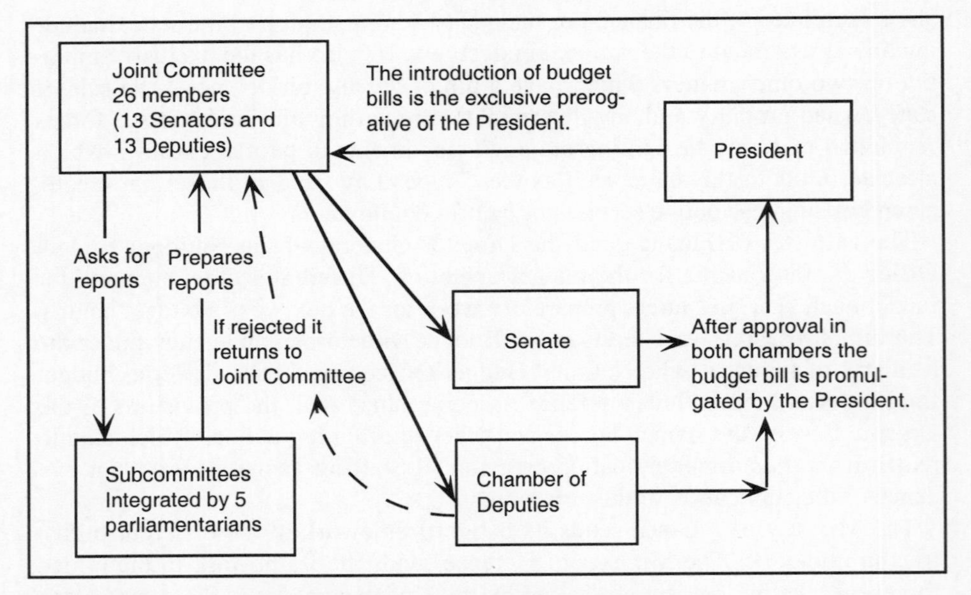

Figure 2.4. The budget law process in the Chilean Congress

days after it was introduced, the Bill submitted by the President of the Republic shall apply'.[34] The Congress may reduce expenses that it judges excessive, as is the general rule in budget issues, but expenses like salaries cannot be reduced and the 'floor' fixed by the organic law has to be respected.

Although the Minister of Finance is responsible for submitting the budget to the Congress, in practice the relevant undersecretaries are responsible for arguing the case before the congressional committees. The Congress establishes a Joint Committee to study the budget, which consists of 13 Senators and 13 Deputies, presided over by a senator. The Joint Committee is divided into subcommittees which study the different ministries' budgets, plan hearings and prepare reports: subcommittee 2 is responsible for studying the budgets of the MOD, the armed forces and other decentralized institutions, and five other important ministries. Basic information is provided to the subcommittees but, if deemed necessary, they may request additional information. In practice they have relied on ad hoc studies requested from non-government agencies.[35] Independent information provided by academic centres is considered particularly valuable.

The Congress must approve the budget within a two-month period (during October and November). In fact, the budget is analysed during a three-week period by the Joint Committee and then for another three weeks by the Senate and the Chamber of Deputies. The effectiveness of the subcommittees is limited

[34] *Constitución Política de la República de Chile* [Political constitution of the Republic of Chile] (Editorial Jurídica de Chile: Santiago, 1994), art. 62.

[35] Gaspar (note 32), p. 3.

by several other factors. They are formed by five members of the Congress and chosen by agreement between the parties once a year, so that the representatives are usually new to the task. The extremely short time available makes accurate study of the budget difficult: in recent years the entire defence budget has been examined in an average of three to four days.[36] The breakdown of the defence budget does not go further than subtitles: several members of the Congress have requested that the information be broken down into programmes for action. Moreover, the defence budget by its nature is not likely to attract the interest of members of the Congress since it will not directly affect their own districts or electorates. Another disincentive to a more exhaustive analysis of defence issues is that, when the defence budget is debated, it is backed by the opposition as a way to support the work carried out by the armed forces and by the parties in government because they cannot radically oppose a directive issued by the executive.

The budget approved by the Congress is published in the official gazette, the *Diario Oficial*, in December and takes effect on 1 January of the following year. Monitoring and follow-up of budget implementation are the responsibility of the Office of the Comptroller General, a completely independent body. Its main function is to monitor administrative integrity, but its auditing is limited to finance and administration and it does not evaluate value for money in arms procurement. It has not been possible to find published information about the level of detail in its analysis or its evaluations of arms acquisitions. It has a multidisciplinary staff which includes lawyers and auditing and finance experts.

The Copper Law

At present arms procurement is fully financed by the Copper Law.[37] Initially the law imposed a 15 per cent tax on the net profit on exports of copper and copper by-products, to be spent on the acquisition of military equipment. A 'floor' of $90 million was also established. These revenues have always been placed at CONSUDENA's disposal, to be distributed between the three services, although originally the law contained no provisions on the allocation of funds between them. In 1973 changes were introduced, many of which are still in force: the tax rate was reduced from 15 per cent to 10 per cent of the income received by CODELCO from copper exports; and the three services were to allocate $3.35 million each per year to CONSUDENA to finance projects involving more than one service. The current practice of allocating the funds to the services in three equal parts began in 1975. In 1985 Copper Law no. 18.445 increased the 'floor' to $180 million, to be adjusted according to the US CPI. In

[36] E.g., in 1994/95. Chilean National Congress, *Boletín de Sesiones, 1994–1995* (Congress Printing Office: Valparaíso, 1996).

[37] See note 26. Between 1974 and 1985, the funds provided for arms procurement under the Copper Law averaged US$128 million per year, varying from $90 million in 1975 to $184 million in 1980 (in current prices). Pattillo (note 18), p. 5. Since the return to democracy in 1990 they have averaged $200 million per year. Fuentes, C., 'Arms supplying and transparency: the case of Chile', SIPRI Arms Procurement Decision Making Project, Working Paper no. 63 (1997), p. [6].

the event of tax to be paid amounting to less than the floor, the government would be required to make up the difference.

Although these changes implied an increase in the funds available to the armed forces, they reinforced the very partial method of analysis of procurement investment. Because resources were allocated individually to each branch, the potential effects of projects on Chile's overall defence system were not considered. Inter-service coordination is not a high priority. CONSUDENA continues to be responsible for approving individual projects, but its proceedings and decisions are secret.

The law ensured a certain amount of funding for arms procurement but by eliminating congressional approval it prevented the elected representatives of the Chilean people from understanding the needs of arms procurement and excluded civilian politicians from consideration of an issue which is essential to national stability. Thus, key decisions pertaining to Chile's strategic military capacities were based on the revenues produced by its copper exports. A considerable gap has been created between the process of automatic allocations for arms procurement and its rationale, since the price of weapon systems is unrelated to the price of copper or to production decisions of the state-owned CODELCO.

The importance of the Copper Law for the armed forces can be judged in several ways. Those who believe that it plays a crucial role in maintaining peace and stability are not entirely wrong, but it is also true that the law was adopted at a very different point in time and requirements are bound to change further in the future. Despite the advent of a democratically elected government, a change to the Copper Law, introducing a different arms procurement method, more neutral in terms of resource allocation and more appropriate to circumstances, is not imminent because of the still transitional character of Chilean politics. Two of the major faults for which the system is criticized are: (*a*) that the method used for making allocations does not reflect the objectives of specific projects; and (*b*) that it does not make it possible to increase programme coordination, rank goals according to priorities, evaluate the cost-effectiveness of each option and optimize investments.

The method of disbursing funds without any coordination between the services is a major problem. It means that defence policy is not conceived on the basis of shared strategic considerations. Although limited coordination exists at the level of the defence general staff (for example, with communication systems,) there is no coordination process for joint projects between the services.

Offsets

No explicit offset policy exists, but in most recent procurement operations offsets have become significant. The goal is not only to purchase specific weaponry but also to ensure that after-sales service, spare parts and other supplies are guaranteed and that a significant part of these functions is reinvested in projects in Chile.

Cost assessment

The primary consideration in deciding on a purchase is the contribution a particular weapon system will make to the capabilities of the purchasing agency.

Assessments are carried out before procurement of the life-cycle costs of systems, their service life and different alternatives. The problem with this assessment is that objective variables other than the concerns of the agency involved are not considered. Procurement projects are therefore not analysed in terms of their overall role in the defence system. Moreover, given that the cost of any major weapon system is substantially larger than the funds allocated annually under the Copper Law, major purchases must be financed on a long-term basis. Interest payments therefore represent a significant share of the total cost of the equipment.

Although initial estimates are made regarding the cost of subsequent operations, according to some analysts[38] projects are usually not assessed after procurement and actual total cost is therefore not determined. They base this judgement on the fact that, since arms purchases are not funded by the normal defence budget, cross-subsidies exist in each military agency which could hide the real cost of any weapon system. There are no pricing systems for internal transfers and no cost determination or follow-up of projects to obtain precise data on actual costs at a given point in time or upon completion of service life.[39]

The first cost analysis is conducted in each branch of the armed forces in study committees, the results being analysed with the highest authorities in the MOD, particularly the minister. Because there are large areas of ambiguity in costing, in this process resources are probably not being used optimally. Since each institution implements its own procurement system and a general defence policy is lacking, resources are wasted which might have been put to better use if there had been an integrated assessment. In particular, if the requirement to allocate identical amounts to each branch of service did not exist, a major allocation to one service would be possible as and when required: the cost of interest payments could in this way be avoided and an efficient utilization of funds could be ensured.

Balancing arms procurement with national socio-economic problems

Arms procurement decisions are completely unrelated to the socio-economic problems that affect Chile. Since the funds are provided by a special law from copper export revenues and not subject to congressional debate, there is no forum for public discussion of the priorities or of investing such funds for other, socially pressing programmes in Chile.

The state is obliged to maintain a minimum level of arms procurement, and laws would have to be amended before these funds could be allocated to other areas. The balance of political power and the quorum required in the Congress

[38] Meneses (note 19).
[39] Pattillo (note 18), p. 12.

for an amendment to the Copper Law mean that this would not be feasible, even if the government were to attempt it. The same applies to the floor established by the Organic Law on the Armed Forces:[40] for political reasons, the executive and the Congress have little freedom of action.

V. Technical and industrial problems

The Chilean arms industry

Since the mid-19th century Chile's slow economic development has been a determining factor in the development of an arms industry and the modernization of military equipment. Domestic industry generally has operated at a level lower than that of the developed countries and the gap has been increasing. This meant that a capability for military R&D was virtually non-existent.

The Chilean defence industry developed to address the military needs as a result of the US embargo on sales to Chile of new weapons and spare parts. In addition to the embargo, tension with two neighbouring countries (Peru in 1974 and Argentina in 1978) encouraged the private sector to invest in repairing and even manufacturing weapons in Chile to supply low-cost arms to the Chilean armed forces. Because of the region's political dependence, lack of strategic involvement and low levels of conflict, the demand for weapons and equipment systems has been irregular and low.

Major differences in policy in arms procurement between the three armed services, which are due to historical reasons[41] but persist to this day, are also reflected in the development of the arms industry.

The Chilean Army continued with its own autarchic supply policy, manufacturing military supplies under licence, through FAMAE, the other military manufacturing facilities or the private sector (mostly civilian industry). In preserving its logistical independence, however, it incurred the cost of relative technological backwardness.

FAMAE has dedicated itself to manufacturing light weapons, munitions and light army vehicles, and to making advances in specific technological sectors rather than all-round industrial modernization. As a consequence it invested in rocket production, the result of this decision being the Rayo. Long-term considerations which influenced the decision included: (*a*) the adoption of internationally competitive standards in engineering and work practices in the defence industry; (*b*) developments in systems and logistic engineering skills, as well as integration of different engineering specializations; (*c*) access to training opportunities, creating a cadre of high-quality professionals, which constitutes the core FAMAE skills for the next century; and (*d*) the identification of technological areas for future development of FAMAE, in human

[40] See note 2.

[41] Traditionally the army was based on the Prussian model and the navy on the British Navy, and the air force drew its standards from the USA.

resources as well as in engineering standards. Key factors are the skills needed to interact efficiently with the industrial and technological centres in Chile.

FAMAE, an important complex of military industries, is geared for the production of arms, munitions and vehicles. A mixed organization was adopted in 1992 as a solution for a better integration of the company's functional capabilities—divisions by product, structured according to specialization in the human resources and plant needed in production. This organizational structure is consistent with international trends, and is at the heart of the new industrial model currently being formulated.

The situation in the Chilean Air Force and Navy is more complex. Both were compelled to accept international technological and logistical dependence because their main systems require considerable technological capacities, which Chile lacks. Maintenance problems were solved by sending military personnel to the United States and to European supplying countries for training. The dependence created by such programmes did not encourage domestic development and discouraged the Chilean private sector from undertaking R&D projects relating to sub-assemblies.

The navy's arms company is ASMAR (Astilleros y Maestranzas de la Armada) and the air force's is ENAER (Empresa Nacional del Aeronautica). ASMAR carries out repair and modernization of ships from different nations. It has delivered the offshore patrol vessel (OPV) *Vigilant* to the Government of Mauritius and its experience in the export of naval vessels opens new possibilities in the future. Its delivery to the navy of the Taitao patrol ships, *Corneta Cabrales* and *Piloto Sibbald*, also indicates the scale of its work in shipbuilding. ASMAR's socio-economic contribution to the regions where its shipyards are located and to the maintenance of the Chilean Navy confirms it as a strategic industry. ENAER's activities are exclusively in the aeronautical sector. It cannot engage in other areas of industrial production, except in special circumstances. In the national and international market it is engaged in: (*a*) the design and fabrication of airships; (*b*) the fabrication of parts and pieces for aeronautical use; (*c*) maintenance and modernization of airships; and (*d*) maintenance and repair of aero-engines and aeronautical components.

The private-sector arms industry in Chile is limited to servicing certain types of systems which have low utility in the military. In 1978, when tensions with Argentina were high, the Chilean Army asked Carlos Cardoen, a businessman and engineer, to develop an arms industry along with other industrialists. Cardoen's industries manufactured landmines and prototypes for armoured cars which were never mass-produced and exported cluster bombs to Iraq during its conflicts with Iran. Cardoen has stopped manufacturing arms in Chile for the time being, but in the event of an increase in demand it would be able to recommence production of military items for which it has the know-how.

The survival of the arms industry has depended on arms exports. Since the mid-1990s it has sought to enhance its export potential by forming partnerships with third parties to produce goods for the civilian and military markets.[42]

Transnational corporations hold a considerable share of the Chilean arms market but do not operate directly in Chile. Chilean state-owned companies continue to dominate as military suppliers and in the maintenance and repair of existing systems. FAMAE, ENAER and ASMAR are the most important arms companies in the country but, with the end of the weapons embargo, they had to adapt their production in order to manufacture dual-purpose goods. They have formed partnerships with other companies in the region and internationally to manufacture spare parts.

R&D and international technical collaboration

The influence of economic considerations on strategic philosophy in Chile led to the maintenance of a large army at levels of technology comparable with those of other countries in the region and a relatively small technology-intensive navy and air force.

The armed forces made significant efforts to attain limited degrees of technological excellence in order to retain operational effectiveness in the region. Policy on military technology was based on adapting military strategies to financial and technological constraints (selection of imported equipment was based on Chile's capacity to maintain the equipment) and on optimizing the capabilities and maintenance of the weapon systems purchased.

Even though R&D policies were never formally defined, they have recognizably existed from 1974 onwards, in two different phases.

First, from 1974 to 1990, R&D in Chile was determined by market-based policies of selection according to the dual criteria of 'product market' and 'product need'. As technical and scientific military capabilities were considerably reduced in 1974, policy at the time was to maintain and if possible improve the weapon systems in operation. The criteria set were cost and quality competitiveness, without discrimination between Chilean and foreign companies. Domestic suppliers were invited to manufacture goods which could not be purchased abroad. Items which private enterprise had no interest in would have to be manufactured by military facilities.

The crises with Peru in 1974 and Argentina in 1978 gave rise to intensive activity in military R&D. This was based on the need to copy and produce spare parts that could not be bought abroad and which were essential to keep defence systems operational. Projects to refurbish existing equipment were developed along the same lines. Throughout this period technology for military use was obtained as a by-product of the process of operating, maintaining and repairing existing systems. This began with the task of specifying the technical parameters for arms procurement and continued with increasing capacity for repairs up to the total reassembling of weapons.

[42] Varas and Fuentes (note 12).

After the early 1980s, development focused on hardware and software in the private-sector electronics industry. An example is SETAC—Sistema de Entrenamiento Tactico Computacional (the Tactical Training Computer System), a joint project initiated in 1982 by the Chilean Army and engineers from the Catholic University of Chile. A program for military decision-making simulation at brigade and division level involving planning, execution and control began to be used in the Academy of War in 1994. It has been sold to other armies in the region, such as Mexico's and El Salvador's. The main components of the projects were manufactured mostly by private-sector industry, both domestic and foreign.

Contrary to expectations, no joint ventures were established with companies from the developed countries to manufacture weapons. The main reasons were Chile's poor economic situation and the experience that these companies had had with investments in defence in other developing countries. Both factors made private companies reluctant to enter into partnerships.

The period after 1991 was marked by the return to democracy and the re-opening of the international arms market to Chile. Although the civilian authorities legitimized the national defence industry,[43] no official guidelines were provided for its subsequent development or for military R&D. Consequently, efforts have been made to maximize domestic advantages to competitively develop technological niches that will eventually become areas of specialization. Technical and economic criteria are being considered, together with strategic and military factors, for developing self-financing, dual-application technologies. The technology sector aims at a balance between the private- and public-sector industry in order to preserve industries seen as essential to national defence. However, there is no independent R&D authority responsible for testing and monitoring the military industry.

The parameters which guide the selection of weapon technology emerge from four different types of agency: (*a*) armed forces logistic support agencies; (*b*) the Armed Forces Defense Staff and technical divisions; (*c*) private enterprise; and (*d*) state and private-sector science and technology research institutes. Communication between these actors is usually informal but well developed because of the small size of the science and technology community. Several technical experts who formerly served in the military are now working either for private industry or in institutions of higher technical education.

To modernize or upgrade weapon systems, foreign companies are involved for the sake of their technological skills. Such involvement usually begins with a collaborative effort for systems integration and frequently extends to the logistical functions of supply, maintenance and repairs. The state-owned defence companies acquire technology by three different methods: (*a*) through their own developments; (*b*) by means of transfers from foreign companies in joint-venture partnerships; and (*c*) by direct purchase. Technological collabora-

[43] This statement was made by the then Minister of Economic Affairs, Carlos Ominami, in a speech on 'Guidelines for a development policy for the military goods industry in Chile' at the Simposio de Industria de Defensa, organized by ASMAR, Valparaíso, 14–16 Nov. 1991.

tion with France, Israel, South Africa, Switzerland and the UK between 1990 and 1994[44] was concentrated almost exclusively in the state-owned companies.

VI. Behavioural and organizational issues

The influence of elite motivations and the legislature

Decision making in Chile is organized along traditional lines, that is, the political elite that controls the state steers opinion and interest groups in the political parties. The recent trend in Chilean politics towards consensual politics actually indicates lack of formal dissent.

The Chilean political regime is characterized by a strong presidency. The 1980 Constitution grants the President broad powers considerably in excess of those of the legislative branch. For this reason, the traditional elite has concentrated its efforts on decision-making processes in the executive branch.

The Congress is essential to the political elite because it facilitates the highest levels of institutional dialogue between them. It is also an important step in the development of their political careers. The power of the Congress has been curtailed, in comparison to the power wielded by the executive branch, because the constitution reaffirmed the power of the President to co-legislate (as he has the right to rule by decrees). It limited the ability of the Congress to introduce bills and the control that could be exercised over government actions.

Although there is no doubt that decision making in defence matters is concentrated in the executive branch, there is an interest in and obligation to discuss defence issues in the Congress, and this led to the creation of the Defense Committees of the Chamber of Deputies and the Senate. Both are permanent (their composition used to change each year, although a representative could remain on the same committee). There are also a few permanent staff members to support their administrative work. They are recruited by the Congress.

The Chamber of Deputies' Defense Committee is made up of 11 deputies, and the Senate's of five senators. According to regulations, the functions of the two committees are basically the same: (*a*) to study all projects related to national defence and security matters; (*b*) to examine all bills and issues in the first or second constitutional stages as well as all remarks made by the President about projects approved by the Congress and other issues which have to be examined by the committee as defined by the regulation; and (*c*) to collect antecedents and study facts of which the Defense Committee judges that the Senate and or the Chamber needs to be informed. It can request the participation of officials to explain elements in the security debates, seek advice of specialists in the subject of study and hear the institutions and persons it thinks relevant. These are typical parliamentary hearings similar to those held in other democracies by their parliaments. The committees can move to any place within the country to exercise their authority if this is agreed by three-quarters of their

[44] Varas and Fuentes (note 12).

members and approved by the Chamber. They are not allowed to represent the Congress or to adopt agreements during these visits.

The responsibility for implementing decisions belongs with the executive. However, in decisions relating to defence matters, particularly arms procurement, the Ministry of Defense does not have much freedom of action. The military recommends and the Ministry concurs. Decisions on the type of weapon systems to be purchased are made basically by the user service of the armed forces.

This situation is changing. The Minister of Defense participates increasingly in arms procurement decisions as a result of a policy of enhancing civilian participation in defence matters. In turn, civil–military relations are being normalized with the departure of General Pinochet as Commander-in-Chief in March 1998. During Edmundo Pérez Yoma's tenure as Minister of Defense, the defence community began to take shape and facilitated a dialogue between the leaders in the executive branch to develop an explicit defence policy. With this change, arms acquisitions became a part of the decision agenda of the Minister of Defense. At the same time, the process of coordination between the MOD and the three armed services that is described in section III above began to emerge, since the executive branch had to authorize specific purchases by issuing corresponding executive orders.

Historically, in Chile the decision-making elite is a group that has led the country from the political, economic and cultural points of view—in essence the leaders of political parties, the military, businessmen and intellectuals. In the case of arms procurement decisions, however, the position has been rather different, for three major reasons.

First, the degree of autonomy the armed forces enjoy and access to procurement funds under the Copper Law allow them to implement their own decisions as to the armaments they need and to remove arms procurement decisions from the political arena.[45] Under the legal system in force, they have ample room for action. The most important decision-making levels are those within the individual armed services. As mentioned above, to change decisions that have been made by the armed forces is politically difficult.

Second, it is difficult for the elites to participate more actively in arms procurement decision making because of the technical nature of the matters involved. This has also prevented active participation by civilians in general, a situation which dates back to the time of the enactment of the Copper Law in 1958 when it was decided to exclude this issue from congressional debate. The issue has only recently begun to be analysed again as an integral part of public policy. This accounts for the establishment of agencies such as the Academia Nacional de Estudios Políticos y Estratégicos (ANEPE, the National Academy for Strategic and Political Studies), attached to the MOD. Other agencies (like FLACSO, the Facultad Latinoamericana de Ciencias Sociales, research units in

[45] Pattillo (note 18), p. 6.

the universities and colleges, and specialized centres in each of the armed services) have also started studying defence issues.

Third, the secret or 'reserved' nature of this field results in a lack of civilian involvement. The current administration is trying to change this without losing sight of the strategic element involved in arms procurement and to make the process more transparent. The *Libro de la Defensa Nacional* is evidence of this.

Furthermore, the political elite has historically paid no real attention to military issues or arms procurement decision making. Under the military government analysis of such issues was restricted to the military so that the new generation of the political elite is not familiar with the subject; the democratic governments have given priority to other issues, such as health and education.

Monitoring processes

Chile is a country with a low level of corruption in the government. Cases of corruption are occasional and there are no major networks of state corruption. To date, the problem has been limited to isolated cases. Instances of corruption in arms procurement have not come to light, probably because access to information is restricted and the arms suppliers are the only actors who can report on corrupt practices. Historically, there has been almost no corruption in the military generally.

There are no constitutional or other provisions to compel the government to provide information on the purchase of weapons or any other related matter to the public or the media. The latter are therefore forced to obtain information directly from the armed forces and depend on the willingness of the agency involved to make information available. The Defense Committees of the Congress have the right to information but in general their meetings are secret. Moreover, if the Congress is to be able to ask the right questions, it needs a permanent advisory expert staff which would constantly monitor national defence and arms procurement issues.

There are no established accounting practices in Chile to scrutinize military matters. The constitution and the law have not established any special agency to monitor expenses incurred specifically by the armed forces differentiated from other state agencies.

The Office of the Comptroller General is responsible for monitoring all accounting aspects of the implementation of the state budget. It can audit the accounts for arms procurement but is not authorized to examine the choice of arms to be procured or whether value for money has been obtained. Nor is the Congress empowered to conduct this type of monitoring. Its only authority, as part of its responsibility for inspecting and monitoring public activities, is to establish a committee to determine whether any corrupt or illegal act has been committed. In early 1999, for example, a deputy asked specific questions about the legitimacy of some payments for 20 Mirage-5 combat aircraft purchased from Belgium in 1995–96. Within the established legal limits, no agency has authority to reject or criticize expenses incurred by the military.

It would not be accurate to suggest that there is ongoing coordination between the Comptroller General and the Congress. Both deal with the inspection of the state powers, but the Comptroller monitors public spending in advance, usually in relation to the budget, while the Congress is limited to intervening *ex post facto* if it suspects for any reason that there have been shortcomings in budget implementation. If this happens it can ask the Comptroller General for all the antecedents it judges to be relevant, which the Comptroller has to supply, or it can ask the Comptroller to carry out a specific investigation.

The armed forces are subject to the same procedures and controls as any other public-sector agency. The use of any funds allocated under the Copper Law must be authorized by an executive order. That is how the executive branch holds the ultimate right to veto the purchase of weaponry. Control over the movement of funds allocated by the law is exercised by the Office of the Comptroller General, but this supervision is purely administrative in nature. Funds from the Copper Law are not included in the general accounting to which the rest of the public sector is subject. Instead, they fall within a parallel system that is never consolidated. Therefore no data on the flows of funds originating from the law are available, nor is information on the commitments undertaken by the armed forces in advance of purchase. However, the final amount allocated each year is published in the CODELCO annual report.

Before 1990 the military government prohibited any examination of arms procurement decisions, and it is almost impossible to conduct a meaningful audit of these expenses as there are no parameters on which to base it. The development of a national defence policy and attempts to define a framework to ensure transparency may make it easier to analyse these matters. However, civil society has not been actively involved in this process.

Developing good governance through monitoring of public spending

Although the concept of good governance is accepted in the legal framework, in terms of military expenditure and particularly in the purchase of weaponry it has not been possible to apply it because of the constraints inherited from the military regime. A prerequisite for the proper allocation of public spending is to reinforce the evaluation system of public investment, placing emphasis on programme budgeting and on linking spending to proposed goals in order to ensure efficiency.[46] At present, these methods are not developed in Chile. There is no possibility to create an agency to formulate a coherent defence policy or coordinate weapons purchasing. The current system is rooted in the unique professional cultures of the three branches of the armed forces, which are compartmentalized in a bureaucratic, centralized decision process that is more suited to the past than to the goals to be achieved.

[46] Lahera, E., 'Políticas públicas: un enfoque integral' [Public policy: an integrated approach], ed. E. Lahera, *Cómo Mejorar la Gestión Pública* [How to improve public administration] (Corporación de Investigaciones Económicas para Latinoamérica (CIEPLAN) and FLACSO: Santiago, 1993).

External control of military expenditures is minimal. Although there are some indicators of control being exercised by agencies external to the military, for instance, by the Office of the Comptroller General on actual budget spending, no specific mechanism is currently in place to control and evaluate how the armed forces spend the funds allocated by the Copper Law. Control is limited to a form of negotiation between the armed forces and the executive branch, represented by the Minister of Defense and the President's authorization of all arms procurement. Since the start of the second administration of the CPD, such executive control as exists has been exercised prior to decision on a purchase by the Minister of Defense and CONSUDENA. The role of the Congress is perfunctory.

Sociology of national decision-making behaviour

Even after Chile became a republic, the decision-making processes were controlled by the political elite, a carry-over from the original landowners who made all the political decisions. In the mid-19th century new political and social groups and in the 20th century the professional middle classes and the state bureaucracy began to participate in national politics.

The return to democracy allowed the development of a system of cooperative alliances of political parties which adopted non-confrontational approaches. The parties are not attempting to change society radically but rather to manage and control the economy in order to reduce social inequality. These political coalitions are a distinct feature that has emerged recently. The political platforms of the first two administrations of the CPD have been similar, the first emphasizing the transition to democracy, the second the modernization of the state and its agencies.

In the current decision-making process, two groups have emerged among the political elite—the traditionalists and the technocrats. The first is made up of politicians who base their actions on political negotiation and the second of professional experts who are also members of a political party. The technocrats first appeared on the scene under the military government and continued to be influential, particularly in economic issues. In the field of defence policy making their influence felt is limited, although their participation is increasing.

There is a public perception that the political parties have distanced themselves from civil society, which increasingly feels less represented. New points of reference will have to be considered in planning policies in the future.[47] A campaign to raise citizens' awareness of their role in managing the affairs of the state will indirectly increase the need to know about national defence issues.

[47] The decrease in active political participation, particularly among younger people, became evident when almost 1 million young people failed to register to vote in the last congressional elections. Navia, P., 'Tendencias de participación electoral en Chile en 1997' [Tendencies in electoral participation in Chile in 1997], *Chile 97: Análisis y Opiniones* (FLACSO: Santiago, 1998), pp. 61–86. In 1993, with a population eligible to vote of 8 925 000, only 8 044 163 were registered. In 1997, with an eligible electoral population of 9 425 000, only 8 069 163 were registered to vote. In the elections of deputies in Dec. 1997, only 5 733 714 voters voted, considerably fewer than in 1993, when 6 738 889 voted.

VII. Conclusions: the ideal decision-making process for arms procurement in Chile

Two related areas can be identified as the basis of the 'ideal type' arms procurement decision-making process for Chile. First, it is necessary to determine the most suitable process for resource allocation, monitoring and evaluation. Second, the current lack of a systematic approach to national defence will have to be addressed.

The current system of resource allocation in Chile needs to be modified. It is not reasonable that the funds for arms procurement are not closely linked to any coherent national strategic plan developed by the national security apparatus but relate to exogenous events, such as the expected exports of CODELCO or the international price of copper, which have no bearing on the price of weapons. The current arms procurement system based on export income from CODELCO should be changed to a consistent method of harmonizing resource allocation priorities with national security perceptions. The amount allocated to procurement should also not be determined by arithmetical proportions but according to objective criteria of national defence priorities.

Funding for arms procurement should be included in the national budget. The funding criteria should be based on medium- and long-term defence policies. If these issues are to be taken into account in the budget, the legislature will need to participate in examining the appropriateness of allocations. Thus, the issue will no longer be subject to confidentiality. However, the Congress will need to build capacities to conduct independent analysis for national strategic assessment and build up data as well as experience.

In order to establish an efficient decision-making process, threat analyses should be based on a comprehensive approach to national defence, in which the three branches of service should participate to coordinate their requirements in a coherent manner. Their decisions and resource allocation methods should be monitored according to objective criteria applied by a higher-level executive entity responsible for coordinating different national security options with overall foreign and security policies. Such overall goals must be included in all analysis conducted by the different branches of service and integrated into the subsequent evaluation made by the senior entity when funding is being sought.

A higher-level agency responsible for analysing and evaluating the different projects from a broad perspective is also required. Such an agency should include experts who are qualified to conduct studies according to the different perspectives and time horizons required by national defence.

General guidelines for arms procurement are currently defined at ministerial level where likely medium- and long-term strategic scenarios are analysed.[48] This demands knowledge of techniques beyond the scope of the training provided to military officers or their professional experience. Since the mid-1960s the armed forces have made considerable efforts to train groups of offi-

[48] These guidelines are restricted documents and are not published.

cers in different aspects of management at both Chilean and foreign universities. At present all three branches of service have large teams for project analysis that are capable of evaluating projects according to the different approaches required. One product of this experience is the development of institutional manuals for project evaluation that are now available in the three branches of the military.

In short, the 'ideal' process requires a global approach to security to include long-term technical and strategic studies; effective coordination between the different agencies involved; and financial forecasting to invest in the best alternatives. Such a process requires specialized skills in the society, which would increase civilian participation and legislative oversight of the decision-making process. The responsibility for final decisions would, of course, remain with the President of the Republic.

3. Greece

*Stelios Alifantis and Christos Kollias**

I. Introduction

Greece, a member of both NATO and the European Union (EU), having joined them in 1952 and 1981, respectively, allocates a substantial part of its national income to defence. Indeed, it is the most militarized of the NATO and EU countries in terms of the human and material resources allocated to defence uses and the military burden.[1] In 1996 Greek military expenditure as a share of gross domestic product (GDP) was more than twice the EU and NATO averages.[2] Similarly, in 1998 and in 1999 the ratio of the Ministry of National Defence (MOD) budget to the total government budget was high.[3] Domestic arms production capabilities are comparatively modest, and according to SIPRI data Greece was the sixth largest importer of major conventional weapons in the five-year period 1994–98.[4]

There have been no comprehensive analyses or systematic studies of arms procurement decision making in Greece, despite the high level of resources allocated to defence. The lack of previous research is a major obstacle to examining this process. The Greek defence planning process and in particular the arms procurement decision-making process are also fairly closed in terms of public accountability, transparency, parliamentary scrutiny, monitoring and oversight.

[1] Kollias, C., 'Country survey VII: military expenditure in Greece', *Defence and Peace Economics*, vol. 6, no. 4 (1995).

[2] In 1996 Greek military expenditure, at 4.5% of gross domestic product (GDP), was more than twice the EU and NATO averages for the same period, which were 2% and 2.2%, respectively. Sköns, E. *et al.*, 'Military expenditure and arms production', *SIPRI Yearbook 1998: Armaments, Disarmament and International Security* (Oxford University Press: Oxford, 1998), pp. 228–29, 232–33.

[3] It was 8.58% in the government budget for 1998 and estimated at about 8.14% for 1999.

[4] Hagelin, B., Wezeman, P. D. and Wezeman, S. T., 'Transfers of major conventional weapons', *SIPRI Yearbook 1999: Armaments, Disarmament and International Security* (Oxford University Press: Oxford, 1999), p. 428.

* The authors wish to thank Dr Theodoros Stathis, MP, for acting as the country adviser in this study and gratefully acknowledge the contributions of Thanos Dokos, Panayotis Tsakonas, Meletis Meletopoulos, Odysseus Narlis, Haralambos Giannias, Christos Valtadoros and Stratis Trilikis, both in their individual research papers and in their contributions during the course of the SIPRI Project on Arms Procurement Decision Making. Their in-depth research made this chapter possible. The constructive suggestions and insightful remarks made by the editor of this volume and an anonymous referee on earlier drafts of this chapter are also gratefully acknowledged. However, the opinions expressed here are the authors' own and any remaining errors, omissions and weaknesses are their exclusive responsibility.

This chapter is based partly on the six working papers prepared by Greek researchers for a workshop held in Athens on 28 February 1998. They are not published but are deposited in the SIPRI Library. Abstracts appear in annexe B in this volume.

Information available publicly on specific weapon acquisitions consists mainly if not exclusively of press reports and articles on the choice of weapon system, rough estimates of their costs, and analyses by defence experts of their operational characteristics and their usefulness for the defence needs of the country. Public knowledge of the arms procurement process is at best sketchy.

There is almost universal consensus in Greece on the need for a strong defence and arms procurement decisions are not often questioned. Any criticism that there is comes from such quarters as defence experts in the media, the academic world, the opposition, and individual members of parliament (MPs) and politicians, and is likely to refer to delays in arms procurement which adversely affect the balance of military strength between Greece and its adversary, Turkey.

This chapter examines the arms procurement decision-making process currently in operation in Greece, the levels of public accountability relating to it, and the barriers to and opportunities for shaping the process to meet the broader needs of Greek society. The remainder of this section presents a historical overview of threat perceptions, linking current security concerns and priorities with the history of the region and in particular the adversarial nature of Greek–Turkish relations. The major changes that have occurred are described and the factors that determine the changes in security perceptions and defence priorities are identified. In section II, national security priorities and the Greek experience of arms procurement in the initial post-World War II period are discussed. Section III examines the current defence planning and decision-making process in the context of the wider strategic environment and the threat assessment and defence priorities of the country. Section IV examines arms procurement decision making, section V the defence budget process, section VI the role of the defence industry in the procurement process, and section VII the strengths and limitations of democratic oversight of arms procurement decision making. Section VIII summarizes and concludes the chapter.

The security environment

Greece is located at the crossroads of three continents in a volatile area of southern Europe—the Balkan Peninsula. Historically, its two major security concerns have been the Balkan Slavs to the north and Turkey to the east.[5] Following the collapse of the cold war bipolarity, the Balkan strategic and security environment has undergone important structural changes. As a member of NATO, during the cold war Greece had borders with Bulgaria, a Warsaw Treaty Organization (WTO) member, and non-aligned Yugoslavia and Albania. However, Greece has long regarded Turkey, another NATO ally, as the main threat to its security interests.[6] Indeed, the consensus across the entire Greek political

[5] Veremis, T., *Greek Security Considerations* (Papazissis Publishers: Athens, 1982) (in English).

[6] 'The central axis of Greece's military strategy is the deterrence of the Turkish threat.' Greek Ministry of National Defence, 'White Paper for the armed forces 1996–97', Dec. 1997, URL <http://www.hri.org/mod/fylladia/bible/e_index.htm>, p. 27. See also Alifantis, S., 'National defence in the

spectrum is that Turkey is the principal and most imminent security concern. This view is shared by the public, the media, politicians and security experts. A 'cold war' prevails in Greek–Turkish bilateral relations.[7] For at least 25 years the domestic security debate has taken the Turkish threat for granted.[8]

The resources allocated to defence reflect the increased security needs of the country. They are also a major obstacle to Greece's achieving economic convergence with the other EU members and joining the Economic and Monetary Union (EMU), which is universally considered to be of paramount importance and of strategic significance.

II. National security priorities and arms procurement experience: a historical overview

External threats to a state's sovereignty and independence can be met by the combined use of two policies which help to deter aggression by a hostile neighbour. The first, known as internal balancing, is the strengthening of the state's military capability through the allocation of resources to defence. The second, external balancing, is participation in international politico-military alliances and coalitions (NATO in the case of Greece) which offer the benefit of reinforcement—political and/or military—in order to balance and deter aggressors. For example, the Greek White Paper for the armed forces for 1996–97 states that among the means used to secure Greece's national interests is the 'maximisation of the advantages from Greece's participation in alliances and collective security organisations . . . . for the protection of its national interests'.[9] Internal balancing is being achieved through the strengthening and modernization of Greece's armed forces. Within this context, the arms and weapon systems procured and held by a state reflect its security concerns and priorities.

Two distinct periods in defence planning and arms procurement can be identified: (*a*) the years up to 1974 and the Turkish invasion of Cyprus; and (*b*) since 1975. A turning point was reached in Greek security concerns and priorities which resulted in a major reappraisal of defence and security policies and therefore arms procurement.

aftermath of the Imia crisis: the concept of "flexible retaliation"', SIPRI Arms Procurement Decision Making Project, Working Paper no. 71 (1998); Meletopoulos, M., 'The sociology of national decision-making behaviour', SIPRI Arms Procurement Decision Making Project, Working Paper no. 74 (1998); Giannias, H. C., 'Arms procurement and foreign dependence', SIPRI Arms Procurement Decision Making Project, Working Paper no. 73 (1998); Dokos, T. and Tsakonas, P., 'Perspectives of different actors in the Greek procurement process', SIPRI Arms Procurement Decision Making Project, Working Paper no. 72 (1998); and Kollias, C., 'The Greek–Turkish conflict and Greek military expenditure 1962–90', *Journal of Peace Research*, vol. 33, no. 2 (1996).

[7] Former Greek Prime Minister Andreas Papandreou once described the state of Greek–Turkish relations as a 'no war' situation, implying that, since peaceful relations were not possible and actual war would be catastrophic for both, the 2 countries had to coexist in a limbo between peace and war.

[8] Dokos and Tsakonas (note 6). The Greek–Turkish dispute is well documented in the literature on international relations. See, e.g., Larrabee, S., 'Instability and change in the Balkans', *Survival*, vol. 34, no. 2 (1992); and Constas, D. (ed.), *The Greek–Turkish Conflict in the 1990s* (Macmillan: London, 1991).

[9] 'White Paper for the armed forces 1996–97' (note 6).

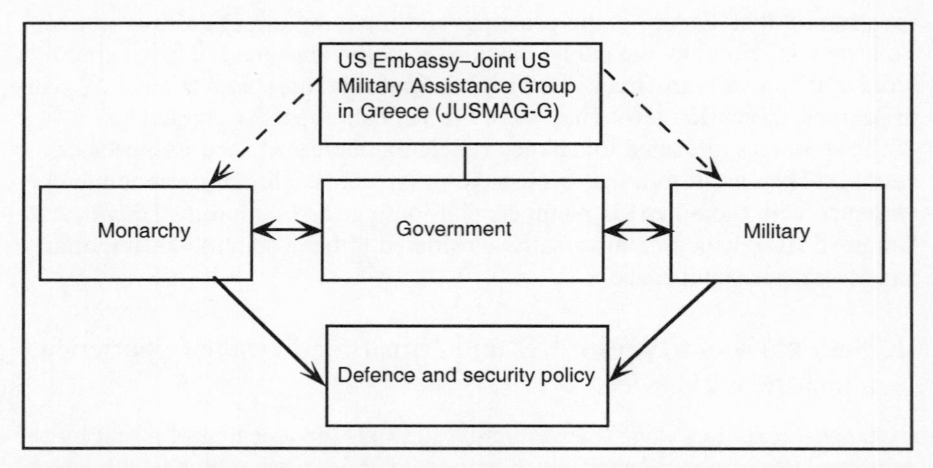

Figure 3.1. US influence on defence and security policy in Greece during the early post-World War II years

Source: Based on Giannias, H., 'Arms procurement and foreign dependence', SIPRI Arms Procurement Decision Making Project, Working paper no. 73 (1998).

After the end of World War II and the subsequent civil war (1945–49), Greece found itself in the Western sphere of influence and a member of NATO. The traumas of the civil war led to profound and extensive political and military dependence on the USA. Up to the early 1960s the main security threats to Greece were thought to emanate from its northern borders and the communists, both externally and internally. The authoritarian state established after the civil war saw NATO and the USA as indispensable for the defence of the country.[10] The structure of the Greek armed forces at this period reflected their mission of maintaining internal security against communist insurgency:[11] the forces were designed primarily to delay a southward push of WTO forces, acting as the tripwire that would set in operation the NATO military machine. The emphasis on internal security also resulted in a poorly developed navy and air force, which in practice meant that the country had an extremely limited capability for independent operations against threats to its national interests. External security rested within NATO's defence planning, which regarded the WTO forces as the only source of external threat.

The period 1949–74 can be characterized as one of almost total political and military dependence on the West, in particular on the USA.[12] Following the victory of the nationalist forces in the civil war, the armed forces were equipped with US weapons (mostly second-hand surplus), reorganized according to US

[10] Giannias (note 6).

[11] Stavrou, N., *Allied Policy and Military Intervention* (Papazissis Publications: Athens, 1976) (in English); and Platias, A., 'Greece's strategic doctrine: in search of autonomy and deterrence', in Constas (note 8).

[12] With the Truman Doctrine Greece passed from the British to the US sphere of influence. The USA played an instrumental role in the assistance provided by the West to the nationalist forces in the civil war with the left in 1945–49.

standards and assigned the role described above.[13] The relative lack of internal legitimacy of the state increased its dependence on the USA. Figure 3.1 shows the channels of US influence on defence and security policy formulation during this period. The US Embassy in Athens, and in the late 1940s the Joint US Military Assistance Group in Greece (JUSMAG-G), directly influenced the decision making of the three dominant institutions in Greece at the time—the government, the monarchy and the military. These formed the three pillars or centres of power of the post-war state.[14] US military advisers were posted at various levels of the command structure of the armed forces such as working groups, committees and councils. They had not only immediate access to the decision-making process but also a direct say and influence in matters of military planning, arms procurement, force structure, operational plans, strategy, military doctrine and so on. The US military mission in Greece was effectively in joint command of the armed forces. Sovereign arms procurement policies and decision making were virtually non-existent since the armed forces depended entirely on the arms and equipment supplied by the USA under its various military assistance programmes.[15] Under military rule (1967–74) Greece began to diversify its weapon acquisition sources because of the arms embargoes imposed by the US Congress and the almost exclusive dependence on US sources was reduced.[16] After 1974, missile boats, AMX-30 tanks, armoured personnel carriers (APCs) and combat aircraft were procured from France; Type-209 submarines, Leopard-1 tanks and fast attack craft from Germany; Kortenaer Class frigates from the Netherlands; and in 1998 and 1999 SA-8 and SA-15 surface-to-air missiles (SAMs) from Russia.[17] Even so, the USA remains the single most important supplier of military equipment and still exercises an appreciable degree of influence over arms procurement decisions through political, diplomatic and military channels.[18]

Following the political changes after 1974 and the drastic reduction of overt US influence in domestic political affairs, the influence of the USA on military affairs in Greece, including force structure and arms procurement, diminished.[19] This does not of course imply that external influences such as alliance policies and commitments are not important in Greek military affairs: they are with all

[13] Giannias (note 6).

[14] Mouzelis, N., *Modern Greece: Facets of Under-Development* (Macmillan: London, 1978); and Mouzelis, N., *Politics in the Semi-Periphery* (Macmillan: London, 1986).

[15] Giannias (note 6).

[16] SIPRI arms transfers database, Apr. 1999. US-made military equipment accounted for almost 81% of all imported equipment in the period 1950–66, about 59% in the period 1967–73, 51% in 1974–89, and 55% in 1990–98. For the entire period 1950–73, US imports accounted for about 75% by value of all imported weapons and for the post-1974 period (1974–98) for about 51%.

[17] SIPRI arms transfers database.

[18] Statements by the US Ambassador in Athens in newspaper and television interviews expressing US interest in the choice of the new long-range anti-aircraft system and the order for new combat aircraft are an example.

[19] In contrast to the period before 1974, US military advisers no longer participate in the committees, working groups or councils of the armed forces. Any points of contact that exist are institutionalized groups and committees provided by bilateral military cooperation agreements, or within the organizational structure of NATO.

NATO members. The difference between Greece and most other NATO members is that Greece feels that its national interests were not, when needed, protected by the alliance. Furthermore, where its current security needs and priorities are concerned, it can hardly rely on NATO for active protection against external aggression if the source of this aggression is Turkey.[20] It must therefore rely entirely on its own military capability as a deterrent.

In 1974, when Turkish forces invaded Cyprus, Greece found itself in a weak position, lacking an independent military deterrent and the capacities to react militarily. A major reappraisal of defence priorities took place. Greater emphasis was placed on strengthening the air force and navy, and a substantial increase in defence expenditure was required. Between 1974 and 1975 it increased in real terms by about 69 per cent and by 1978 it almost doubled. As a share of GDP it jumped from 4.1 per cent in 1973 to 7 per cent by 1977.[21] As a result, it was necessary to allocate substantial resources to building up and modernizing military equipment and infrastructure, especially in the Aegean islands near the Turkish mainland. The major requirements of the Greek military were to install, upgrade and modernize its command, control and communications (C^3) systems and to revise its force structure, military plans and geographical distribution of forces. Emphasis was given to qualitative improvements of the military through the procurement of technologically advanced weapon systems and 'smart' weapons.[22] The deployment of capital-intensive, better-equipped and better-trained armed forces is intended to counterbalance the Turkish superiority in numbers.

III. Defence planning

The two distinct periods in defence planning and arms procurement policies, reflecting changes in threat perceptions and defence priorities, coincide with important changes in domestic politics and the economy. Except for one brief interlude, the pre-1974 period was a period of authoritarian rule in which the army and the monarchy played a major role in political affairs and the military

[20] NATO is obliged by treaty to give military assistance to its members if they are attacked by a 3rd force but not if this aggression emanates from another NATO member.

[21] 'World military expenditure', *World Armaments and Disarmament: SIPRI Yearbook 1979* (Taylor & Francis: London, 1979), pp. 35, 37.

[22] The Minister of National Defence, Akis Tsochatzopoulos, said in a statement to Parliament in Nov. 1996: 'Considering the dimensions of our country, the condition of our economy and the demographic problem, quantitative armaments competition with any hostile power would constitute a particularly costly effort for Greece with an uncertain outcome. Emphasis, therefore, should be put on quality, by adopting a modern strategic and operational doctrine (with emphasis on combined/joint operations), improving personnel training, restructuring combat units (with the aim of successfully carrying out defensive operations, but also to transfer operations on enemy territory), obtaining the necessary modern weapon systems (smart weapons and especially force multipliers) and rapidly integrating them in our Armed Forces. The main element of our defence planning is the achievement of maximum cost-effectiveness.' Dokos and Tsakonas (note 6), p. 7. According to the US Arms Control and Disarmament Agency (ACDA), the size of the Turkish armed forces in 1995 was in the region of 805 000 while the Greek armed forces numbered about 213 000. US Arms Control and Disarmament Agency, *World Military Expenditures and Arms Transfers 1996* (US Government Printing Office: Washington, DC, 1996), pp. 72, 94.

as an institutional group had a decisive say in matters of internal security.[23] Since the collapse of the military government in 1974 and the establishment of a fully functional liberal democracy, the military has been under firm civilian control and has had no political power.

Similar structural changes can be seen in the performance of the economy. Growth was fast in the pre-1974 period, but then slowed, and the period since 1975 has been characterized by serious and persistent economic problems such as rising inflation, increasing public deficits and debt, and in more recent years rising unemployment.[24] Undoubtedly high defence spending presents an additional obstacle to efforts to reduce the budget deficits, inflation and government borrowing. Participation in the EMU is considered to be of paramount economic importance, and a long-term commitment to maintaining a strong defence makes joining much more difficult. In a broader context, Greek military expenditure throughout the post-1974 period has played a role in retarding growth and has used resources which, if allocated to areas such as health care, education and infrastructure, would have contributed to the development and modernization efforts of the country. Although there is broad consensus across the political spectrum that the national defence must be strengthened, it is also recognized that high defence budgets are a heavy burden on the weak economy, especially at a time when successive governments have been implementing strict austerity programmes to meet the EMU convergence criteria.[25]

The defence planning bodies

According to the Greek Constitution (adopted in 1974 and partially revised in 1985), the President of the Republic is the Supreme Commander of the armed forces, but his powers are largely symbolic. The Prime Minister and the Cabinet determine national defence policy, exercise command over the armed forces and make all defence-related decisions.[26]

The Government Council on Foreign Affairs and National Defence (Kivernitiko Simboulio Exoterikon kai Aminas, KYSEA), which usually convenes on an ad hoc basis, is the main decision-making body on issues of national defence and security. KYSEA is chaired by the Prime Minister and its members include the ministers of foreign affairs, defence, the national economy, the interior, public order, and public administration and decentralization, and the Chief of the Hellenic National Defence General Staff (HNDGS).[27] It formulates defence and foreign policy, appoints the Chief of the HNDGS and the Chiefs of

[23] Mouzelis, N., *Modern Greece: Facets of Under-Development* (note 14); and *Politics in the Semi-Periphery* (note 14).

[24] Alogoskoufis, G., 'The two faces of Janus: institutions, policy regimes and macroeconomic performance in Greece', *Economic Policy*, no. 20 (Apr. 1995), pp. 149–92.

[25] See, e.g., Tsakiris, G. and Koronaios, P., *Elefterotipia,* 25 July 1999.

[26] All the government ministers and deputy ministers participate in the Cabinet.

[27] Other ministers may participate on an ad hoc basis if deemed necessary.

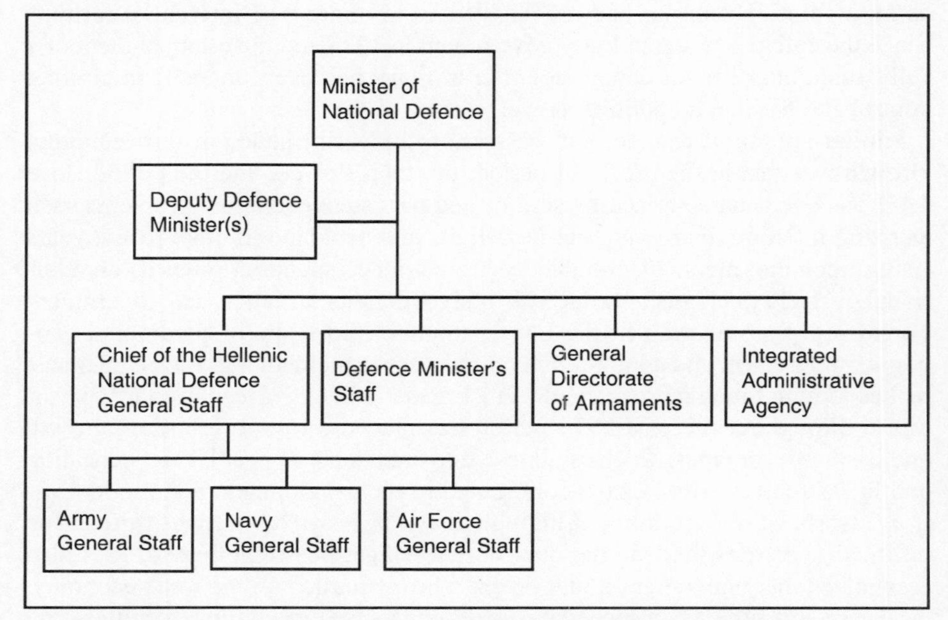

Figure 3.2. The organization of the Greek Ministry of National Defence

Source: Greek Ministry of National Defence, 'White Paper for the armed forces 1996–97', Dec. 1997, URL <http://www.hri.org/mod/fylladia/bible/e_index.htm> (in English).

Staff of the Hellenic Army General Staff (HAGS), the Hellenic Navy General Staff (HNGS) and the Hellenic Air Force General Staff (HAFGS), and decides on the procurement of all major weapon systems.

As described in the White Paper for the armed forces for 1996–97, KYSEA is responsible for: (*a*) formulating national defence policy within the broader context of national strategy and on the basis of long-term evaluations and assessments of security, foreign affairs and relevant international developments; (*b*) deciding on and approving long- and medium-term development programmes for Greece's defence capabilities, and for all major arms procurement, on the basis of national threat assessments; (*c*) deciding on all issues of national defence, particularly those requiring coordination with other ministries; (*d*) deciding to impose or lift national security alert measures and advising the President on the need for partial or general mobilization or the declaration of war; (*e*) selecting the Chief of the HNDGS and the Chiefs of Staff of the other services, following recommendations made by the Minister of National Defence; and (*f*) deciding on the assignment of forces for international operations in line with the international obligations of the country.

The MOD and its subordinate armed forces are responsible for the implementation of national defence policy in line with the general defence and security policy guidelines decided on and formulated by KYSEA. The Minister of National Defence heads and directs the Ministry of Defence Staff, the HNDGS and the three branches of the armed forces through their respective chiefs and coordinates their functions through the office of the Chief of the HNDGS. Within the general framework of defence policy formulated by KYSEA, the minister approves and authorizes national military strategy, military evaluation and assessment, and the general directions of defence planning. He proposes to KYSEA the major changes required in force structure, authorizes the annual budget of the three branches, and coordinates and approves arms procurement programmes. He decides on the required annual reviews of the medium-term defence planning programme, recommends policies for the development and modernization of the defence industry to KYSEA, and submits to Parliament an annual report on the main activities of the armed forces.[28]

The structure of the MOD is shown in figure 3.2. The tasks and jurisdiction of the Deputy Minister(s) of National Defence are decided on jointly by the Prime Minister and the Minister of National Defence. The main agencies and bodies that make up the ministry are:

1. *The Defence Council*. This consists of the Minister of National Defence, the Deputy Minister(s) of National Defence, the chiefs of staff, and if deemed necessary officials from other ministries on an ad hoc basis, such as diplomats from the Ministry of Foreign Affairs. The Defence Council is the highest advisory body to the Minister of National Defence for incident and situation assessment, issues of force structure and arms requirements, budget issues and research and development (R&D) programmes. It promotes a broader understanding of national security in order: (*a*) to develop a wider perspective on issues of defence and foreign policy through the assessment of international developments that could influence national security; (*b*) to improve coordination and communication between the various agencies of the MOD; and (*c*) to act as an internal think-tank submitting policy proposals to government bodies such as KYSEA. The Joint Council of the Chiefs of the General Staff is responsible for military decisions while the Defence Council is responsible for political–military analyses.

2. *The Joint Council of the Chiefs of the General Staff*. This is made up of the Chief of the HNDGS and the Chiefs of Staff of the three branches of the armed forces. Its duties and responsibilities include submitting policy proposals to the Minister of National Defence on issues such as the direction of defence planning, force structure, military strategy, military readiness, and military assessment of incidents and situations.

3. *The Defence Minister's Staff*. Formed in 1996, it includes civilian as well as military personnel with specialist training and knowledge, experience of budgeting, personnel management, R&D, military technology, international relations, national and international law, and so on. Its function is to provide the Minister of National Defence with immediate, specialist information on defence planning, defence policy, foreign relations, and technical and financial issues.

[28] 'White Paper for the armed forces 1996–97' (note 6).

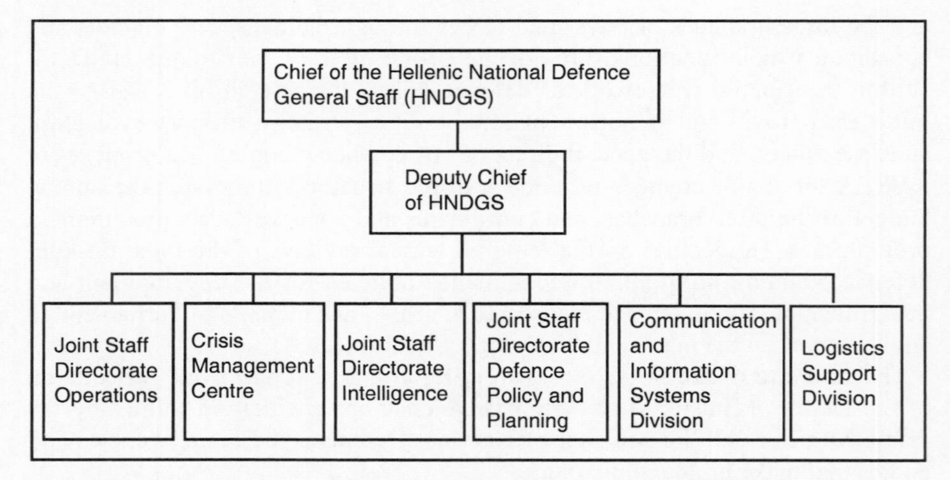

Figure 3.3. The organization of the Hellenic National Defence General Staff

Source: Greek Ministry of National Defence, 'White Paper for the armed forces 1996–97', Dec. 1997, URL <http://www.hri.org/mod/fylladia/bible/e_index.htm> (in English).

4. *The Chief of the HNDGS*, who is the Supreme Military Commander of the armed forces in times of crisis or war. (In peacetime the Chiefs of Staff of the three branches report directly to the Minister of National Defence.) Figure 3.3 shows the staff organization of the HNDGS. The post of the Chief of the HNDGS alternates on an almost regular basis every two years between officers of the three branches. The Chief of the HNDGS is selected by KYSEA among the lieutenants-general, vice-admirals and air force lieutenants-general and is appointed by presidential decree. The two-year period of service can be extended for one more year before the officer is retired. The three Chiefs of Staff, who are also selected by KYSEA, serve for a two-year term, although this can be extended if deemed necessary. The Joint Staff Directorates of Operations and Defence Policy and Planning in the HNDGS are directly involved in the arms procurement process.

5. *The Supreme Council of each of the three branches of the armed forces*. These three councils are responsible for the force and organizational structure of each branch, operational doctrines, identifying and listing arms procurement requirements for each branch, budgeting and so on.

6. *The General Directorate of Armaments (GDA)*. As well as being responsible for the implementation of procurement of major arms and equipment, the GDA coordinates the equipment needs of the three branches of the armed forces and executes the procurement programmes decided on by KYSEA. Established in 1995 by Presidential Decree 438/1995 and operational since 1996, it represents the MOD in international arms procurement negotiations and formulates recommendations on military technology issues. It also coordinates and oversees the domestic defence industry. It is discussed further in sections V and VI below.

The threat assessment process

In general, security and threat assessments are based on information gathered by intelligence organizations such as the Ethniki Ypiresia Pliroforion (National Intelligence Service, EYP), the intelligence branches of the three armed services, the HNDGS and other sources, and the Ministry of Foreign Affairs through its own sources and channels. Intelligence gathered includes information about the military potential of adversaries, changes in their deployment and force structure, their arms procurement plans and agreements with other countries that could affect the balance of power (local and/or regional), and economic and political information which could help in evaluating the overall strengths and weaknesses of foreign powers.

Despite the almost universal agreement on the main principles of national security and threat assessment, there are differences of view among the various actors involved in security policy making as to the best mix of policies to balance the external threat and other challenges to national interests. For example, policy differences often exist between the ministries of defence and foreign affairs as to the best mix of internal and external balancing.

Coordination with foreign policy making

The two departments within the Ministry of Foreign Affairs responsible for threat assessment and security are the Centre of Analysis and Planning and the Permanent Mixed Crisis Management Group. The former is mainly a research group which studies international relations issues across the whole spectrum and, following comprehensive analyses, submits proposals on the conduct of foreign policy and diplomacy. It is headed by an ambassador and staffed by Ministry of Foreign Affairs personnel but, if deemed necessary, can include specialists from academic, research or other institutions. An ambassador also heads the Permanent Mixed Crisis Management Group, and it includes the head of the Centre of Analysis and Planning, representatives from the MOD and the ministries of the national economy and public order, representatives from the press and the media, and personnel from the EYP. Its main task is the formulation of the procedural framework necessary for crisis management in line with the analyses carried out by the Centre of Analysis and Planning and the periodical conduct of simulated crisis management exercises.

Coordination between the MOD and the Ministry of Foreign Affairs on security issues is achieved through the posting of an army officer (usually a colonel) in the latter and of an ambassador in the Defence Minister's Staff. In practice this coordination is not always very effective and is not fully utilized, as recent cases have shown. Examples are the Imia incident involving Turkey in 1996[29] and Cyprus' procurement in 1998 of the S-300PMU-1 SAM system

[29] Turkish troops landed on the uninhabited Greek island of Imia. See, e.g., 'Greece and Turkey in stand-off over island', *Daily Telegraph*, 31 Jan. 1996.

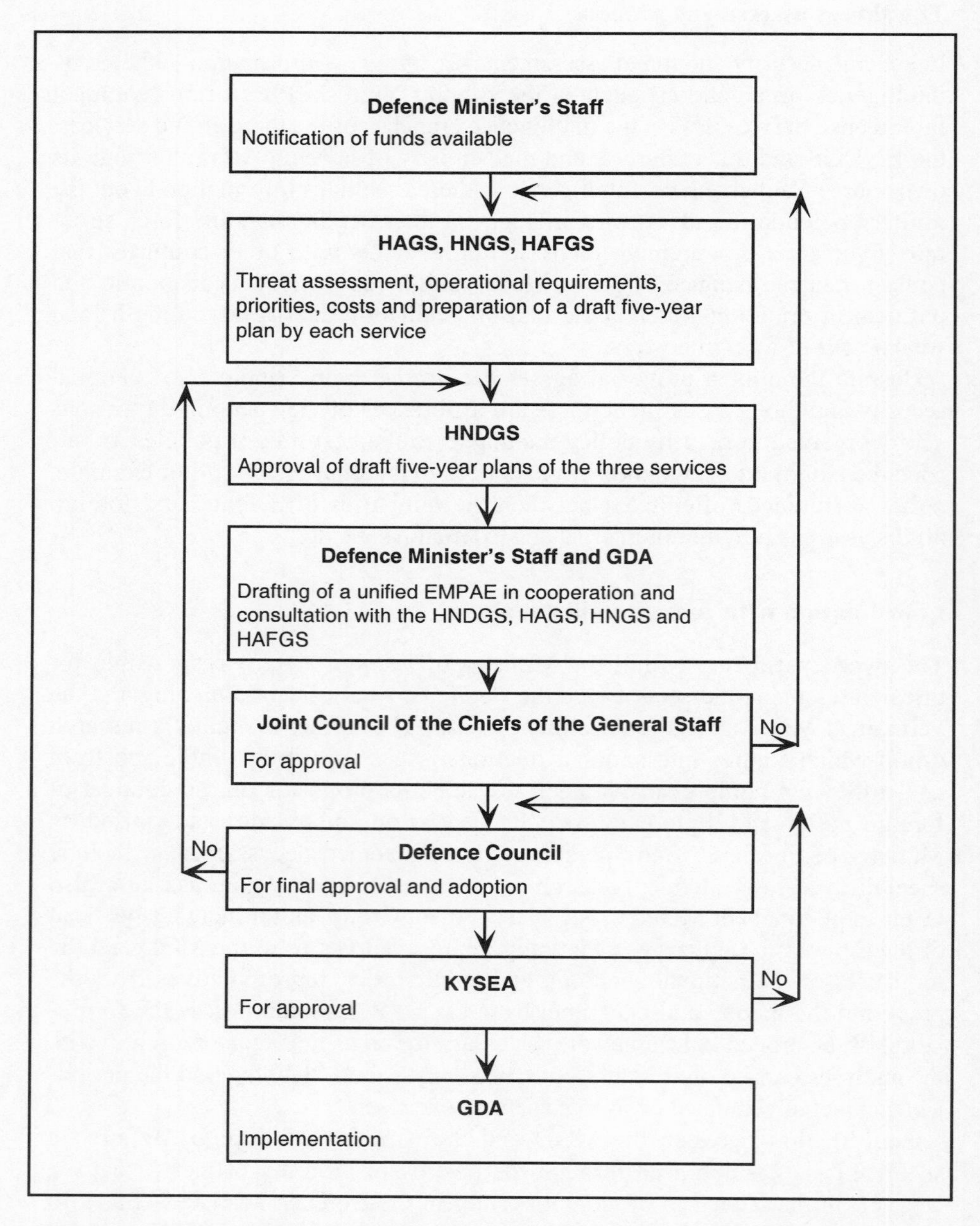

Figure 3.4. The drafting of the Greek medium-term arms procurement programme

Note: HAGS = Hellenic Army General Staff; HNGS = Hellenic Navy General Staff; HAFGS = Hellenic Air Force General Staff; HNDGS = Hellenic National Defence General Staff; EMPAE = Medium-term Programmes of Development and Modernization; KYSEA = Government Council on Foreign Affairs and National Defence; GDA = General Directorate of Armaments.

from Russia in close cooperation with the Greek MOD.[30] In the first case the Ministry of Foreign Affairs failed to notify the military in good time of the impending crisis and the resulting military escalation brought the navies of Greece and Turkey to the brink of war. In the S-300 case, in a closed hearing of the Parliamentary Committee on Foreign Affairs and Defence in February 1999 the Foreign Minister stated that the decision to procure the system was wrong, while the Defence Minister stated that the decision was correct but the political and diplomatic handling of the affair by both Greece and Cyprus was at fault.[31] It has been openly argued that better institutionalization of the coordination and flow of information, which interministerial rivalry has hitherto discouraged, could have prevented the incident. There is substantial room for improvement if effective coordination is to be achieved between the apparatus of the foreign and defence ministries both on the institutional level and on the functional level. As a result, the coherence and effectiveness of national security decision making and policy are impaired.

IV. Arms procurement decision making

The stages in arms procurement planning are shown in figure 3.4. They are: (*a*) preparation and approval by the staff of each branch of service (army, navy and air force) of a draft five-year plan which includes the weapon systems required, an indication of priorities and the estimated budgets; (*b*) approval of the draft five-year plan by the supreme council of each branch of service; (*c*) approval of the draft five-year plans by the HNDGS: these are then incorporated into the Medium-term Programmes of Development and Modernization (EMPAE); (*d*) submission to the GDA and to the Defence Minister's Staff. At this stage the final drafting takes place, taking into consideration operational priorities and co-production capabilities or industrial offset requirements; (*e*) submission to KYSEA; and (*f*) if the plan is approved, execution of the procurement programme by the GDA.

The ultimate decision on the type of weapons and numbers to be procured rests with KYSEA. In practice the Minister of National Defence and his staff (military and civilian personnel as well as political advisers) set the scope and limits of weapon system requirements, according to the available information, through the existing lines of command in the MOD. The original procurement programmes prepared by the staff officers of the three branches may be substantially changed and revised if changes in the geo-strategic environment require it or as a result of financial and budgetary constraints.

Following evaluation of available threat assessments, broad outlines of military strategy and arms procurement policy are proposed. The three branches of the armed forces assess the current and projected capabilities of the forces of foreign powers, their force structure and planned armaments programmes. They identify their military needs and make recommendations for the procurement of

[30] Hagelin *et al.* (note 4), pp. 431–36.

[31] See, e.g., Karaiosifides, F., 'Cyprus', *Ptisi*, Special Edition 1999, 'Balance of power', p. 172.

the necessary weapon systems and equipment to their respective Supreme Councils. The programmes list technical characteristics, operational capabilities and priorities of the weapon systems that are required to meet external military threats. These recommendations are then passed on to the Joint Council of the Chiefs of the General Staff, which finalizes the proposals for procurement, coordinating the needs of the three branches and identifying the weapon systems that satisfy the operational and technical specifications.

Arms procurement decisions are in this sense more often than not reactive, aiming to offset the effects of weapon acquisition by other powers in the region. Since there is national agreement on the primary external threat, threat assessment mostly takes the form of a periodic evaluation of possible changes in its military capability. There is no comprehensive or coherent long-term strategy for meeting current or potential challenges to national interests and threats to national security. This also often results in short-term reactive responses to changes in the security environment which tend to emphasize the military aspect of national security policy.

For obvious reasons the military plays the most significant role in the arms procurement process. Through the three supreme councils and the HNDGS it heavily influences threat assessment, on the basis of which national defence needs and requirements are defined. Defence experts in the academic world and outside government often publish views and opinions on the military strength and long-term strategy of foreign powers or make recommendations and proposals to counterbalance them.[32]

More often than not, an important criterion in the final decision of KYSEA—apart from the obvious financial and budgetary constraints which affect the numbers ordered—is the political leverage that can be gained ('external balancing') by placing the order with one major supplier or another.[33] Questions of long-term procurement needs, in terms of suitability and compatibility of the system and financial issues such as life-cycle costs, often take second place.

The decision in 1985 to opt for two different third-generation combat aircraft, the US F-16 and the French Mirage 2000E, was indicative. It was clearly a political decision to divide the procurement 'pie' between US and French producers. As a result, the numbers ordered (40 of each) were less than the military had recommended (100–120) and the opportunity to enter into a co-production agreement was forfeited since it was not economically viable. It soon became apparent that the newly acquired fighters were not enough to meet operational needs and a further group of 40 F-16s was ordered in 1993. Before all the units of this second group had been delivered the air force was once again preparing to procure a further batch of combat aircraft. The F-16, Mirage-2000-5 and

[32] E.g., retired military personnel—mostly senior officers such as former heads of the HNDGS, HAGS, HNGS and HAFGS—may write journal and newspaper articles on issues of procurement requirements, force structure changes, military doctrine and so on.

[33] In an interview for a Greek newspaper, Minister of National Defence Tsochatzopoulos has stated: 'If the government buying (the weapons) does not at the same time ask for something in return, for support on a governmental level on the basis of its country's needs, then the buyer is not utilizing (and benefiting from) its defence procurement policy'. *Elefterotipia*, 13 July 1999.

EF-2000 Typhoon were selected in 1999. It is now widely accepted that the balance of power in the air tilted against Greece as a result of the 1985 decision to procure from different sources.

Political criteria weigh heavily in the current EMPAE. Contracts for the procurement of new combat aircraft, surface vessels, tanks and long-range anti-aircraft systems are expected to be allocated primarily on the basis of political criteria rather than operational suitability, long-term defence planning or financial terms offered.[34] It is generally expected that the arms procurement 'pie' will be divided in such a way as to include orders for US, European and Russian military hardware.[35] The KYSEA decision of October 1998 to buy the US Patriot long-range air-defence missile system, as well as two medium-range SAM systems from France and Russia, is an instance of large military contracts being used as an instrument of foreign policy.[36] In other words, the arms procurement budget is used as a means of external balancing: large defence contracts are expected to earn Greece a more favourable stance on the part of the supplying countries on issues of interest to Greece.

The current EMPAE (1995–2000) is for a total estimated cost of 4 trillion drachmas (*c*. $14 billion) over five years.[37] It is based on: (*a*) an assessment of the military threat, strength and capability of Greece's main adversary, Turkey; (*b*) the latter's current and planned armament and force modernization programmes; (*c*) projections of how the balance of force between the two countries may be affected; (*d*) an assessment of other sources of potential threat; and (*e*) an assessment of the military needs that stem from Greece's alliance obligations or from other international commitments such as participation in peace-keeping operations, which also influence the armaments programme since they create specific operational needs.

[34] The decisions are often so overtly political that press reports have questioned the need to spend millions of drachmas on committees evaluating technical characteristics and operational capabilities and on testing of candidate weapons since the KYSEA decision is not likely to be based on their reports but rather on political and diplomatic considerations.

[35] 'The execution of the defence procurement programme was based on three main axes . . . the second axis is not to create imbalances between the countries participating as suppliers in procurement programme . . . and the third axis are the political returns that we get for placing the order with the one or the other supplier.' Interview with Minister of National Defence Tsochatzopoulos (note 33).

[36] It was widely expected that Greece would opt for the Patriot rather than the Russian S-300. However, following a diplomatic dispute with the USA over the latter's position on the Cyprus problem, KYSEA postponed the decision in order to put pressure on the USA to change its statements on Cyprus.

[37] It includes the acquisition of *c*. 60 combat aircraft—the US F-16C/D, the French Mirage 2000-5 and the EF-2000 Typhoon were selected—the modernization of 39 Phantom F-4E combat aircraft, transport aircraft, helicopters, attack helicopters, air defence systems (the US-made Patriot and the Russian-made S-300 system were the 2 contenders, with the former getting the contract), new tanks, multiple rocket launchers (MRLs), short-range air defence systems (SHORADS), frigates, corvettes, submarines, smart weapons and munitions, and so on. *Ptisi*, Special Edition 1999, 'Balance of power', pp. 119–21.

Factors influencing arms procurement[38]

Clearly, like any other process in an open society, arms procurement is subject to a number of external influences from sources such as institutional and social groups—political parties and politicians, think-tanks, pressure groups, interest groups or simply the general public and the media—as well as to internal influence from the various actors directly involved, such as the military, the government and its various departments and agencies, and interest groups in the supplying countries. They are of course not independent of one another and there is a considerable degree of reciprocal influence and feedback between them. Furthermore, their relative weights differ substantially and change over time. The influence of the military was much greater in the period before 1974 than it is now. The media exercise more influence in an open society.

For purely economic reasons, the Greek Government can also be subjected to pressure from other governments that wish to see lucrative defence contracts awarded to their national defence industries. Clearly, this tug-of-war has intensified given the shrinking of the international arms market caused by the defence budget cuts in many countries in recent years. Competition among producers has intensified and manufacturers are prone to use all means at their disposal, from large offsets and/or co-production agreements to gentle arm-twisting by their respective governments, which often act as brokers for their national industries.[39] Thus, statements and/or visits by ministers of defence, high-ranking diplomats and other officials of the countries interested are fairly common when a KYSEA decision on a major procurement programme is due.

To this may be added the more covert and unethical means of persuasion that arms producers, both domestic and foreign through local representatives, can use in order to tilt the balance in their favour.[40] This of course raises serious questions about the accountability and transparency of the arms procurement process. As Dokos and Tsakonas observe, such questions are becoming an issue.[41] For example, the decision in 1985 to procure two different types of third-generation combat aircraft instead of one raised many questions and there were a number of accusations of 'foul play' and bribery. There is intense competition between firms for defence contracts. This can take several forms, ranging from price competition to attractive offset programmes and co-production agreements, but can also be more covert—the use of connections and acquaintances in the various MOD departments, for instance (retired senior

[38] This section is based on the research contribution of Dokos and Tsakonas (note 6).

[39] 'It is the governments that exercise pressures, express their wishes, intervene and appeal to the government that wishes to procure weapons systems.' Interview with Minister of National Defence Tsochatzopoulos (note 33).

[40] The Litton case is one example. The Public Prosecutor was called in following reports in the *New York Times* that, in connection with procurement of electronic protection equipment for the Greek F-16s, the US company Litton in 1993 paid bribes of $12 million to tilt the decision in favour of its systems. *Elefterotipia,* 22 June and 3 July 1999.

[41] Dokos and Tsakonas (note 6). Fafoutis, K., *Kathimerini*, 25 July 1999, p. 9 gives an account of the 'war' between weapon manufacturers over the lucrative Greek contracts for new fighters.

officers are often instrumental) to influence decisions by officials, often using bribes or other forms of pressure.[42] However, the importance of pressure from the industry should not be overstated, especially in the case of the domestic arms producers, since the large defence industries are state-owned.

Implementation and procurement procedure

Once the decision is taken on procurement of a weapon system, implementation starts with an international tender. The offers of potential suppliers are submitted to the ministry in sealed envelopes. They include technical specifications, operational capabilities and characteristics, costs, financing, subsystems included, delivery times, offsets and co-production deals.

Following this, specialist committees made up of experts, both military and civilian, with diverse backgrounds and expertise from departments within the MOD evaluate the various offers.[43] Among the aspects evaluated are: technical specifications and operational capabilities as set out in the original call for tender; costs and terms of financing; offsets; the addition of value through co-production with foreign companies; levels of technology to be transferred or made available to domestic producers; supply of spare parts; and the possibility of upgrading. When possible or desired, testing on the ground under simulated conditions may also take place. This gives staff officers and the evaluating committees the opportunity to view performance in action and test the operational capabilities and characteristics of the candidate weapons, and thus compare their performances before final reports are compiled.[44] Once this stage is completed the committees through their respective general staffs (army, navy and air force) and the HNDGS submit their recommendations to the minister, who in turn takes the shortlist of candidate systems to KYSEA for final decision. The reports are not published or made available outside the MOD, but the final ranking may on occasion be reported in the press.

The GDA is then responsible for the execution of acquisition programmes for the signing of the relevant contracts.

[42] Another channel of influence exploited by companies is to offer executive positions to retired senior officers who can influence decisions through their contacts in the MOD.

[43] The evaluating committees are made up of serving officers with different backgrounds and technical expertise. They assess the technical and operational characteristics of the candidate weapons. A report is then drafted in which the pros and cons of each weapon system are set out and on this basis the various systems are ranked.

[44] The ground testing by the army of the contenders for the contract for the new main battle tank recently received particular publicity in the media. This may be viewed not only as a public-relations exercise but also as an attempt by the MOD to emphasize the impartiality of the assessment process and the rigour with which weapons are tested before selection in order to maximize value for taxpayers' money.

V. The defence budget process

The defence budget is part of the annual government budget submitted to Parliament. It is also the only part of the budget that is approved by all parties (with the exception of the left) even if the rest of the budget is rejected.[45]

The budget process is an integral part of defence planning. Until recently, the three general staffs, following a meeting of the Joint Council of the Chiefs of the General Staff, submitted their annual budgets directly to the Finance Ministry to be incorporated in the government budget. Since the Defence Minister's Staff was established in 1996, the MOD budget has been drafted there. The Defence Minister's Staff conveys to the HNDGS and the three General Staffs directives concerning the size and aims of the budget, and they proceed in turn with drafting their preliminary budgets. These are brought together by the Defence Minister's Staff and submitted to the Defence Council for approval. If it is accepted, the MOD budget is then passed to the Ministry of the National Economy and the Ministry of Finance, to be incorporated into the overall budget and submitted for approval to Parliament. If rejected by the Defence Council, it is returned to the Defence Minister's Staff.

The MOD budget can be presented in broad terms or divided into three main categories of expenditure—salaries, operating costs and development expenditure. Funds for procurement come into the third category. In 1998, operating costs (including personnel) accounted for 72.2 per cent of military expenditure, procurement for 24.4 per cent, construction for 3 per cent and R&D for 0.3 per cent. (The figures for 1997 were 75.8 per cent, 21.3 per cent, 2.6 per cent and 0.19 per cent, respectively.)[46]

The HNDGS and General Staffs implement the budget while the Defence Minister's Staff supervises implementation. Public accounting procedures require each of the General Staffs to have accounting offices to supervise and audit expenses. The Defence Minister's Staff is also responsible for the budgeting for and drafting of the five-year procurement plan.

Offset policies and priorities

The Offset Benefits Directorate in the GDA is responsible for the negotiation and implementation of offsets offered in the major defence contracts awarded to foreign companies. A principal aim of offsets is that each major defence contract should achieve the participation of local producers in the execution of each pro-

[45] The Communist Party's view is that procurement decisions reflect not the actual defence needs of the country but rather NATO's requirements. It also argues for a fundamental diversification of the sources of supply in order to reduce dependence on the West and on the USA in particular. The 2 major parties, the social democratic PASOK and the conservative New Democracy, when in opposition reject the annual budgets submitted but always vote in favour of the defence budget. Any criticisms raised usually concern the officer corps' pay scale and delays in the execution of armaments programmes.

[46] Instrument for standardized international reporting of military expenditures, for 1997, UN document A/53/218, 4 Aug. 1998, and 1998, UN document A/54/298, 17 May 1999.

gramme. Offset policy is aimed at co-production of defence materials which could be produced by the Greek defence industry without requiring heavy expenditure for expansion or changes in infrastructure.

The beginnings of the Greek offset policy can be traced to the directives issued in 1985 in connection with the procurement of 40 Mirage 2000 combat aircraft from France and 40 F-16Cs from the USA. According to these directives, if the procurement value was more than 250 million drachmas (about $2 million at 1985 exchange rates), the foreign firm must agree to offsets which were divided into three categories. Category I included work to be undertaken by foreign firms in Greece or for export and use in similar armament systems. Category II included other products of the Greek defence industry which the foreign firms agreed to purchase. Category III included products for exports from Greek agriculture and industry and promotion of foreign tourism.[47] Categories I and II were weighted with a base factor of 2 or 3, and Category III with a base factor of 18. This means that the amount spent by a foreign firm is divided by the corresponding base factor to count towards the firm's offset obligation.[48]

To implement the offsets policy, offices were set up at the ministries of defence, commerce and industry, energy and technology. The Ministry of the National Economy implements Category III agreements.

Although the indications are that offsets may contribute significantly to the development of the Greek defence industry through co-production programmes and exports, there are significant problems and delays. For instance: (*a*) smaller private corporations are unable to take advantage of offsets as they have had to compete with large public corporations; (*b*) there is a lack of coordination between the offices responsible for implementing offsets and the interested manufacturing entities; (*c*) penalty clauses have not been included in the offset agreements to provide for obligations not being fulfilled; (*d*) French companies enforced lower prices for the parts of the Mirage 2000 made in Greece and ordered smaller numbers; and (*e*) the lack of technological infrastructure, specialized personnel, quality control systems and correct programming has impaired the successful absorption of technology under offset agreements by small and medium-sized companies.[49]

[47] Antonakis, N., 'Offset benefits in Greek defence procurement policy: developments and some empirical evidence', ed. S. Martin, *The Economics of Offsets: Defence Procurement and Countertrade* (Harwood: Amsterdam 1996), p. 168.

[48] Offset agreements made under these directives include: (*a*) the agreement for procurement of the Mirages, which required the French suppliers to provide within 15 years offsets worth up to 60% of the purchase price of the aircraft; (*b*) the agreement for purchase of the Meko 200 frigate from Germany in 1988 which obliged the seller to provide offsets up to 45% of the contract value in Categories I and II, and up to 55% of contract value for Category III; and (*c*) the contract for upgrading of the Harpoon guided missiles, which included offsets worth 70% more than contract value, mainly in terms of technology transfer. Antonakis (note 48), pp. 169–72.

[49] Antonakis (note 47), pp. 173, 174.

The audit process

The State Audit Council (Elentiko Sinedrio) oversees and audits government spending and the execution of the budget. It produces an annual report which is submitted to Parliament. Public accounting procedures require that payment orders issued by each of the three armed forces staffs (the HAGS, HNGS and HAFGS) are submitted to a reporter of the State Audit Council, which in theory has the authority to reject them if they present problems.

This procedure is not applicable to a number of types of expenditure that relate directly to the defence capability of the country or to defence procurement. However, procurement agreements signed by the MOD with suppliers are checked by the State Audit Council. There is also a process of internal auditing by the Army Inspector General. This is an entirely internal process and little information about it ever becomes public. The competence and effectiveness of such internal auditing can be questioned; nevertheless it has a deterrent effect and can safeguard against misuse of resources and malpractice.

VI. The defence industrial and technology base

Greece is a net importer of arms but since the mid-1970s has also attempted to partially replace imports with domestically produced arms and equipment. In the past two decades arms imports have on average accounted for about 4.2 per cent of total imports, reaching an all-time high of 12.4 per cent in 1989.[50] Most indigenous arms production in Greece started as joint ventures with foreign companies such as Lockheed, Westinghouse, Steyr, and Heckler & Koch, and/or licensed production from imported systems and subsystems. All companies that were set up as joint ventures in the 1970s later came under state control through nationalization programmes. With marginal indigenous technological capabilities, the Greek defence industry still relies heavily on imported technology and know-how.

The Greek defence industry consists mainly of five state-owned companies which have played a prime role in the effort for import substitution, plus a number of small and medium-sized private enterprises engaged in the production of components. The state-owned companies are: the Greek Powder and Cartridge Company (PYRKAL, founded in 1874 and state-controlled since 1982); Hellenic Shipyards (founded in 1957); Hellenic Vehicle Industry (ELBO, 1972); Hellenic Aerospace Industry (EAB, 1975); and Hellenic Arms Industry (EBO, 1977). In addition there are private-sector producers and a number of army factories under the various corps of the armed forces, which primarily maintain, repair and modernize army hardware such as tanks, as the recent upgrading of M-48 tanks to M-48A5 level indicates.

[50] US Arms Control and Disarmament Agency, *World Military Expenditures and Arms Transfers 1990* (US Government Printing Office: Washington, DC, 1990), p. 106.

SEKPY, the Hellenic Manufacturers of Defence Materials Association, was founded in 1982, initially with 20 member companies employing about 2000 people. There are currently about 110 member companies, together employing 19 500 people, about 7000 of them in EBO, PYRKAL, ELBO and EAB.

The GDA oversees and coordinates the domestic defence industry. It aims to increase the participation of the domestic industry in the procurement of arms and equipment. Since it was established in 1995 it has implemented a programme of structural changes in the state-owned defence industries to reduce losses and accumulated debts and make them profitable and competitive.[51] It has signed a number of defence *matériel* orders with improved offsets in terms of local manufacturing of components and technology transfer,[52] and international defence production and technology cooperation agreements. It has also initiated a process of improving the procurement system for the acquisition of secondary hardware (spare parts, auxiliary equipment and so on) and other *matériel* in order to nationalize orders for such *matériel* and equipment, aiming to maximize domestic value added in the production of such secondary inputs.

Figure 3.5 shows the structure of the GDA. A Defence Industry Directorate (DID) which was set up in 1977 to oversee and coordinate the state-owned arms industries (EAB, ELBO, EBO, PYRKAL and so on) has come under the GDA since the latter was established in 1995. It is responsible for the development of the domestic defence industry, for participation in co-production consortia and for the continuous monitoring of the local defence industry's manufacturing capabilities. The Armaments Programmes Directorate (APD) is responsible for the execution of the programmes of military equipment acquisition as these are decided by KYSEA. The Technological R&D Directorate is in charge of military R&D policy, the supervision and coordination of the research centres of the ministry and the branches of the armed forces, and local and international R&D contracts. In particular, as Narlis writes, it includes the Department of Research Centres and the Department of Scientific and Technological Cooperation.[53] The former is responsible for the coordination and supervision of the three research centres belonging to the MOD—Kentro Technologikon Erevnon Stratou (KETES), Kentro Technologias Aeroporias (KETA) and Kentro Technologikon Erevnon Naftikou (KETEN). The latter is also responsible for Greece's participation in international defence organizations such as NATO/RTO (Research and Technology Organization) and Panel II of the Western European Armaments Group (WEAG) of the WEU.

Figure 3.6 is a flow-chart of the methodology used to examine the different procurement options in order to maximize the participation of domestic defence producers and achieve the maximum possible technology transfer.[54]

[51] 'White Paper for the armed forces 1996–97' (note 6).

[52] Two examples are the co-development and production of the Hermes communication system between Siemens and EAB and the participation of EAB as a partner in the production of the EF-2000 Typhoon combat aircraft which is to be procured by the air force.

[53] Narlis, E., 'Arms development and defence R&D growth in the Hellenic Republic', SIPRI Arms Procurement Decision Making Project, Working Paper no. 75 [1998].

[54] Narlis (note 53).

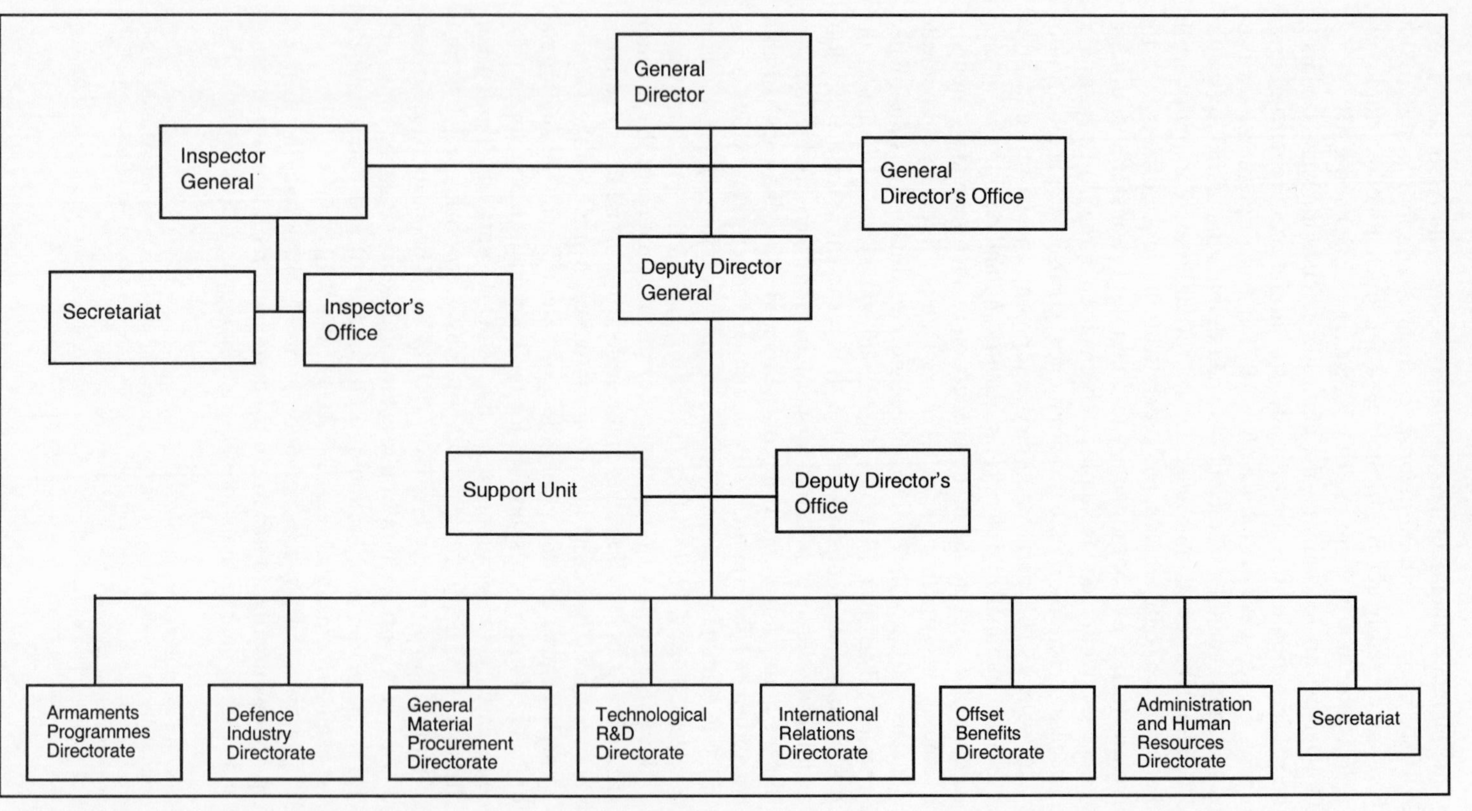

Figure 3.5. The organization of the Greek General Directorate of Armaments

Source: Greek Ministry of National Defence, 'White Paper for the armed forces 1996–97', Dec. 1997, URL <http://www.hri.org/mod/fylladia/bible/e_index.htm> (in English).

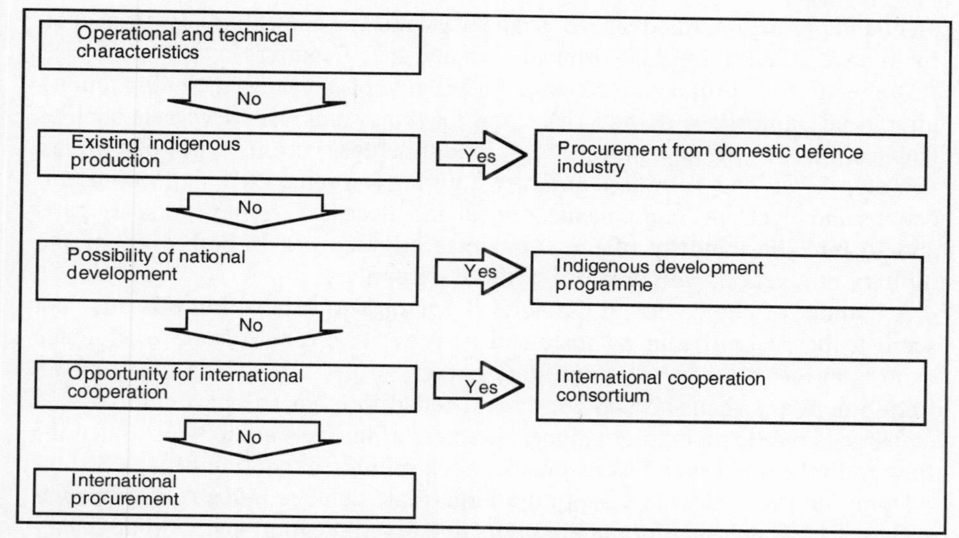

Figure 3.6. Procurement decision making in the Greek Armaments Programmes Directorate

Given the procurement programmes and the required operational and technical characteristics, the options for every procurement proposal are analysed by a team made up of representatives of the APD, the DID, the Technological R&D Directorate and the Offset Benefits Directorate. This team compiles a study in which the advantages and possible disadvantages of the proposed procurement are described. The analysis of the local development capability (i.e., which equipment or which module of a weapon system will be developed locally and which technology will be requested as part of an offset programme) is fully integrated into the decision flow-chart.[55]

Under the EMPAE the target is to achieve about 15 per cent domestic participation in the weapons procured—a very optimistic target considering that currently this figure does not exceed 4–5 per cent.[56] The long-term target is to meet defence procurement needs by 37 per cent local development, 33 per cent from participation in international development/co-production consortia, 20 per cent by co-production and only 10 per cent by direct imports.[57] Clearly, given that Greece's technological capabilities are marginal and its industrial base weak, and considering the technological sophistication of modern weapon systems and the huge R&D costs involved, this represents more a wish-list than a feasible outcome. A typical illustration of Greece's technological weakness is the fact that even advanced upgrading of existing systems in operation with the armed forces, such as CH-47 Chinook helicopters and the F-4E Phantom-2 combat aircraft, is contracted to foreign companies. Past attempts at domestic develop-

[55] Narlis (note 53).
[56] 'White Paper for the armed forces 1996–97' (note 6).
[57] Narlis (note 53).

ment and production of advanced weapon systems have been largely unsuccessful or have fallen short of the original aims and specifications.[58]

Domestic arms production covers a wide range of *matériel,* including ammunition and explosives, assault rifles and machine-guns, naval vessels such as frigates, missile and fast patrol boats, landing ships, various types of military vehicle, APCs and armoured infantry fighting vehicles (AIFVs), communications and electronic components, optical and electrical equipment, spare parts and so on. The majority of the companies are engaged in both civilian and military production, with only a handful of exceptions.

A number of companies in the defence sector also export, but this does not seem to be a significant or sustained activity. A sizeable share of defence exports in the past was 'grey' military exports, especially during the 1980–88 Iraq–Iran War, when the two countries needed to bypass export controls and embargoes by the main international suppliers of military equipment. When the flow from the main suppliers resumed, Greek military exports fell.[59]

Domestic demand cannot support a large-scale defence industry. Companies such as ELBO are therefore rapidly pursuing diversification into civilian manufacturing, such as the assembly of buses, coaches and fire-fighting vehicles. Participation in international development and co-production projects as well as in indigenous production of parts and sub-assemblies of imported equipment through the various offset programmes negotiated does appear to be a viable solution for the Greek defence sector.

VII. Democratic oversight

The arms procurement decision-making process is not open to outside scrutiny from other bodies or agencies such as Parliament. If there is any parliamentary involvement it takes the form of retrospectively questioning the correctness of specific decisions.

The details of the budget are rarely debated by Parliament, and when arms programmes do attract attention MPs tend to concentrate on raising issues concerning their implementation rather than the processes, costs and finances involved. In general, in the Greek political system parliamentary committees do not possess any real power. The Parliamentary Committee on Foreign Affairs and Defence holds hearings on various defence and foreign policy issues but lacks any real authority, for example, to veto procurement projects, influence armaments programmes or review and monitor the decision-making processes[60]

[58] An example is the case of the Artemis-30 air defence system developed by EBO, originally intended to be a technologically advanced short-range air defence system against low-flying targets. It was mainly developed through the domestic integration of subsystems from various foreign systems already in operation. Development has so far cost an estimated 110 billion drachmas and the final product falls far short of specification. The major problems were encountered at the integration stage of the various subsystems, indicating EBO's lack of technology and know-how.

[59] Based on US Arms Control and Disarmament Agency, *World Military Expenditure and Arms Transfers* (US Government Printing Office: Washington, DC), various issues.

[60] Dokos and Tsakonas (note 6), p. 3.

and its activities are limited to briefings by the foreign and defence ministers. It is not divided into specialized subcommittees that could examine and scrutinize arms procurement programmes and contracts. What power or influence it has springs only from the weight that the publicly expressed opinions of its members may carry. This reflects the fact that in Greek politics the executive branch to a great extent allows the legislative branch a symbolic role only. From the point of view of public accountability, empowerment of this committee to review the arms acquisition process will only come about as a result of a wider improvement and extension of Parliament's involvement in monitoring the public policy and decision making generally.

Nor does Parliament carry out regular review of procurement decisions after the event in terms of the suitability of weapons acquired or of whether the best possible deal was struck and whether the contract was awarded to the supplier that offered the best financial and/or co-production deal. The true ownership costs of the weapon systems over their entire life cycle are rarely if ever given any attention. In any case, for Parliament to address such issues it would require not only the advice of experts but also access to details and financial data on operational and life-cycle costs of the various weapon systems. Such information is not publicly available.[61] There is a clear lack of legislative oversight of the arms procurement process, either *ex post facto* or *ex ante*.

For the military, accountability and transparency are issues of less significance than the weapons' technical characteristics, operational capabilities and delivery times. In any case, for major procurement projects the final decision rests entirely with the politicians—KYSEA—and from the military's perspective its own role is limited to that of a technocrat offering his expert opinion. Accountability is an issue for the politicians.

There is no institutionalized process whereby procurement decisions can be seriously questioned and where the expert opinion of serving officers can be called upon (for instance, by the Parliamentary Committee on Foreign Affairs and Defence). Accusations and/or media reports of malpractice are dealt with through internal official inquiries[62] commissioned by the minister, but findings are rarely published.[63] Solidarity among fellow officers naturally hinders the process of such inquiries. Nevertheless, individual officers do on occasion leak information to the press and/or politicians on specific cases of malpractice, waste, fraud or abuse of power.

There is in fact some accountability and transparency in the arms procurement process where large defence contracts, which can be the source of political

[61] The media on occasions carry reports of how much a flying hour costs for the different combat aircraft operated by the air force, but such sketchy and perhaps unreliable information is no basis for scientific analysis of operational and life-cycle costs.

[62] Cynics would point out that an internal inquiry is often the best way to stall, to obstruct justice and eventually to cover up scandals.

[63] The Litton affair cited above (note 40) and the case of the air force being overcharged by an estimated 150 million drachmas (*c.* $500 000 at 1997 exchange rates) for the supply of ground equipment for the F-16s (*Elefterotipia,* 7 July 1999) are 2 cases in which the judicial system is currently involved. They may signal a change towards more openness and accountability in procurement.

tension and accusations of foul play, are concerned. There is the call for public tenders; technical and financial evaluation committees are made up of military as well as civilian experts; and there are oral briefings and written reports to Parliament and the Parliamentary Committee on Foreign Affairs and Defence.

The absence of an institutionalized process of scrutiny and accountability in arms procurement to a certain extent reflects the dominant view that confidentiality and secrecy are important in matters of national defence and security, but it is also indicative of the workings of the Greek political system, which is centralized in the Office of the Prime Minister with its extensive powers.

Apart from institutional influence in the weapons acquisition process, the influence of public opinion, the media, various think-tanks and the defence manufacturers must be allowed for.

Public opinion broadly supports the strengthening of Greece's military capabilities. Even so, questions on the choice of weapons, priorities and resource allocation within the three branches of the armed forces as well as on issues of professional assessment and integrity in the decision-making process have on occasion been raised.

The media occasionally play a role in promoting transparency and accountability by revealing possible wrongdoings and by criticizing specific procurement decisions.[64] The influence of the press should not be overemphasized: it is perhaps mainly a deterrent against malpractice, and criticisms in the media are not always based on a sound knowledge of the capabilities and technical characteristics of specific weapon systems. However, such analyses can be found in specialized defence journals as well as the general press. Newspapers and magazines frequently carry articles from academics and specialists from the two main think-tanks and research institutes in Greece—the Institouto Diethon Scheseon (IDIS, the Institute of International Relations) and the Elleniko Idruma Europaikes kai Exoterikes Politikes (ELIAMEP, the Hellenic Foundation for European and Foreign Policy). Furthermore, the influence of such institutions and specialists is not limited to publicly stated views and opinions, since many of them act as advisers to ministers and government agencies, and studies are also often commissioned by the MOD and the Ministry of Foreign Affairs on issues affecting defence and foreign policy.

Barriers to and opportunities for transparency and accountability

In recent years there have been efforts to improve the accountability and to a lesser extent the transparency of the procurement process. These include the publication of two defence White Papers (in 1995 and 1997) which aim to provide more information to decision makers such as MPs, the media and the public on current defence issues, defence policy, the principles of military strategy, force structure, arms acquisition programmes and other aspects of the

[64] The procurement of the 3rd-generation fighter aircraft is a case in point. See also, e.g., *Nafteboriki*, 30 Dec. 1997; *Kathimerini*, 10 Nov. 1996 and 25 July 1999; and *Elefterotipia,* 22 June and 7 July 1999.

activities of the MOD and the armed forces. Similarly, an Annual Defence Report to Parliament (prepared by the MOD since 1996) aims to improve the channels of communication between the MOD and Parliament and to allow the possibility of greater parliamentary scrutiny of the activities of the MOD.

Undoubtedly further steps towards greater accountability and perhaps transparency are possible within the constraints of the Greek political system. The role of the Parliamentary Committee on Foreign Affairs and Defence in the procurement process could be upgraded, giving the committee greater involvement in monitoring and reviewing the arms procurement process.[65] This could, to start with, take the form of advisory reports and recommendations and even eventually the power to veto specific decisions. This presupposes that expertise is available to the legislative branch that will improve its ability to scrutinize arms procurement decisions.

Given the almost universal agreement on broad issues of threat assessment and defence priorities, a case can be made for the appointment of a Deputy Defence Minister, a fixed-term permanent under-secretary, to be responsible for the long-term armaments requirements planning and execution programmes and appointed by Parliament with reinforced majority (for example, with two-thirds or more of the votes) for a term longer than the maximum four years that a government can stay in office. The time horizon of such requirements and the execution times involved with major weapon acquisition programmes often outlast both defence ministers and governments. The person selected could also be accountable to Parliament through regular reports and closed hearings of the Parliamentary Committee on Foreign Affairs and Defence. The creation of such a post would have a number of advantages and disadvantages. One clear advantage would be continuity in weapon acquisition planning and procurement policies, resulting in cost savings, greater accountability and stronger defence. However, further examination of this possibility is beyond the scope of this chapter.

Clearly, greater transparency in the various stages of the arms procurement process and accountability for the resulting choices and deals struck are in the public interest and can not only safeguard against the possibility of malpractice, waste and fraud but also lead to a better allocation and use of the scarce resources that the country invests in its national defence. However, as has been seen, the elected representatives of the public presently play an extremely limited role in oversight and monitoring of defence decision making and their influence on and ability to scrutinize and review arms procurement are virtually non-existent. Public opinion, stirred up by reports of wrongdoing and foul play in the media, is at present the only potential influence that can be used to promote the accountability of the military for their decisions.

[65] Dokos and Tsakonas (note 6), p. 3.

VIII. Conclusions

This chapter has identified the two distinct periods which reflect the major changes that have occurred in threat perceptions, security needs and defence priorities in Greece. Greece's allocation of a substantial part of its national income to defence, which is the result of the strategic instability of the region and in particular the continuing tensions with Turkey, conflicts with its socio-economic priorities, in particular its efforts to join the EMU. To meet the main external threat Greece relies on a mix of internal and external balancing.

Since the possibility of military confrontation cannot be ruled out, the country's current arms acquisition emphasizes the accomplishment of a defence that is strong in terms of quality, using lucrative defence contracts to gain political and diplomatic leverage from the supplying countries. Issues of public accountability, legislative monitoring and oversight in this process are of secondary importance since emphasis is given to the technical characteristics and operational capabilities of the weapons procured for building strong deterrence. Indeed, the current system is a fairly closed one, which does not allow a great degree of parliamentary scrutiny.

Since recent years have seen a significant and adverse change in the balance of power between Greece and its main antagonist in the region, the short-term emphasis will continue to be on the speedy acquisition of weapons that will help in the maintenance of a minimum balance of power. Within this context efforts are being made to achieve improved offsets and co-production deals in order to minimize the negative economic effects of defence spending. At the same time Greece tries to maintain a balance in the sources of weapons in order to reduce dependence and to gain diplomatic benefits from more than one major supplier.

The resources allocated to defence are undoubtedly a heavy burden for the weak economy of the country. Clearly, a greater degree of accountability, transparency, scrutiny and parliamentary oversight of the weapons procurement process is in the public interest and can lead to better defence at a lower cost. This point is further strengthened by the fact that programmes to acquire modern weapon systems, from the initial stages of identification of need and planning by the staff officers to actual procurement and acquisition, usually outlast any government and/or defence minister. Greater involvement of other institutions such as Parliament would ensure greater bipartisan agreement and thus long-term planning and consistency in the modernization programme of the armed forces.

4. Malaysia

*Dagmar Hellmann-Rajanayagam**

I. Introduction

Malaysia has become one of the major political players in the South-East Asian region with increasing economic weight. Even after the economic crisis of 1997–98, despite defence budgets having been slashed, the country is still determined to continue to modernize and upgrade its armed forces.

Malaysia grappled with the communist insurgency between 1948 and 1962. It is a democracy with a strong government, marked by ethnic imbalances and affirmative policies, strict controls on public debate and a nascent civil society. Arms procurement is dominated by the military. Public apathy and indifference towards defence matters have been a noticeable feature of the society. Public opinion has disregarded the fact that arms procurement decision making is an element of public policy making as a whole, not only restricted to decisions relating to military security. An examination of the country's defence policy-making processes is overdue.

This chapter inquires into the role, methods and processes of arms procurement decision making as an element of Malaysian security policy and the public policy-making process. It emphasizes the need to focus on questions of public accountability rather than transparency, as transparency is not a neutral value: in many countries it is perceived as making a country more vulnerable.[1] It is up

[1] Ball, D., 'Arms and affluence: military acquisitions in the Asia–Pacific region', eds M. Brown *et al.*, *East Asian Security* (MIT Press: Cambridge, Mass., 1996), p. 106.

* The author gratefully acknowledges the help of a number of people in putting this study together. Eleven working papers were prepared by Malaysian researchers in connection with the workshop on arms procurement decision making organized jointly by the Universiti Kebangsaan Malaysia (UKM) and SIPRI at the UKM on 18 Aug. 1997. The authors of working papers also followed up their papers in discussions. People in government service, private industry, the military and academia, most of whom wish to remain anonymous, gave generously of their time to help clear up doubts, uncertainties and confusion, in particular, Puan Siti Azizah Abod, formerly Under-Secretary for Policy, Ministry of Defence, Malaysia; Mr Rajayah, also from the Ministry of Defence; Dato' Brig.-Gen. Richard Robless, who enlightened the author on various points connected with budget planning and decision making; Puan Faridah Jalil from the Faculty of Law, UKM; and D. E. Dasberg, German Naval Group, who clarified the working of offsets and the tender and negotiation process from the viewpoint of private industry. Prof. Zakaria Haji Ahmad, Prof. Baladas Ghoshal and anonymous reviewers from SIPRI commented on earlier drafts. All their comments and advice helped to make this a better chapter. Puan Faezah saw to the administrative glitches and drew some of the figures. Not least, thanks are due to UKM for funding the UKM–SIPRI workshop and for facilitating the research.

The working papers are not published but are deposited in the SIPRI Library. Abstracts appear in annexe B in this volume.

to the informed public to push for greater accountability in order to make the processes less wasteful and more focused on public priorities.

A study of Malaysian defence policy making and arms procurement decision-making process should also identify the problems associated with the tension between the public's right to know and military's 'need-to-know' policies. Only if a balance could be achieved between public accountability, confidentiality and efficiency can restraints on arms acquisitions be introduced voluntarily (in contrast to the international arms control initiatives which have not apparently been able to control the conditions that fuel the arms race).

The chapter examines the arms acquisition processes in Malaysia from the perspectives of the different agencies and interests involved to understand the rationale behind the decisions. This is done in the context of the kind of democratic oversight of arms procurement decision making that would be desirable, involving the informed public and its elected representatives. Conversely, the means available to and used by the government to restrict the flow of information in the interests of confidentiality and security are investigated. The study describes the current situation and not the one that should be. However, where there are obvious inconsistencies or even redundancies in the system, attempts are made to suggest possible remedies.

Section II of this chapter describes the actors involved in defence decision making in Malaysia and section III the national defence policy generally. Section IV examines the links between defence policy making and arms procurement planning, section V budgeting, financial planning and auditing, and section VI the defence industrial aspects of arms acquisition. Section VII examines accountability in arms procurement decision making and section IX summarizes the deficiencies in the process and sets out conclusions.

II. The structure of the defence organizations[2]

In order to determine in what ways the official channels are used, modified or perhaps simply overridden it is important to understand the organizational and management structure of national defence decision-making processes within the Malaysian security system.

1. The highest body to discuss questions of internal and external security is the National Security Council (NSC), chaired by the Prime Minister. Other permanent members are the Deputy Prime Minister, the ministers of information, defence and home affairs, the Chief Secretary to the Cabinet, the Chief of the Defence Forces (CODF), the Inspector General of Police and the Director General of the Department of National Security. In attendance are the Attorney General and the secretaries general of defence, home affairs and foreign affairs. Representatives of other agencies may participate depending on the issues being

[2] The following description is based on Siti Azizah Abod, 'Decision making process of arms acquisition in the Ministry of Defence, Malaysia', SIPRI Arms Procurement Decision Making Project, Working Paper no. 85 (revised version, Sep. 1998).

discussed.[3] The NSC normally meets once a year to discuss national security priorities. In the Prime Minister's Department there is a National Security Division which functions as the secretariat to the NSC.[4] Its decisions are confidential. The NSC gives guidelines and instructions, on which the ministries develop position papers and which they have to implement.

2. The deployment of the armed forces is under the control of the government, with execution authorized by the Cabinet through the Ministry of Defence (MOD). (The King is technically the Supreme Commander of the Malaysian Armed Forces—the MAF—but his role is only ceremonial.) In the present Malaysian context, the Prime Minister is most influential in the overall decision-making process and has a major influence on the direction of the military's expenditure and its expansion programmes.[5]

3. At the functional apex of the defence establishment is the MOD, headed by the Defence Minister who provides the political leadership and is responsible to the Cabinet. He is a civilian and is assisted by the Secretary General in charge of general policy, finance, external defence relations, human resources, infrastructure, defence science and industry. The CODF advises the minister on operational matters and the implementation of defence policy.

The MOD determines strategic interests within the parameters of national interest in general and foreign policy as formulated by the Ministry of Foreign Affairs. The political division in the MOD takes the lead in formulating the draft paper on strategic policy, which is the basis of the national defence policy,[6] assisted by other divisions in the ministry and the armed forces. The paper is endorsed by the ministry and presented to the Cabinet and other ministries involved through the NSC.

Figure 4.1 shows the structure of the MOD. It consists officially of two halves—one civilian and the other military—responsible for planning and for implementation and operational matters, respectively. On the basis of the defined defence policy, the military's operational arm examines the capabilities required to implement it and the types and scale of weapons and equipment, human resources, and infrastructure required. The civilian staff in the MOD maintains contacts with the Ministry of Foreign Affairs and other departments at ministerial level.

4. Various committees have specific areas of responsibility. Important among them is the Armed Forces Council (AFC), established by Article 137 of the constitution and chaired by the Minister of Defence. It delegates its powers to its members and thus to the Defence Minister. It consists of the Secretary General of the MOD, who acts as the secretary of the AFC, a representative

[3] Personal communication from Siti Azizah Abod, Under-Secretary, Malaysian MOD.

[4] Balakrishnan, K. S., 'Examine the institutionalisation of decision-making processes based on the principles of good governance: problems, apprehensions and barriers in building public awareness, public interest, transparency and accountability', SIPRI Arms Procurement Decision Making Project, Working Paper no. 78 (1998), p. 4.

[5] Balakrishnan (note 4), p. 3.

[6] This paper is produced annually but is classified.

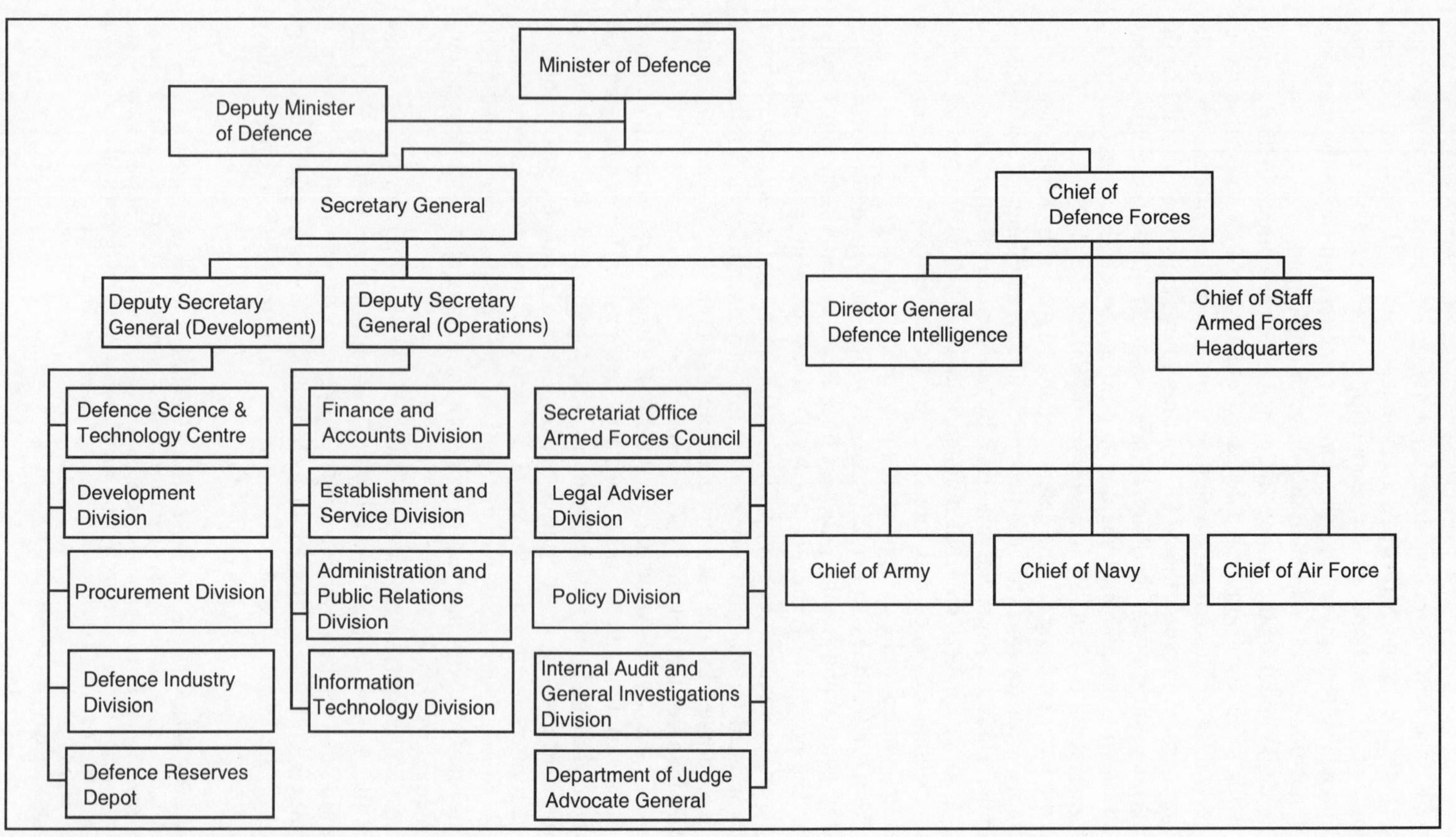

Figure 4.1. The Malaysian Ministry of Defence

Source: Malaysian Ministry of Defence, URL <http://www.mod.gov.my/btmk/mindef/english/eng_str_org_carta_link.htm>.

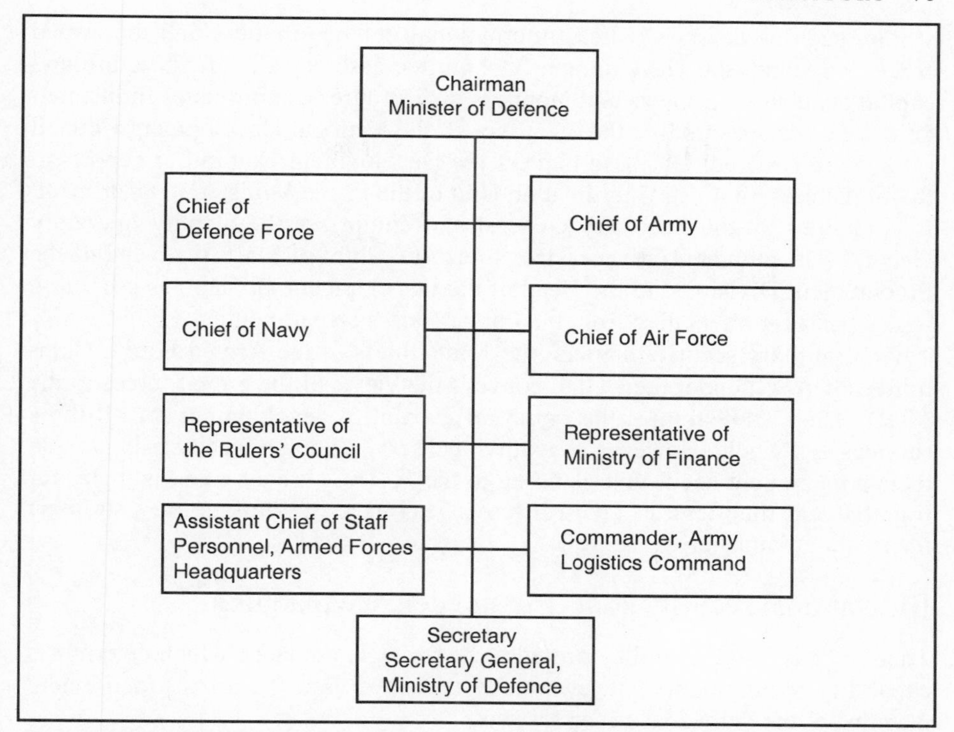

Figure 4.2. The organization of the Malaysian Armed Forces Council
Source: Asian Defence Yearbook 1998/99 (Syed Hussein Publications: Kuala Lumpur, 1999), p. 102.

from the Rulers' Council,[7] the CODF, all service chiefs, the Armed Forces Logistics Commander and the Chief of Staff Personnel. Other members are the Deputy Defence Minister, the Deputy Secretaries General of Development and Operations, and the Chief of Staff of the Armed Forces.[8] The AFC discusses and advises on matters of service, discipline and administration. It has corporate responsibility for the management of the armed forces, although operational responsibilities, apart from logistics, equipment and so on, are outside its purview.

5. There are two other committees besides the AFC that to some extent decide policy relating to arms procurement. First, the Defence Minister's Board is the highest policy-making body in the MOD, comprising the Minister as chairman, the Deputy Minister, the Secretary General, the CODF, the service chiefs and heads of the divisions. It has two functions: (*a*) to discuss policy on security issues; and (*b*) to decide on policy issues affecting the ministry as a

[7] Composed of the rulers of 9 of the 13 states in the federation.
[8] Sharifah Munirah Alatas, 'Government–military relations and the role of civil society in arms procurement decision-making processes in Malaysia', SIPRI Arms Procurement Decision Making Project, Working Paper (draft, 1998), p. 6.

whole, such as defence policy, international defence matters and the armed forces.[9] Second, the Development Committee assesses all infrastructure and capital acquisition projects and monitors their progress, coordinates implementation and makes sure that the directives of the National Development Council (NDC) are realized. On these matters the Development Committee reports to the Implementation and Coordination Unit of the Prime Minister's Department. It is chaired by the Secretary General and comprises the Deputy Secretary General (Development), the service chiefs, the Chief of Staff, the Head of the Procurement Division and the Head of the Development Division as secretary. It also has a representative from the Public Works Department.

6. On military security matters, the Joint Chiefs of the Armed Forces Committee (JCAFC) under the CODF conveys the views of the armed forces to the MOD. The CODF frames the policy of capability development for all three services as a whole. However, the individual services do not necessarily pass on their procurement needs and planning to the JCAFC. Procurement is their prerogative and they tend to guard it rather jealously and only convey strategic ideas and opinions to the JCAFC.

III. National security concerns and defence planning

Theoretically, defence policy provides the strategic guidance which determines capability requirements and eventually translates into the arms procurement planning of the individual armed services.

Malaysian defence policy is politically driven and is so acknowledged by the armed forces. There is no White Paper as such outlining the context of the country's security concerns. A document detailing the defence policy was laid before the NSC in 1987 and endorsed by the Cabinet in 1990[10] and in late 1997 a publication of the MOD described its organizational structure and strategic perspectives—'protection of its national strategic interests and the preservation of national security'.[11] It emphasized that Malaysia's security cannot be seen in isolation from that of other countries of the Association of South-East Asian Nations (ASEAN).[12] This was an attempt at greater transparency, but is not considered as a White Paper by the MOD or by outside defence experts since it does not follow the guidelines generally expected of White Papers in the Asia–Pacific region,[13] nor was it tabled and debated in Parliament as a White Paper would be.

The approach to defence policy is pragmatic rather than idealistic and the MAF have translated it operationally into concepts of defensive defence, com-

[9] Siti Azizah Abod (note 2), p. 1.

[10] Personal communication from Lt-Col Abdul Rahman Adam (ret.).

[11] Malaysian Ministry of Defence, *Malaysian Defence: Towards Defence Self-Reliance* (Ministry of Defence: Kuala Lumpur, 1997), p. 21.

[12] *Malaysian Defence: Towards Defence Self-Reliance* (note 11), p. 22. The members of ASEAN are Brunei, Cambodia, Indonesia, Laos, Malaysia, Myanmar, the Philippines, Singapore, Thailand and Viet Nam.

[13] Kang Choi, 'The approach to a common form of defence White Paper', *Korean Journal of Defence Analysis*, vol. 18, no. 1 (1996), pp. 205–21.

prising; (*a*) deterrence; (*b*) forward defence; and (*c*) total defence. Deterrence in this context means deterring potential enemies from the use of force. This necessitates credible armed forces with conventional war-fighting capacity, with regard both to military hardware and to manpower. Forward defence in the Malaysian context is understood as ensuring that a war is not fought on Malaysian soil. The armed forces are stationed and kept in a state of readiness towards this aim. It is claimed, however, that the capability for this does not exist in Malaysia, as the force structure was until recently not balanced and the air assets needed to make forward defence possible were lacking.[14] Total defence means the full support of the nation for the armed forces and integrated efforts by all actors, both within and outside the government, to defend the nation.

The MAF are apolitical—although top military leaders are often picked for their empathy with the ruling elite as much as for their leadership qualities[15]—and would not protest even if they disagreed with political measures or policies regarding their role. They acknowledge that defence policy making has to be guided by the political leadership, both in the field of foreign and security policy and in that of arms acquisition.[16]

The defence policy guidelines are threefold: (*a*) self-reliance with regard both to the internal security of the country and to external security in its immediate surroundings; (*b*) regional cooperation, which is very significant for geographical reasons and entails cooperation within ASEAN; and (*c*) external assistance to complement self-reliance and regional cooperation. The Five-Power Defence Arrangement (FPDA) of 1971 with Australia, New Zealand, Singapore and the UK is Malaysia's only multilateral defence arrangement with other countries.[17]

The apparent absence of immediate threats to Malaysian security has enabled arms acquisition to be spread over a longer period, incorporating the concept of self-sufficiency under Plan 2000 for the armed forces.[18] Attempts to downsize the armed forces led to demands for modern equipment with greater firepower and mobility. The army was restructured by reorganizing infantry battalions into specialized units and mechanized, support and parachute units. In 1995 the Army Air Corps was established and a Rapid Deployment Force was set up in 1996.[19] The priorities for equipment therefore changed: the emphasis on light equipment suitable for jungle warfare was replaced by the capability to protect the country from external threats.

[14] Abdul Rahman Adam (Lt-Col, ret.), 'Dynamics of force planning: the Malaysian experience', SIPRI Arms Procurement Decision Making Project, Working Paper no. 77 (revised version, Oct. 1997), pp. 4–5.

[15] Mak, J. N., 'Security perceptions, transparency and confidence-building: an analysis of the Malaysian arms acquisition process', SIPRI Arms Procurement Decision Making Project, Working Paper no. 82 (revised version, Sep. 1997), p. 5.

[16] E.g., the decision to take part in UN peacekeeping exercises was a political decision, taken although the MAF had absolutely no training for mountain or desert operations and were therefore unsure about the mission. Abdul Rahman Adam (note 14), p. 15.

[17] Siti Azizah Abod (note 2), p. 2.

[18] Abdul Rahman Adam (note 14), p. 9. Plan 2000 is part of the 6th (1991–95) and 7th (1996–2000) national plans.

[19] Abdul Rahman Adam (note 14), pp. 9–10.

Emphasis was placed on privatization of some maintenance functions, reductions in manpower and acquisition of new technologies with special attention to command, control and communications (C^3I) capabilities. Greater emphasis was placed on joint exercises and an integrated Joint Operational HQ under the command of the CODF was created.[20] The navy's Year 2010 Plan is based on capability building instead of threat perceptions, and thus on the acquisition of air and anti-submarine warfare (ASW) armament.[21] The air force has acquired state-of-the-art aircraft, especially for maritime operations. It has moved away from air support in counter-insurgency warfare and is turning into an air defence-capable force. It has also upgraded its air defence radar system.[22]

The relationship between foreign and defence policies

It is acknowledged that defence policy should be in line with foreign policy, and that its strategic guidance should also determine the capability and eventually arms procurement requirements of the armed services. Foreign policy provides the wider framework within which arms procurement decisions are also made. This perception has been strengthened since 1989.[23] This does not exclude the possibility of foreign and defence policy diverging or the trajectories of the political and the military establishments differing.[24]

Foreign policy is seen as the prerogative of the Prime Minister and the United Malay National Organization (UMNO), the strongest party in the ruling alliance. This means that foreign policy initiatives often originate from the Prime Minister's Department rather than the Foreign Ministry.[25] Chandran Jeshurun outlines measures taken by the government to keep the military informed of foreign policy initiatives by involving senior military officials as part of the Prime Minister's delegations on official visits,[26] although not the formulation of decisions. The MOD follows this lead, since it sees no external threat.[27]

This latter view is not uncontested. Even though since the end of the cold war the focus of defence policy has changed from threat perception to protection of strategic interests,[28] it cannot be said that there are no outside threats, given regional rivalries, notably the uncertainty in the Spratly Islands and over

[20] Sengupta, P., 'The MAF and force modernisation challenges in the post-cold war era', *Asian Defence Journal*, no, 4 (1998), pp. 16–17, 21, 26–28.

[21] Dantes, E., 'RMN's force modernisation plans', *Asian Defence Journal*, no. 12 (1997), pp. 14–21.

[22] Abdul Rahman Adam (note 14), p. 11.

[23] Unless otherwise indicated, the discussion in this section follows Chandran Jeshurun, presentation at the UKM–SIPRI workshop on arms procurement decision making, 18 Aug. 1997.

[24] Mak (note 15), p. 1 and *passim*.

[25] Zakaria Haji Ahmad, 'Change and adaptation in foreign policy: the case of Malaysia's Foreign Ministry', ed. B. Hocking, *Foreign Ministries: Change and Adaptation* (Macmillan: London, 1999), p. 7.

[26] According to the MOD, military personnel participate in delegations only when the Prime Minister visits troops on peacekeeping missions, e.g., in Bosnia or Somalia. Personal communication from Siti Azizah Abod. The contradiction has to stand unresolved.

[27] Zakaria Haji Ahmad (note 25).

[28] Siti Azizah Abod (note 2), p. 3.

China's role in general.[29] A change has, however occurred in the MAF's perception of their role from the time of counter-insurgency warfare, when they saw themselves as mainly aiding the police forces to ensure continued Malay political dominance in general, and UMNO pre-eminence in particular.[30] The MAF move cautiously in military planning and capacity building.

Attempts at improving communication notwithstanding, the perception gaps between the MOD and the Foreign Ministry have not yet been satisfactorily resolved.[31] According to MOD experts, however, there is consultation between them, the goals and directions of the political and military establishment are identical, and defence policy is subordinate to and guided by foreign policy.[32]

Political and economic considerations, including technology transfer,[33] outweigh purely technical or military considerations where defence procurement is concerned. Procurement takes place within the framework of the foreign and development policies of the state as a whole. The Economic Planning Unit (EPU) and the Treasury (the finance ministry) have the final say in procurement matters—something that is resented by some in the military. Political agencies take decisions on the basis of considerations other than military. These can involve trade-offs, as in the famous 'arms for aid' case with the UK in 1994, which generated intense irritation among the military.[34] The arms procurement requirements of the MAF are subordinated to foreign policy considerations where choice of suppliers is concerned. For instance, when the MiG-29S combat aircraft was purchased from Russia in 1995, although it was considered a good aircraft, there were apprehensions as to its maintainability and sustainability and criticism of the quality of Russian training for an air force used to Western equipment, but the recommendations of the armed forces were overruled in the final stages of decision making and a politically motivated deal with Russia was pushed through instead.[35] The British Aerospace (BAe) Hawk combat aircraft was bought in 1990 although it was seen by the air force as inferior to the Tornado (which the UK had refused to supply with the state-of-the-art electronics) because the deal involved an offset agreement with BAe to buy back local manufacture and establish the necessary training and support infrastructure.[36] This instance also makes it clear that counter-trade and offset programmes are not uniformly seen as beneficial by the armed forces.

[29] This point came up in the UKM–SIPRI workshop on arms procurement decision making, 18 Aug. 1997.

[30] This within the context of the riots of 1969.

[31] Chandran Jeshurun, 'Malaysian defence policy revisited: modernization and rationalization in the post-cold war era', *Journal of Southeast Asian Affairs*, vol. 16 (1994), p. 203.

[32] Personal communications with Siti Azizah Abod.

[33] Offsets and technology transfer are considered further in section VI in this chapter.

[34] Sharifah Munirah Alatas, 'Government–military relations and the role of civil society in arms procurement decision-making processes in Malaysia', SIPRI Arms Procurement Decision Making Project, Working Paper no. 84 (1998), p. 43.

[35] Yap Pak Choy, 'Air power development: the Royal Malaysian Air Force experience', Air Power Studies Centre, Australia, and Faculty of Humanities and Social Sciences, UKM, Bangi, 1997, pp. 45–46. The MiG-29 deal also included $200 million in offsets out of an overall value of $381 million.

[36] Yap Pak Choy (note 35), pp. 43ff.

The armed forces contend that for reasons that are not transparent there have been occasions when the government 'strongly suggests' weapon purchases that are not planned by the forces. Such decisions will distort the expenditure planned for in the Perspective Plan.[37] According to this view, procurement is not the joint process that it is meant to be. On the other hand, political considerations are not necessarily a drawback or harmful, and in situations where other things—that is, quality and standards—are equal they are the only ones applicable. This aspect is sometimes overlooked by the armed forces.

In the end, as the content and direction of Malaysia's defence policy are not publicly accessible, it is difficult to evaluate arms procurement decisions or to know whether defence policy fits in with the political leadership's perceptions of strategic planning in a comprehensive and systematic fashion.[38]

IV. Arms procurement planning

According to Mak, the Malaysian political elite still perceive challenges to the regime and its legitimacy as the greatest threat, followed by internal security threats, external security threats being a weak third. For this reason Malaysia was able for some time to avoid investing in building up conventional military forces for external defence. The end of insurgency therefore spelt the beginnings of a divergence of interpretations between the political and military elite.[39] The concept of 'self-reliance' defined in the 1987 national defence plan concentrates on comprehensive (internal and external) security, whereas MAF arms procurement planning appears to focus on acquiring conventional military power for the deterrence of external threats. Such inconsistencies impinge on force development and arms acquisition planning.

Force planning

Until the late 1980s, when the end of the insurgency spelt a shift in threat perceptions that significantly influenced force planning, the army was privileged over the other two services in terms of arms procurement. The withdrawal of the British east of Suez in 1971 and the end of the Viet Nam War in 1975 led to the PERISTA (Perancangan Istimewa Angkatan Tentera, Special Armed Forces) plan of 1979–83. It was designed to facilitate the development of the armed forces' capability through arms acquisition, with a decided naval emphasis. The end of the communist insurgency freed the armed forces from domestic concerns. The air force came to be seen as an indispensable element for self-reliance in military security. In the 1990s, arms procurement priorities focused on naval and air force requirements, leading the army to feel neglected in the acquisition

[37] Sharifah Munirah Alatas (note 34), p. 41.

[38] Zakaria Haji Ahmad, 'Defence industrialisation in Malaysia', Paper presented at the Chatham House Conference on European Defence Industry in the Global Market: Competition or Co-operation?, London, 20–21 May 1996, p. 3.

[39] Mak (note 15), pp. 8–9.

of new technology.[40] The navy and air force have moved away from supporting army operations and assumed a more independent role, which for the navy was the protection the exclusive economic zone (EEZ).[41] With the extension of the EEZ the navy had to shed its image of a coastal force, and with the problems developing in the South China Sea, the US withdrawal from the Philippines in 1993, piracy, smuggling and illegal immigration, new security concerns have come to the top of the naval agenda.[42]

The individual services and the Armed Forces Staff Headquarters set up in 1993 should logically be working on joint operational doctrines in the light of the changes in the region.[43] While joint operations and exercises have been successful, they have revealed problems, often deriving from the fact that the cooperation only extended to the operational levels and is not integrated at a higher level, as is indicated by the absence of any joint doctrine on acquisition.[44]

Arms procurement planning

Arms procurement planning starts with the respective armed services preparing their capability requirements, which are submitted to Armed Forces Headquarters. These are the basis on which the MAF prepare a Perspective Plan. These plans are consolidated at a JCAFC meeting and sent to the Development Division of the MOD, which in turn examines and integrates the plans, and are then sent to the Development Committee.[45] The Development Division follows the guidelines given by the EPU in the Prime Minister's office, in which defence issues are not necessarily the main focus.[46] Economic and development concerns take priority over the needs of the military.

The EPU examines the plans for affordability and feasibility and sets a spending ceiling. The MOD receives the plans back, re-examines them and passes them back again through the same stages.

After consultation with the services requesting procurement, priorities are identified or changed and the process is repeated up to the EPU, from where the plans are sent to the NDC, which is the central planning body in terms of capital outlay and functions under the Chief Secretary to the Cabinet. The plans are then presented to the Cabinet, which examines them and passes them back, and then laid before Parliament. The Cabinet does not directly intervene in procurement decisions made by the ministry; it merely passes the annual budget as a whole together with whatever arms purchases are included.[47]

[40] On the relative weights of the 3 services in the 1990s, see Thananthan, S. and Sengupta, P., 'Articulating Malaysia's total defence capability: Syed Hamid gives his views', *Asian Defence and Diplomacy*, Mar./Apr. 1996, pp. 5–7.

[41] Siti Azizah Abod (note 2), p. 3.

[42] *Malaysian Defence: Towards Defence Self-Reliance* (note 11), p. 19.

[43] Mak (note 15), p. 11.

[44] Chandran Jeshurun (note 31), pp. 199, 202.

[45] *Malaysian Defence: Towards Defence Self-Reliance* (note 11), p. 25.

[46] The EPU is discussed further in section V of this chapter.

[47] Faridah Jalil and Noor Aziah Hj. Mohd. Awal, 'Control over decision-making process in arms procurement: Malaysia', SIPRI Arms Procurement Decision Making Project, Working Paper no. 81 (1997), p. 5.

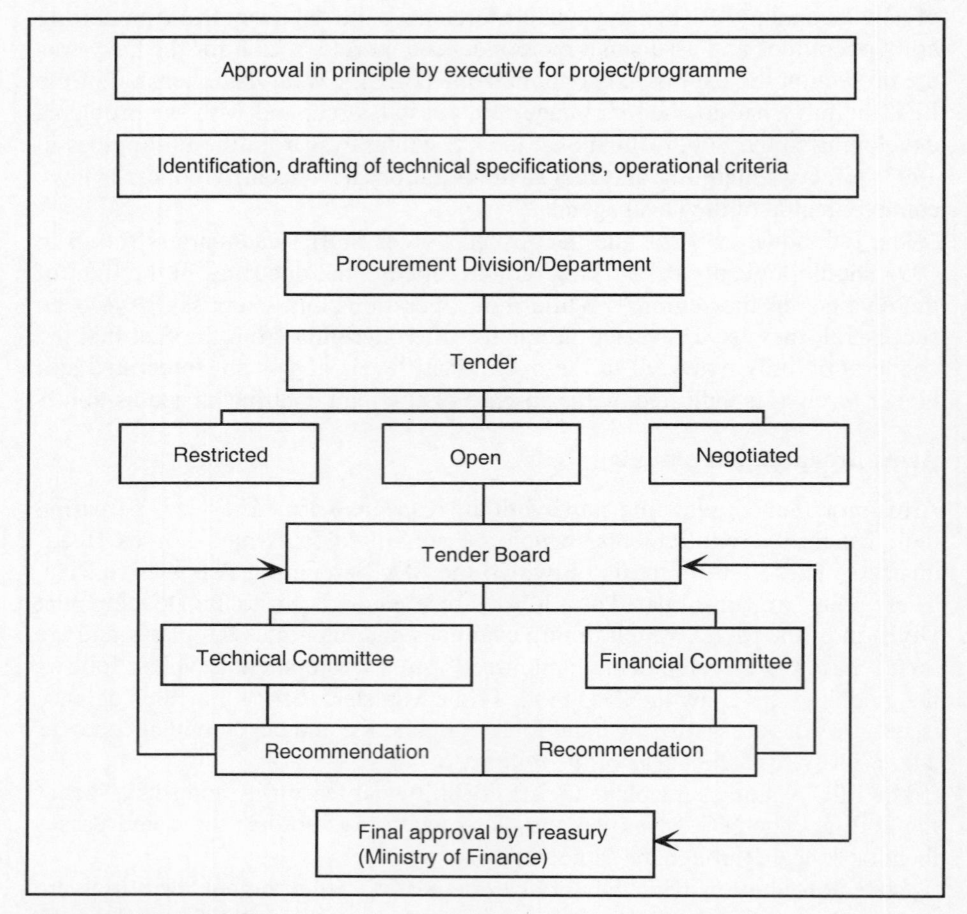

Figure 4.3. The Malaysian tender and procurement process

Source: Mak, J. N., 'Security perceptions, transparency and confidence-building: an analysis of the Malaysian arms acquisitions process', SIPRI Arms Procurement Decision Making Project, Working Paper no. 82 (1997), appendix.

Stages in the arms procurement process

There are seven steps in the process:

1. The General Staff Requirements (GSR) at armed forces level for single services are generally for the purchase of equipment off the shelf.

2. For capital items made to order, a specification committee tests viability and local content. There is a specification committee for each service as well as one for the three services jointly. The members are drawn from the different equipment departments of the services according to requirements. For example, in the air force the actual users are members of the committee. A tri-service

specification committee whose members are determined by the forces themselves is established, for example, for the specification of major items to be used by all the services, such as C³I equipment. The committee suggests the type of equipment preferred and methods of procurement.

3. The deputy heads of the services coordinate the recommendations and pass them to the Procurement Division (formerly known as the Supply Division) of the MOD. The division is headed by a Secretary who reports to the Deputy Secretary General (Development) and through him to the Secretary General. Its members are drawn from the executive offices of the MOD and from the diplomatic and administrative services.[48] It decides on the method of procurement and the type of tender. The MOD handles procurement of equipment below 5 million ringgits (about $1.3 million); proposals for items costing more than that must be approved by the Treasury. Thereafter proposals are evaluated by the Technical Committee of the Procurement Division.[49]

4. A technical evaluation committee carries out technical evaluation and field tests for the suitability of the equipment in terms of specifications and user requirements. It also examines the life-cycle costs, local content, infrastructure and other logistical requirements. It comprises end-users and technical experts from the relevant MOD departments, such as the Defence Science and Technology Centre (DSTC), the Defence Industry Division of the MOD and the Information Technology Division. Its membership is determined by the MOD and can, but rarely does, involve experts from outside the government. Debate over government purchases is usually confined to the technical committee established for the particular tender, whose composition varies according to the type of equipment to be purchased.

5. The Procurement Division decides on the method of procurement and type of tender: (*a*) open tender; (*b*) restricted tender; and (*c*) direct negotiation (negotiated tender). This stage is shown in figure 4.3. In the open tender procedure, the bidders are required to meet basic criteria. In cases where open tender is not considered suitable, restricted tender is designed to save time when potential suppliers are few because the equipment involved is highly specialized: for instance, builders of conventional submarines may be approached directly to submit tenders. In a negotiated tender a supplier has been identified as the only one offering the equipment that meets the specific requirements of a user agency, for instance, for spare parts for vehicles that are not available from any other source. As the name implies, negotiations are carried out to establish price, delivery dates, support and so on. A negotiated tender can also apply in the case of government-to-government purchase.[50]

Tenders for programmes costing below 5 million ringgits are managed by the General Secretary of the MOD, and for items above 5 million ringgits by the Treasury. It issues its own tender, although the end-user or the specific armed service initiates the tender process by identifying and writing the technical

[48] Interview by the author with Siti Azizah Abod and Mr Rajayah, MOD, Sep. 1998.
[49] Siti Azizah Abod (note 2), p. 5.
[50] Information provided by MOD staff in discussions, Oct. and Nov. 1998.

specifications and operational criteria. These are passed on to the agency's procurement division or department.

6. The next stage is tender evaluation. Depending on financial limits, the Procurement Division of the MOD or the Treasury forms an evaluation committee with technical experts from the services, the DSTC and the Information Technology Division, which submit tender briefs to the Tender Board. A financial evaluation committee also evaluates the financial merits of proposals such as industrial offsets, financial packages including modes and payment schedule, and other cost-related criteria. This committee is practically the same for all ministries.[51]

7. The Tender Board, chaired by the Secretary General of the MOD and comprising the Deputy Secretary General for Development, representatives from the Armed Forces HQ, the services and the Treasury, considers the tender brief and either approves or rejects the recommendations, or calls for a re-tender. The Treasury has the right to accept or reject any or all proposals against the recommendations of the Tender Board and the two tender sub-committees (for technical and financial evaluation). For tenders called by the Treasury, the MOD will forward the technical evaluation report directly to the Treasury.[52] If the Treasury handles the procurement of certain high-value equipment, a special committee will be appointed to look into the commercial proposal before the Request For Proposals (RFP) is made. The committee looks into delivery, costs and terms of payment, warranty and such aspects as offsets or counter-trade, transfer of technology and local content.

At the level of the individual armed services the process is similar, with minor variations. For the army, GSRs are examined in the Army Operational Equipment Committee, which consists of the Deputy Chief of the Army and the heads of the relevant departments such as logistics, equipment, and mechanical and other specializations. The GSRs are examined in line with army doctrine, operational factors and training requirements and then passed to the Procurement Division of the MOD to be processed by a technical evaluation committee. The air force procedure involves the Technical Specification Committee, which passes the Air Staff Requirements to the Air Specification Committee and then to the MOD Procurement Division. Thereafter it follows the same procedure as described for the army. In the navy the GSRs are evaluated by the Chief of the Navy Committee, which passes them on to the Procurement Division in the MOD. The detailed working of the tender process in the navy is given in a judgement in an appeal case in 1979 initiated by Lim Kit Siang, the then leader of the opposition in Parliament.[53]

[51] Mak (note 15), pp. 20–21.
[52] Siti Azizah Abod (note 2), p. 6; and Mak (note 15), p. 22.
[53] Public Prosecutor v. Lim Kit Siang [1979] 2 *Malaysian Law Journal* (MLJ) 37, p. 294.

V. The defence budget and financial planning[54]

The defence budget planning and arms procurement processes are part of more general budget making by other ministries and agencies. They go through similar mechanisms for financial vetting and control.[55] However, there is no detectable public process of examining the question how much is enough and allocating a fair share of the national budget to the social sector at the cost of security, and the question of opportunity costs remains unexamined.

During the period 1991–95, most capital-intensive and sophisticated weapon acquisitions programmes were approved. Despite frequent changes in defence policy, defence spending in Malaysia has remained at around 4 per cent of gross national product (GNP) for much of the past two decades, which is just above the average of 3.3 per cent for the region.[56]

The MOD imposes a five-year planning structure which establishes manpower levels, equipment requirements and financial ceilings to guide the formulation of annual budgets. The MOD and the Treasury discuss expenditure within these parameters, and the latter determines the annual budget allocations accordingly. They are divided into two headings in the five-year plan: (*a*) capital procurement, part of which is infrastructure; and (*b*) operational expenditure. The capital budget is reviewed half-way through the planning period, which provides an opportunity to apply for new or additional requirements or an increased allocation.[57] The operational budget is revised annually. The five-year budget is broken down into annual allocations for the departments of the MOD.

The budget cycle

The budget cycle is standardized for all government departments. It is shown in figure 4.4. The defence budget is a two-tiered exercise: the overall allocation for defence in the national budget is set in the medium term and then distributed between the individual services.[58]

In February of each year the initial internal screening and consolidating by the end-users (Logistics, Medical, and so on) are carried out, prior to a circular being sent out by the Finance Division of the Treasury in March/April asking for requirements for the financial year. Requirements are submitted by May. The MOD Finance Division examines them and submits them to the Budget Division of the Treasury by June. Here they are examined and consolidated. By July, the allocations are known informally to the MOD and end-users, and

[54] Unless otherwise indicated, this section follows Robless, R. (Brig.-Gen.), 'Harmonizing arms procurement with national socioeconomic imperatives', SIPRI Arms Procurement Decision Making Project, Working Paper no. 83 (1997).

[55] Siti Azizah Abod, personal communication.

[56] Robless (note 54), p. 16.

[57] Siti Azizah Abod (note 2), p. 4.

[58] Robless (note 54), para. 5.

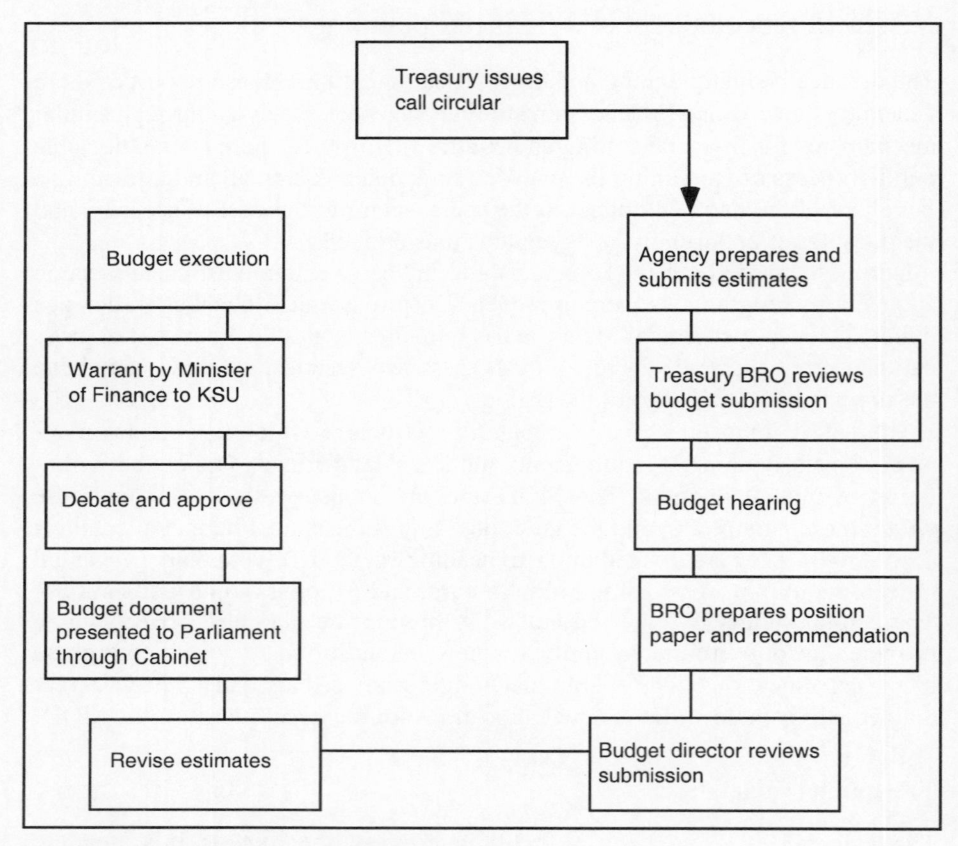

Figure 4.4. The Malaysian annual budget cycle

KSU = Secretary-General of a ministry. BRO = Budget Review Office.

Source: Supian Ali, 'Harmonizing national security with economic and technology development in Malaysia', SIPRI Arms Procurement Decision Making Project, Working Paper no. 87 (1997), p. 38.

planning for the financial year's procurement begins at user level.[59] The Treasury goes through its own process to evaluate requirements, but the MOD and services are represented to answer technical questions. While the Treasury has the power of decision, the MOD identifies its operational expenditure and capital procurement. It is here that the services compete for funds, and competition can become quite severe in the absence of any joint doctrine that identifies priorities or missions.[60]

The five-year plan is the primary planning instrument which indirectly determines spending priorities among the services. The annual estimates concentrate

[59] The author thanks Mr Arunasalam, Managing Editor, *Asian Defence Journal,* for clarifying this part of the budget cycle.

[60] Abdul Rahman Adam (note 14), p. 16.

on service expenditures, which are submitted to Parliament.[61] Recording expenditure is a crucial part of defence budgeting. Operating expenditure cannot be carried over into the next financial year. Expenditure on capital projects, however, can be spread over several years, so that a large part of each year's capital budget, or of each forthcoming five-year-plan, is often committed in advance and this eats into funding for new acquisitions and capability development.[62] Figure 4.5 shows the system of financial planning under the Malaysian five-year plan.

In 1992 the government introduced the Modified Budgeting System (MBS) in order to manage the budget and align expenditure with stated policies. The MBS requires spending agencies to commit themselves to objectives and targets against which their performance is then measured. It is meant to look comprehensively at input, throughput and output (in this regard it is similar to the old Programme Analysis Review used in the UK), to make the armed forces more goal-driven and accountable and to avoid incremental budgets.[63] The application and use of funds have become more stringent and streamlined since the MBS was introduced, because underspending can lead to cuts in allocations.[64]

While requirements for the medium and short term (five years and one year, respectively) are fairly clearly identified, longer-term needs are less clear. There is a lack of long-term financial planning in the defence sector, which is particularly important for the capital-intensive procurement programmes. The fact that the decision-making process is long-drawn-out can lead to technological, operational and financial problems in the sense that inordinate time delays may involve eventually paying higher prices. It was for this reason that the Minister of Defence, Datuk Syed Hamid Albar, has called for a 15- to 20-year financial planning procedure to be introduced.[65]

Without a long-term strategic perspective, expenditure planning and budget allocations are dependent on threats perceived over the short or medium term and can thus fluctuate considerably from year to year.[66] This leads to a tendency for margins to be increased in times of perceived threat, as in the 1970s during the Viet Nam War or in the 1960s when threats were perceived from Indonesia, and a disproportionate scaling down when the threat passes.

[61] Robless (note 54), para. 6.

[62] Sharifah Munirah Alatas (note 34), p. 41.

[63] Supian Ali, 'Harmonizing national security with economic and technology development in Malaysia', SIPRI Arms Procurement Decision Making Project, Working Paper no. 87 (1997), p. 30.

[64] Robless (note 54), para. 7. The phenomenon of 'Christmas shopping' at the end of the budget year occurs in order to use up the allocation. Supian puts this down to bad planning and budget control. Personal communications with Supian Ali.

[65] Balakrishnan, K. S., 'Arms procurement budget planning process: influence of cost and supply source of alternative systems, procurement negotiations and methodologies, offset policies, contracting process and the issue of transparency', SIPRI Arms Procurement Decision Making Project, Working Paper no. 79 (1998), p. 3; and *Asian Defence Journal,* no. 4 (1998), p. 3.

[66] Robless (note 54), para. 22.

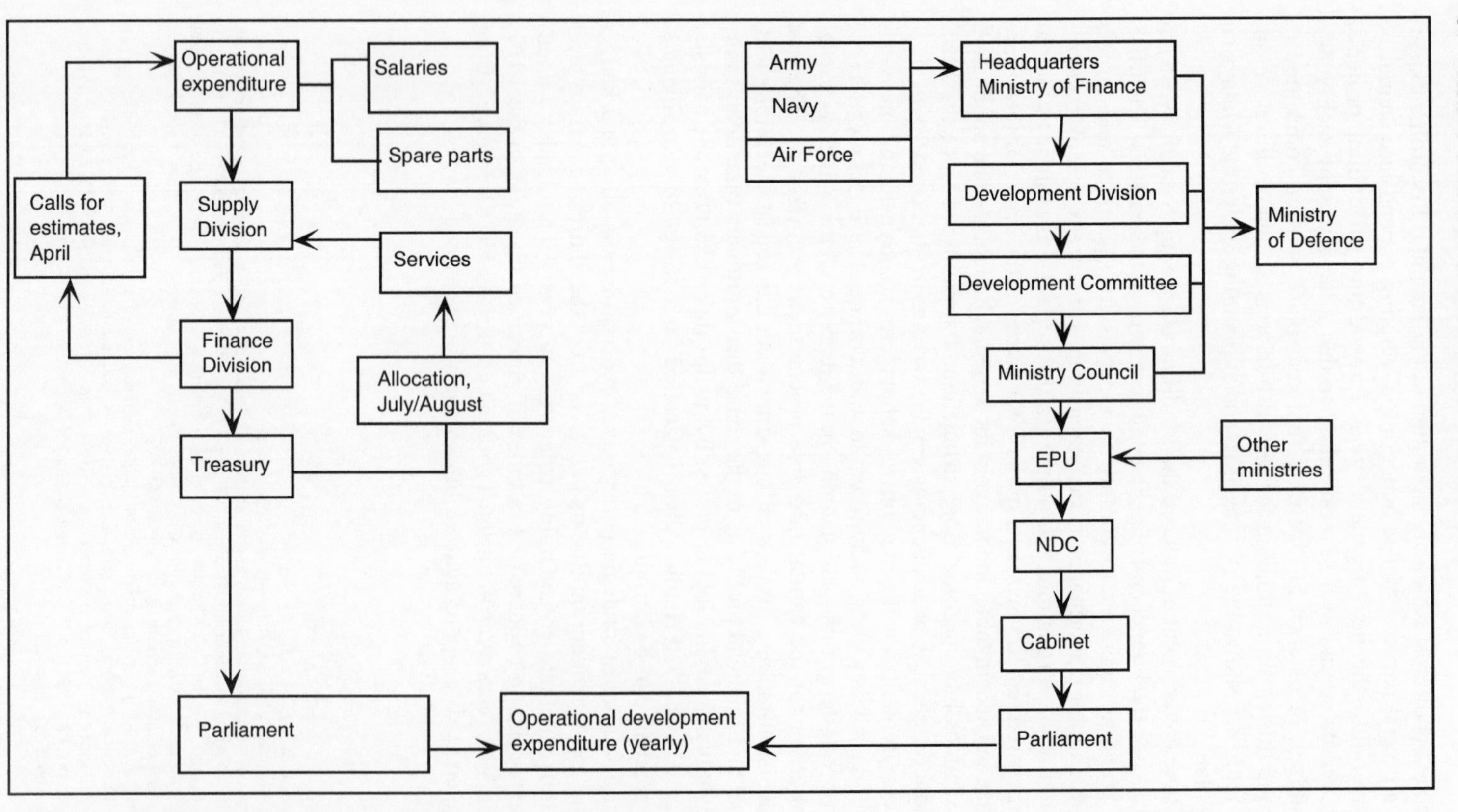

Figure 4.5. Malaysian financial planning under the five-year plan

Source: Compiled by the author on the basis of information from the Malaysian Ministry of Defence.

The financial audit processes[67]

Internal audit

Internal audit aims to avoid or correct any mistakes or deficiencies before the Federal Audit Department does the external audit. The MOD's internal audit department assists financial management of the armed forces. It reports on financial compliance and makes recommendations to the Secretary General of the MOD on the actual state of affairs to ensure that management decisions are implemented in accordance with pre-defined rules and procedures. The internal audit reports also check productivity levels and target attainment, and provide management consultancy services within the department. The Secretary General, as the Controlling Officer under the Financial Procedure Act 1957, upon receiving written complaints can also order the internal audit department to investigate any allegations of misappropriations and abuse of power and take disciplinary and/or corrective measures against the staff responsible.

Equipment audit in terms of life-cycle costing and maintainability is the responsibility of the individual services. The army has its inspection and evaluation division, the air force and navy their inspectorates general. They compare the performance of equipment to the financial efficiencies of a particular system. The reports of their investigations are presented to the service chiefs who report to the minister when deficiencies are found. These inspectorates are separate from the internal audit department, which is responsible to the Secretary General, the civilian arm of the ministry, since usage relates to the operational arm.

Statutory audit

The Federal Audit Department is a statutory organization under the Auditor General, answerable not to the Government but to the Public Accounts Committee of Parliament, made up of MPs of the ruling and opposition parties. It can audit the internal audit departments as well. The department carries out performance auditing after transactions have taken place in terms of verifying financial objectives, outlay, performance, maintenance and reliability. The staff are government servants but it is a 'closed service', drawn from accountants and bookkeepers who receive additional training in the service. The audit teams do not have military or technical backgrounds since accounting and auditing procedures are standard for all government departments.

The annual audit report presented to Parliament and the Public Accounts Committee is accessible to the public,[68] and on that basis the public and Parliament can question individual ministries and departments.[69] The Public Accounts

[67] It was almost impossible to get any but the sketchiest information on auditing in the MOD. The following information is based on general sources and on personal interviews.

[68] E.g., *Laporan Ketua Audit Negara* [Report of the Federal Auditor General for the year 1995] (Jabatan Audit Negara Malaysia, 1996).

[69] Personal communication with Mr Arunasalam, Managing Editor, *Asian Defence Journal* The discussions are laid down in Hansard, but the subsequent action is not made public.

Committee (although not the Federal Audit Department) can demand action on issues not clarified. This may be disciplinary action against the persons responsible; this is not made public, but the Committee has to be kept informed. Inspections continue until a satisfactory explanation is accepted or a problem that has arisen is solved by the department concerned or a decision is made by the Public Accounts Committee. The Federal Audit Department can question the Finance and Accounts Division of the MOD on certain items, and it in turn will ask the relevant departments—in the case of arms procurement the Procurement Division—to reply. If the reply is not satisfactory, the problem can be taken up with the Deputy Secretary of Finance in the MOD and may go to the Public Accounts Committee.

The Federal Audit reports are received by the King, the Sultans of the states, Parliament, the state legislatures and the government authorities and other bodies concerned. The necessary action is taken by the Treasury, heads of departments, the agency concerned and the Public Accounts Committee after the report has been tabled in Parliament. The methods and procedure of the Federal Audit Department are at the discretion of the Auditor General.

A financial review is carried out at mid-term of the current plan or year or if and when the need arises, for instance, during an economic crisis. Spot checks are also done if deemed necessary. On these occasions it can be decided to increase or reduce the expenditure for the armed forces.

VI. Defence industrial aspects

Decisions to make or buy military equipment have been characterized by a cautious, pragmatic and gradualist approach. The accent remains on buying weapons rather than developing them. The building up of the defence industry seems to be a secondary concern of the government and is not, as it is in some countries, seen as the spearhead of the drive for industrial capacity and capability. Defence industrialization is still regarded as belonging more in the realm of defence policy than in that of industrial or economic development policy.[70]

Defence research and development[71]

Responsibility for defence research and development (R&D) belongs to the DSTC, which was founded in 1968. Its role is to give scientific and technological advice to the MOD and the MAF in meeting their capability requirements and to carry out R&D to promote local defence production. In addition, it should identify key technologies and post-evaluate military manoeuvres.[72] It reports to the Deputy Secretary of Defence for Development in the MOD and

[70] Zakaria Haji Ahmad (note 38), p. 2.

[71] Unless otherwise indicated, this section is based on Sukumaran, K., 'Defence research and development (R&D) and arms procurement decision making', SIPRI Arms Procurement Decision Making Project, Working Paper no. 86 (1998).

[72] *Malaysian Defence: Towards Defence Self-Reliance* (note 11), p. 57.

consists of four units: R&D, Quality Assurance, Technical Support and Administration.

The DSTC has not kept up with other areas of national research. It suffers from inadequate funding and therefore inadequate manpower and equipment. For instance, out of 234 positions approved in the DSTC, only 195 were filled on 1 May 1997 and of these only 54 were professional researchers, the rest being support staff. The R&D Unit, the most important unit, has only 18 researchers. The ratio of research staff to auxiliary and technical support staff is about 1 : 4. The DSTC's activities remain ad hoc because it does not have a long-term plan for staff requirements. Defence R&D is confined to applied research in areas such as maintenance and training requirements. No basic research is carried out and the R&D which is done has little impact: the organization is not geared to conducting meaningful military R&D and the quality of its research is poor.

The DSTC's funding is meagre even compared to that of other sectors of the military. Under the Fifth Malaysia Plan 9 million ringgits ($2.4 million) were allocated, and under the Sixth Plan 10 million ringgits ($2.6 million)—just 0.1 per cent of the total capital expenditure of the MOD or 1.73 per cent of the national R&D allocation. Under the Seventh Plan (1996–2000) 30 million ringgits ($7.9 million) was requested for the DSTC; only 6 million ringgits was allocated.[73] Funds for the DSTC are handled by the Development Division of the MOD, which also handles the fund for asset acquisition for the whole ministry; thus the DSTC's funds are not kept separately from the MOD funds. If costly equipment is acquired, this also eats into the DSTC's funds (even if it has not used all its allocated amount thus far). The low funding and priority for the DSTC might be due to a perception that defence R&D has no useful spin-offs for industry and is not conducive to overall development.

Generally it can be said that any meaningful R&D is carried out either by the private sector (training facilities for aerospace engineering with the Mara Institute of Technology, ITM) or together with government agencies other than DSTC (e.g., rapid prototyping together with SIRIM, the Standards and Industrial Research Institute Malaysia, which helps in product development). The Second Industrial Master Plan (1996–2005) produced by the Ministry of International Trade and Industry spelt out the role of the private sector in R&D in general.[74] However, among the eight key areas or industry groups identified as priorities only one is linked with the defence sector—aerospace. Here the plan mentions explicitly desired cooperation between the Royal Malaysian Air Force (RMAF) and Malaysian Airlines System (MAS) to develop local expertise and technological know-how.[75] Other elements of this approach include: (*a*) modifi-

[73] This can probably be explained by the fact that out of 10 million ringgits ($2.6 million) allocated under the 6th Malaysia Plan, only 6 million ringgits were spent.

[74] Sukumaran, K., 'The role and function of the DSTC, Ministry of Defence Malaysia', thesis for the MA degree, National University of Malaysia, Bangi, 1997, p. 14 (unpublished).

[75] Sukumaran (note 74), pp. 15ff.

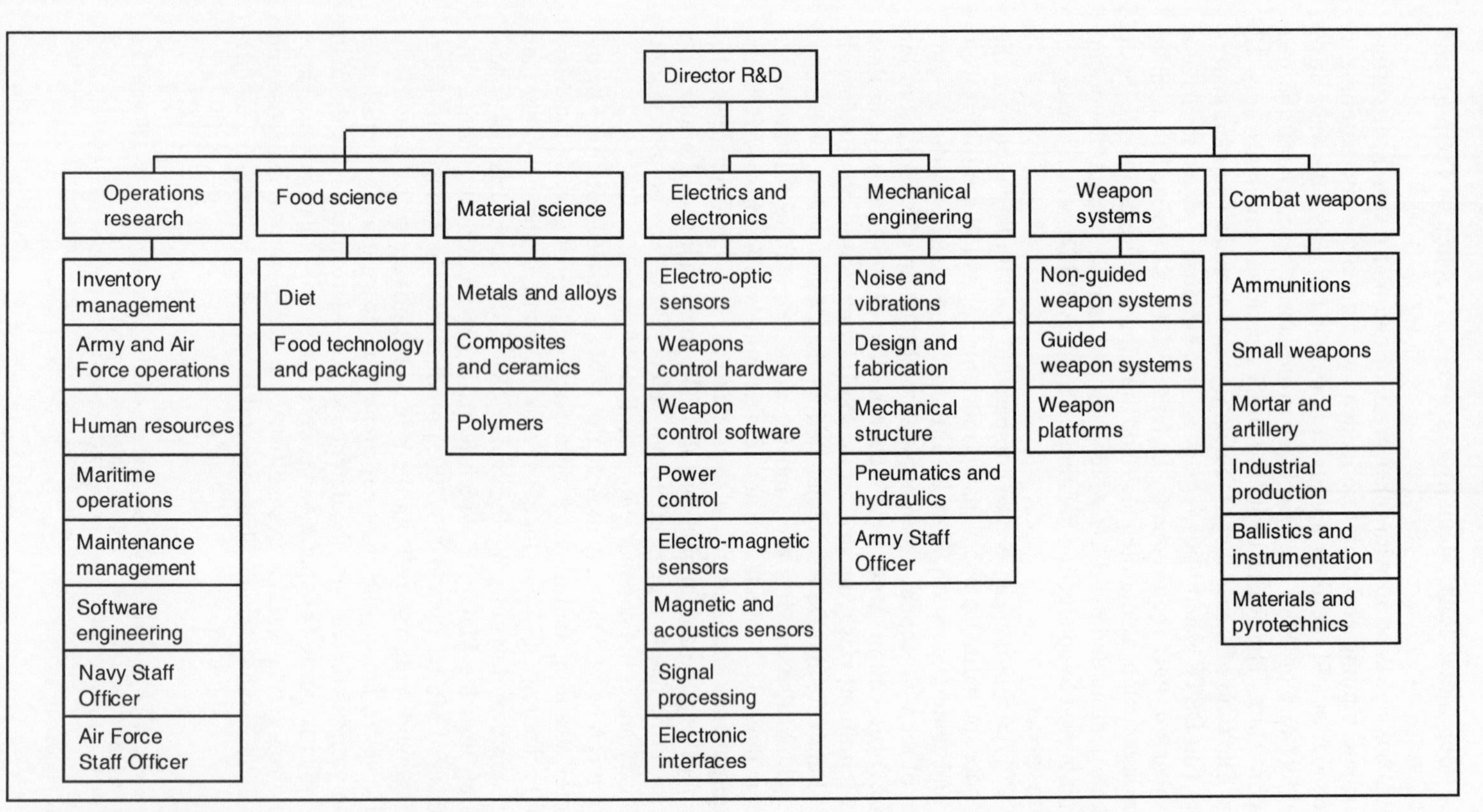

Figure 4.6. The Research and Development Unit of the Malaysian Defence Science and Technology Centre

Source: Sukumaran, K., 'Defence research and development (R&D) and arms procurement decision making', SIPRI Arms Procurement Decision Making Project, Working Paper no. 86 (1998), p. [21].

cation and conversion activities together with the RMAF, which at a later stage could benefit other sectors of the economy; and (*b*) enhancing Malaysian capability for building light aircraft for both the domestic and export markets.

Areas that may have applications in the defence sector, such as the electrical, electronics or materials industries, are left to the private sector and it has not developed cooperation with public-sector defence R&D. At a seminar on defence R&D at the DSTC in 1995 it was revealed that out of 57 R&D projects undertaken by the DSTC in 1995 none was in cooperation with the private sector or other institutions of higher learning. Even relatively successful privatized industrial establishments like SIRIM, Airod and the Lumut Naval Dockyard concentrate to a great extent on applied research.

As certain sectors of defence R&D are handed over to the private sector without the DSTC having any say or participation and the DSTC has not come up with research project proposals,[76] interest from private industry in cooperation with the DSTC is low. Evidence of technology spin-offs between defence and civil R&D is not enough to support any definite conclusions.[77] Defence research is seen as 'lost research'. 'R&D efforts have not borne commercial success.'[78]

However, since 1998 the DSTC has been developing bilateral relations with defence science and technology institutions abroad as well as links with industry, academia and research institutions.[79] Foremost among these is the UK's Defence Evaluation and Research Agency (DERA), which held a joint seminar with the DSTC during the DSA Exhibition in April 1998.[80]

Defence industrial production

After 1986 Malaysia encouraged private-sector involvement in defence production to the extent of privatizing some government concerns. Privatization in the defence industry mainly took the form of what were called Non-Financial Public Enterprises, which kept a measure of government control while encouraging technology absorption by the private sector. In 1982 the Defence Industry Division formulated a National Defence Production Policy (NDPP) in which defence items were classified into three categories—'strategic', 'essential' and 'non-strategic'. The NDPP 'recognized the need to be self-reliant in the production of strategic items and leaving the non-strategic items to be produced by semi-government agencies and in the private sector'.[81]

In 1990 a Malaysian Defence Industry Council (MDIC), a private-sector initiative, was established as an umbrella organization to promote defence industrialization. Its establishment signalled an awareness of the information gap on defence needs in the country—the lack of doctrine or a well-identified defence

[76] Sukumaran (note 71), pp. 15, 16.
[77] On spin-offs, see Supian Ali (note 63), pp. 17–18.
[78] *Malaysian Defence: Towards Defence Self-Reliance* (note 11), p. 61.
[79] Sukumaran (note 74), p. 59.
[80] 'DSA 98 today', supplement to *Asian Defence Journal*, 22 Apr. 1998, p. 5.
[81] Zakaria Haji Ahmad (note 38), p. 4. As a result the cooperation between SME and Steyr-Mannlicher to produce assault rifles was undertaken in 1989. Supian Ali (note 63), pp. 18–19.

policy, leading to improvisation on the part of the MOD, and excessive political control of defence planning and procurement which led to several undesirable consequences.[82] The MDIC identified four priority areas for defence industrialization—ordnance, aerospace, shipbuilding and communications—to gear long-range defence industrial planning more closely to actual needs.[83] It envisaged an integrated pattern of defence industrial production. According to Zakaria, the MDIC is considered defunct.[84]

Because economies scale are unlikely to be developed in private-sector defence industries, demand from the MAF being low, and because a large percentage of the defence budget had always gone into importing military equipment, the Defence Minister in 1997 recommended developing the industry through more collaborative arrangements with industries in the ASEAN countries and the West.[85] The government tried to broaden the supplier base and increase the technology transfer element of trade agreements. These efforts led to a number of moderately successful technology transfer agreements in defence production. Other enterprises were encouraged to engage in dual-purpose technology to provide a cushion against a potential drop in demand.[86]

The strategies identified to develop Malaysia's defence industry were: (*a*) a focus on key industries such as the automotive industry, aerospace, shipbuilding, electronics, arms and ammunition, and advanced materials; (*b*) privatization; (*c*) offsets; and (*d*) incentives. The incentives deemed crucial by the government are: (*a*) terms and conditions similar to those offered in other industries; (*b*) direct negotiation with potential suppliers; (*c*) the assurance of sustained purchases from a certain supplier over a certain period; (*d*) economic volumes of production; and (*e*) long-term contracts to enable companies to recover their investment.[87]

Joint ventures

Defence production through private-sector companies in joint ventures between the government and a foreign partner has progressed considerably. However, most of these joint ventures remain under some form of government control through share holdings. A variety of defence products are offered by MMC Engineering; DRB-HICOM together with FMC-Nurol and related companies; Airod with Aérospatiale and OFEMA; MOFAZ and MOS with South African Advanced Technologies and Engineering Co.; Syarikat Malaysia Explosives

[82] 'The Malaysian Armed Forces: an exclusive interview with Chief of Defence Forces Gen. Ismail Omar', *Asian Defence and Diplomacy*, Dec. 1996, p. 20.

[83] Supian Ali (note 63), p. 20.

[84] Zakaria Haji Ahmad (note 38), p. 4.

[85] *Military Technology*, Dec. 1997, p. 19.

[86] Zakaria Haji Ahmad (note 38), p. 19. The process began with a joint venture between Dynamit Nobel, the Swiss company Oerlikon and the Malaysian Government which is today known as Syarikat Malaysia Explosives (SME), producing small arms and munitions and, since 1989, the Steyr assault rifle. Three spin-offs were SME Tools, SME-Trading and Tenaga Kimia, with Nitro Nobel, which manufactured explosives.

[87] Zakaria Haji Ahmad (note 38), p. 6.

(SME); the Lumut Naval Dockyard; ATSC, a joint venture between the Malaysian Government and Russia for maintenance support of Malaysia's MiG-29 aircraft; Sapura; Malaysia Shipyard; and Hong Leong Lurssen.

The Defence Industry Division of the MOD oversees collaboration with foreign partners and technology transfer to local companies. Lately some joint ventures in the telecommunications sector have come on-stream, such as those between Marconi and LTAT and between Sapura, Siemens Plessey and CRL Ltd. The purchase of Russian MiG aircraft in 1995 brought about cooperation between the Russian state corporation Rosvooruzheniye and Universiti Sains Malaysia for an aerospace school and a technological research centre.[88]

Offsets

In the 1980s counter-purchase was the predominant form of offset in Malaysia (in some areas it still is) because the local industry was not considered ready for offset programmes. As the increased prosperity of the early 1990s led to more arms being bought, Malaysia began to exert more leverage over the conditions of arms contracts, spread its sources and insist on transfer of technology along with the finished weapons imported. Arrangements included barter,[89] counter-purchase, technology transfer and local content.[90] The prevailing view in the government is that any weapon purchase should allow for 'offsets', preferably technology transfer and training.[91]

Offset packages have now become the most crucial element in the procurement decision-making process. The most common kinds of offsets are: (*a*) technology transfer to maintain and modify equipment or manufacture components; (*b*) training; and (*c*) technical assistance. Government policy suggests that all arms purchases should allow technology transfer and training arrangements as offsets.[92] Appendix 4A illustrates the cooperation over offsets between the MOD, the government and the private sector. In this instance the purchase of naval patrol vessels was almost incidental in the sense that technology and price were less important than the economic and development benefits to be derived.

VII. Checks and balances

Checks and balances exist in the arms procurement processes as for all public procurement decisions, and financial checks and balances are quite stringent. It remains to be seen whether they are efficient or meet the standards required as

[88] Robless (note 54), para. 20.

[89] In 1993 Russia undertook to buy 237.5 million ringgits-worth ($61 million) of palm oil over 5 years as part payment for 18 Russian MiG-29s. Abdul Rahman Adam (note 14), p. 14.

[90] In the British Aerospace (BAe) Hawk agreement, some parts for the aircraft were made locally. This had commercial advantages later when the UK imported the parts and sold them to all Hawk users worldwide. Yap Pak Choy (note 35), p. 45. The offset content of the contract for technical support for the McDonnell Douglas F/A-18D in 1993 was worth 677.5 million ringgits ($178 million) out of a contract value of 1.4 billion ringgits ($368 million). Balakrishnan (note 65), p. 8.

[91] Zakaria Haji Ahmad (note 38), pp. 7–8.

[92] Zakaria Haji Ahmad (note 38), pp. 6–8.

regards legal accountability. It should also be borne in mind that civilian control of arms procurement does not necessarily mean parliamentary control or political accountability.

Legal and political accountability

Malaysia is a constitutional monarchy with a bicameral Parliament on the lines of the Westminster model and a Cabinet of which the Prime Minister is the head. However, its style of functioning is very different and one would look in vain for the same kind of adversarial debate. Consensus is sought in debate: controversial discussion is not considered acceptable in Malaysian civil society when social or political superiors are involved. Such norms have resulted in 'strong government', with the Prime Minister at the apex of the system in a position which is more presidential than prime ministerial.[93] He gives the direction for others to follow. In a dispute, his word decides. Attitudes of deference to authority and above all loyalty are also crucial virtues in Malay culture.

In the case of arms procurement decisions, the Cabinet assesses the annual budget as a package and normally does not interfere with the MOD's recommendations. In the normal course of events, Parliament is not informed of arms procurement decisions, nor does it exert pressure on other decisions regarding the armed forces. Decisions are recorded as minutes by the Cabinet staff. The Cabinet meetings are secret; its papers are classified and even the minutes do not reveal the views of individual ministers.[94]

The executive is collectively responsible to Parliament. Ministerial responsibility, collective and individual, can be invoked during Question Time in the Parliament, in debates and in parliamentary committees.[95] The individual responsibility of a minister is acknowledged: when questions are addressed to a minister at Question Time he is obliged to respond. After the first question, a second, related question is allowed, and this is sometimes the only way to get controversial topics discussed in Parliament. For instance, after a question on the budget the second question might be on a controversial arms procurement decision.[96] Responsibility lies with the minister, not with the office, and a minister cannot plead ignorance of matters within his competence.[97] However, no Malaysian minister has yet resigned over misdemeanours of his ministry.

[93] *Far Eastern Economic Review*, 4 Sep. 1997, pp. 18f. See also an interview with Prime Minister Mahathir in *Far Eastern Economic Review*, 2 July 1998, pp. 15–17.

[94] Hickling, R., *An Introduction to the Federal Constitution* (Malaysian Law Publishers: Kuala Lumpur, 1982), p. 22.

[95] Faridah Jalil and Noor Aziah Hj. Mohd. Awal (note 47), p. 1 This section follows these authors unless otherwise indicated.

[96] Faridah Jalil and Noor Aziah explained the procedure at Question Time to the author. See also Hickling (note 94).

[97] E.g., in Dec. 1993 Dr Tan Seng Giaw, MP, from the Democratic Action Party asked whether the decision to buy the MiG-29 had been influenced by the promise to build a factory in Malaysia to manufacture spare parts and the assurance that Malaysia would not need to pay if there was any defect within 20 to 30 years of purchase. Faridah Jalil and Noor Aziah Hj. Mohd. Awal (note 47), p. 13fn.

The potential of parliamentary questions was made clear in a judgment in 1978:

> Members of Parliament do raise questions in the Parliament pertaining to the information received and . . . introduce accusations that may not have any basis whatever. They are entitled to ask questions, raise the issues in the debates or even introduce motion on any such matter in parliament . . . The information contained in the disclosure may be discussed or deliberated in the proceedings in Parliament even though the disclosure may otherwise be held to be a breach of the Official Secrets Act if it is done outside Parliament. The protection is derived from the privilege that Members of Parliament enjoy in regard to freedom of speech and debate or proceedings in the Houses of Parliament.[98]

However, the right to ask questions does not imply the right to get answers in all cases. Under the Official Secrets Act of 1972 (amended in 1986), decisions concerning national security, defence and international relations are classified and confidential by definition.[99] A questioner must not seek information about something that is by its nature secret.[100] The minister can therefore refuse to answer any questions on defence matters (although he is in principle entitled to declassify military information).[101]

Parliament has the power to review ministers' decisions and if required to initiate legal action against their decisions, but this is rarely done. Court action is the last resort against decisions to procure arms or allocate money, but *mala fides* would have to be proved for a conviction to be possible. The Malaysian democratic system has persistent remnants of the deification of political leadership, so that it is difficult to question decision makers in the government for fear of losing face or being accused of not knowing one's place.[102]

The Public Accounts Committee can question the MAF on their expenditure and in some cases a special parliamentary committee is set up to enquire into the matter, although no such a committee has ever been established for an arms procurement case.[103] A parliamentary committee is set up but not controlled by Parliament, nor are its proceedings public. While a minister has to answer its questions, the committee has no power to change decisions or policies.

In principle there are quite broad entitlements to review and seek information, but these opportunities are limited to the environment of Parliament. If the public or the media seek information, several laws can restrain them. The constitution, although embodying the right to freedom of expression, does not explicitly guarantee the right to information on what the government is doing.

[98] Public Prosecutor v. Lim Kit Siang, 1978, MLJ 1979, pp. 44–45.

[99] *International Law Book Series*, Act 88, Official Secrets Act, Section 2, 1972, p. 227.

[100] *Parliamentary Debate, Dewan Rakyat*, 1 Dec. 1993, p. 58.

[101] *International Law Book Series*, Act 88 (note 99), p. 207.

[102] Sharifah Munirah Alatas (note 34), pp. 43–44.

[103] One such committee was demanded and a White Paper asked for by members of Parliament after 3 crashes of Nuri helicopters occurred in quick succession in 1996 and 1997. Eighteen airmen were killed. Since being commissioned in 1968, the Nuri has been involved in 14 accidents. Statement by the Parliamentary opposition leader, DAP Secretary-General and MP for Tanjong, Lim Kit Siang, on 19 Oct. 1996 and 20 Mar. 1997. Lim Kit Siang, URL <http://www.lks.tm.my>.

Chief Justice Azlan Shah (later HRH Sultan Azlan Shah of Perak) defended the right of the state to withhold information: 'In deciding how much information the state may withhold from the public and how much may be disclosed, a balance has to be drawn between two main principles; on the one hand the disclosure of certain kind of information may hinder the sufficient functioning of the executive and administrative machinery, whilst on the other, the rights of the public may be restricted if access to certain information is withheld from them'.[104]

The Internal Security Act of 1960, reframed in 1988, and the Official Secrets Act could theoretically be used against persons seeking classified information. While appeals against decisions under the latter are possible in the courts, no appeal can be entered against decisions under the Internal Security Act. There are safeguards against misuse of the Internal Security Act, but these have been diluted since 1988.[105] The laws that effectively discourage the media from inquiring into matters that are deemed sensitive are the Societies Act and the Printing Presses and Publications Act, both of 1984. Newspapers and journals are required to obtain a permit to publish. The Sedition Act of 1948 (revised in 1969)[106] and the Police Act of 1967 can also be used against the media and individual journalists, thus further undermining their inclination to 'wash dirty linen in public'. The use of these acts limiting the right to information runs counter to the professed goal of an informed, democratic society in Malaysia.

There have been instances of corruption in arms procurement. Until 1976, the MAF were free to make their own arms procurement decisions once funds were allocated. Then the 'Cuckoo's Nest Scandal' of the early 1970s, involving the purchase of jet aircraft from the USA, identified kickbacks and commissions paid to individual officers.[107] A restructuring took place which required foreign contractors to deal with the government through their local agents and not through the MOD any longer.

Corruption and maladministration are dealt with by the Public Complaints Bureau (PCB), established by the Cabinet in 1995. It can receive complaints from the public and has a permanent committee to take action where necessary. However, because it was not established by Act of Parliament, it has no powers of sanction or subpoena; it can only pass on its findings to the departments concerned and recommend action to be taken. Its function resembles that of an ombudsman. The members of the PCB committee include the Anti-Corruption Agency, the Prime Minister's Department, the heads of public service departments and the Malaysian Administrative Modernization and Management Planning Unit (MAMPU).[108] Informally, the PCB is quite powerful, since its head is the Cabinet Secretary.[109] It seems to have produced some positive

[104] HRH Sultan Azlan Shah, quoted by Faridah Jalil and Noor Aziah Hj. Mohd. Awal (note 47), p. 8.

[105] Personal communications with Faridah Jalil and Noor Aziah Hj. Mohd. Awal.

[106] Act no. 15 (1948), introduced at the beginning of the Emergency and revised 20 Nov. 1969.

[107] Supian Ali (note 63), p. 29.

[108] Balakrishnan (note 4), p. 7.

[109] Information provided by Faridah Jalil, 26 Feb. 1999.

results in dealing with corruption-related problems and improvements in governance.[110] Its success is also due to the fact that over the past few years the investigative profile of the media has improved across the board. Reports of corruption in high places have multiplied.[111]

Controls over the decision-making process are in theory stringent, but, once again, the sensitivity of the topic makes it difficult to assess how and how well they actually work.

Technical inspection

Although the DSTC is responsible for testing equipment before acquisition, the recommendations of its investigation reports can be enforced by directives and inspections after the event. However, problems often arise as maintenance is the responsibility of the armed services.[112] Even in the armed forces a 'maintenance culture' has not developed; wastage, accidents and overspending are the result.[113]

The inspectorates of the three armed forces have been mentioned above. Their tasks are clearly defined, but implementation is sometimes another matter as they report to the service chiefs. They are separate from the internal audit (which is responsible to the Secretary General, the civilian arm of the ministry). The Inspector General is a lower-ranking officer, but is called on to inspect the services and thus the service chiefs themselves, so that objectivity may sometimes be difficult to achieve. It is difficult, moreover, for the Inspector General to get dedicated staff from the services, since they would be required to question the decisions of their own chiefs; yet qualified staff are needed from the fields of logistics, engineering and so on.

The drawbacks of this process become clear when accidents occur. The Inspector General forms an independent investigation team, drawn not only from his department but from all parts of the armed services. This team reports to the Inspector General, who in turn reports to the service chief.

VIII. Conclusions and recommendations

Problems in the acquisition process

The problems identified in the existing arms procurement process are both structural and political, and are in a way summed up as 'rationalized decision rather than rational decisionmaking'.[114] The following can be identified: (*a*) the

[110] *New Straits Times*, 1 May 1997. According to the PCB annual report for 1995, the majority of complaints were against the Home Ministry, the Finance Ministry and the Prime Minister's Department: the Ministry of Defence came off rather well in comparison, with only 14 complaints from the public in 1995, as against 28 complaints in 1994. The full report was not published.

[111] Balakrishnan (note 4), p. 5.

[112] Personal communication from the MOD, Nov. 1998.

[113] The Prime Minister has criticized a lack of maintenance culture in the country. *New Straits Times*, 6 Jan. 1999, p. 2.

[114] Abdul Rahman Adam (note 14), p. 17.

lack of an explicit national defence policy and a joint operations doctrine, which in turn inhibits proper definition of arms procurement needs; (*b*) a less than stringent budget review which is liable to political intervention; (*c*) external intervention in arms procurement decisions; (*d*) a lack of transparency and predictability in the acquisition processes and decisions; (*e*) insufficient political and public accountability (in contrast to financial accountability); and (*e*) the indifference of Parliament to proper oversight over procurement matters, which is the result of general public indifference to military and defence matters.

Absence of a national defence policy

Apart from the principles stated in *Malaysian Defence*,[115] no long-term defence planning document exists which would make the procurement process rational and more efficient. Decisions on weapon systems are changed without professional rationale and often unbeknown to the armed forces.[116] The military's requests over the past several decades for a White Paper on defence policy have not been heeded sufficiently and *Malaysian Defence* did not adequately address the need for joint planning.[117] As late as the 1980s there was an 'absence of a well-conceived and co-ordinated joint operational military doctrine. This has resulted in confusion over the acquisition of appropriate weapons systems due to financial considerations, and also the inevitable inter-service squabbles'.[118]

Inter-service rivalry is also indicated by the army's attempts to retain its preponderance of the past four decades, despite the changed circumstances, and keep its share of budget at the cost of the other services. This has led to the three services developing their own versions of operational doctrine, which have not yet been made public. Without a joint operational doctrine, arms procurement decisions cannot address Malaysia's security needs in a comprehensive manner.[119] With an integrated planning process, inter-service rivalry would be likely to decrease.[120]

The budget process

A comprehensive budget review process is lacking in Malaysia, which relies on the MBS.[121] This can lead to external interference in the budget review, push up costs and lead to questionable practices which could provide opportunities for

[115] Malaysian Ministry of Defence (note 11).

[116] A piquant example appeared in the *New Straits Times* and *New Sunday Times* on 11 and 12 Oct. 1997, p. 1 and p. 2, respectively. It was reported that Australia claimed that the order for some vessels for the Royal Malaysian Navy (RMN) had gone to a German consortium rather than to Australia in spite of competitive bidding because former Prime Minister Paul Keating had called Malaysian Prime Minister Mahathir 'recalcitrant' over APEC (the Asia–Pacific Economic Cooperation forum). No independent confirmation of that claim was forthcoming.

[117] Chandran Jeshurun (note 31), p. 200.

[118] Chandran Jeshurun (note 31), p. 202.

[119] Abdul Rahman Adam (note 14), p. 11.

[120] Two army training bases that were to have been closed were therefore restored when the Tornado deal fell through. Yap Pak Choy (note 35), pp. 44, 50.

[121] See section V of this chapter.

corruption. The MBS was an improvement over the earlier Programme Performance Budget System (PPBS) because it looks at all stages of spending and at results against stated objectives, but it does not provide for programme evaluation or alternative costings. Thus incremental budgeting, which the MBS should have prevented, creeps into the process in the drive to reach targets.

Discussions in budget committees in the MOD, at regular budget reviews and in the Public Accounts Committee often fail to scrutinize projects where very large sums of public money are involved.[122] 'Soft' issues and increases at the margin are thus discussed in great detail whereas major items are passed easily or requests simply carried forward from year to year. Past expenditures concealing ever-higher life-cycle costs are treated as continuing commitments.

External influences

The military in general resents two kinds of outside influence in the arms procurement process: (*a*) by political and civilian agencies whose priorities might differ from its own; and (*b*) by external agencies pursuing commercial or personal interests.

Where political agencies are concerned, the services complain of 'ad-hocism' and muddled competences. The Treasury holds the 'purse-strings'. The services are asked what they require technically and professionally and advise accordingly, but because neither the Treasury staff nor the auditors have, by their own admission, the necessary military or technical expertise they go by cost and what is on offer. The resultant equipment is therefore often below optimum; this was the case with the four Assad Class corvettes ordered from Italy in 1995 because they happened to be available cheaply. The Cabinet has at times awarded tenders without informing senior civilian and military officials of the MOD, which the latter acknowledges is the Cabinet's prerogative but is not considered courteous.[123] When the Cabinet decided to award a contract to the German Naval Group (GNG)[124] the MOD received this information from the GNG.[125] The decision to buy the MiG-29S was made for valid economic and political considerations, which, however, totally disregarded the operational merits of the aircraft.[126] Such instances can create tension between the military and the civilian agencies which control the process of arms acquisition.

It must be said, however, that it is doubtful whether the military by itself could make informed choices with regard either to defence policy or to technology assessment. In the absence of basic and applied R&D in the country, it is not certain that even it has the necessary resources to test equipment.

[122] The well-known 'bicycle-shed' syndrome occurs: not much is known about sophisticated equipment involving large sums of money and hardly anybody feels directly affected by these purchases or equipped to evaluate them, but the discussion on building a cheap bicycle shed is hotly contested.

[123] Abdul Rahman Adam (note 14), p. 13; and Personal communication from MOD staff, Nov. 1998.

[124] See appendix 4A in this volume.

[125] Personal communication with Mr Arunasalam, Managing Editor, *Asian Defence Journal*.

[126] Yap Pak Choy (note 35), pp. 45–46.

Robless argues that the problem lies also in the character of the military itself: it is not geared to handle the sophisticated requirements of specification writing, comparative financial evaluations, project handling, lobbying and so on. Cost estimates especially are done badly, leading to cost overruns, reviews and delays.[127] An observation by the retiring RMAF Chief Lieutenant-General, Datuk Seri Abdul Ghani Aziz, in 1996 made it clear that problems of unsuitable equipment being selected might not necessarily be avoided even if the military had a greater say in the matter. He alleged that former RMAF officers working for defence equipment suppliers had frequently compromised the safety of their former colleagues and the operational readiness of the RMAF by selling 'unsuitable and obsolete equipment'.[128] He demanded that 'profit-driven' arms dealers should not take advantage of their connections with officials in the armed forces in order to sell them inferior equipment.[129] These charges were refuted by the Defence Minister, who stated that he had not received any reports about air force equipment not being up to the required standards. A day later a Nuri helicopter crashed.[130] Instances like these reinforce the bureaucratic attitude that sees little sense in defence spending, given a low threat perception and economic difficulties, and will delay or stifle it through complex procedures and over-zealous screening.

The problem of influence being used for commercial or personal interests has less to do with the process of arms procurement than with transparency and accountability. Alatas sees the decision-making process as sometimes dominated more by the interests of individuals in securing interpersonal relationships within the hierarchy than by the efficacy of the items purchased. Even though confidentiality is needed in many areas, a better-informed public would reduce errors of judgement and the influence of the interest groups.[131]

Lack of administrative transparency

'Transparency takes place only after all decisions are made and when the top officials are ready to announce their decisions.'[132] The balance between confidentiality relating to the effects of arms procurement and the deployment of weapons, on the one hand, and the public's right to information to evaluate defence expenditure and procurement decisions in order to avoid waste and abuse in the system, on the other, has remained unaddressed.

The Official Secrets Act is available to be used at the discretion of the executive, which determines what is secret information.[133] Even the courts have little leeway to question the executive, although they can entertain appeals against

[127] Robless (note 54), para. 18.
[128] *New Straits Times*, 9 Aug. 1996, p. 4.
[129] Lim Kit Siang, URL <http://www.lks.tm.my>, 11 Aug. 1996.
[130] See note 103.
[131] Sharifah Munirah Alatas (note 8), pp. 24–26.
[132] Balakrishnan (note 65), p. 10.
[133] Faridah Jalil and Noor Aziah Hj. Mohd. Awal (note 47), p. 8 One participant in the UKM–SIPRI workshop cited an case in which the racial breakdown of prostitutes in Malakka in 1948 was deemed to be information falling under the Official Secrets Act and therefore not accessible to a researcher.

decisions relating to penalties.[134] Even this is rarely done, because the courts are reluctant to interfere with ministerial decisions.[135] Very few political leaders and senior officials are aware of all the decisions made at every step of the procurement process. This is not at all unusual, given the need for confidentiality during any tender process to prevent undue influence-peddling.

There seems to be a resistance to greater transparency on the part of both government and military. It has been said that information is not even fully shared between government agencies: the MAF have their own long-term security and acquisition plans, but these may not necessarily be known to the officials engaged in defence budget making.[136] The government is secretive not only in arms procurement processes but in the majority of its transactions with the private sector as well. It is thus a problem of organizational behaviour that is not easily amenable to solution in the security sphere alone.

The arms procurement process has been described by Mak as transparent at the macro level and opaque at the micro level, partly because there is no tradition of open debate about military procurement.[137]

Here more than in any other public procurement process there is a natural tension between the desire for accountability and the need for confidentiality. The question must be asked whether it is transparency in the decision-making process that could be harmful to security or transparency of the eventual decisions. Transparency is not, of course, the same thing as accountability. Nor does it necessarily lead to greater restraint in arms acquisition or to more rational choices.[138]

Lack of public accountability

Public accountability works primarily in the financial and administrative realm. In principle it is quite highly developed, with every level of the executive or the military being accountable to that above, but at the highest political level of decision making there is no insightful parliamentary control over arms procurement—although Lim Kit Siang has been vocal in demanding greater transparency in defence matters over the past 20 years. He has repeatedly drawn attention to the fact that billions of ringgits have been spent for defence equipment that might actually endanger security personnel: 'Something is very wrong with the arms procurement process'.[139]

It might be argued that public accountability is evaded because of executive control on information. The principle of collective responsibility reduces the chances of wrong decisions being corrected. A strong government reinforces secrecy in the decision-making process. It prevents disagreements between depart-

[134] Faridah Jalil and Noor Aziah Hj. Mohd. Awal (note 47), pp. 9–10.
[135] Information from Faridah Jalil, 26 Feb. 1999.
[136] Personal communications with Dato' Richard Robless, Zakaria Haji Ahmad, J. N. Mak and others.
[137] Mak (note 15), p. 3.
[138] Mak (note 15), p. 4.
[139] Lim Kit Siang, URL <http://www.lks.tm.my>, 11 Aug. 1996.

ments coming out into the open and uncomfortable questions being asked. It is difficult for the public in general and even for the elected representatives to locate responsibility and accountability for any decision. Here accountability ties in with transparency.

Ultimately, the political elite has the final word. The Minister of Defence is accountable for arms procurement expenditure to the Public Accounts Committee and, given the principle of collective responsibility, cannot off-load this onto the Treasury publicly. There is little possibility for decisions to be scrutinized before they are made and the executive does not sufficiently realize that it is answerable to Parliament. The accountability of the executive to Parliament is in effect a myth, both because of the subservience of party members to their leaders and because of the weakness of civil society. This weakness leads to public apathy.[140]

The weakness of civil society

Could the absence of political accountability be attributed to indifference in civil society, which even if it has the means to acquire information is not interested to do so?[141]

The government proclaims itself 'open, liberal and responsible', but more often hands out information it thinks the public should have instead of information the public really ought to have. The overwhelming dominance of the ruling coalition in Parliament makes it difficult to question the government. The public seems to refrain from asking questions or seeking information. Fear of losing their licences prevents newspaper publishers and journalists from voicing strong criticism or investigating defence matters too closely.

It is true that immense difficulties exist not only in accessing information but also in verifying it. Researchers are often not allowed to quote documents and papers consulted or are denied access even to documents tabled as Cabinet papers. Public documents can often be accessed by researchers only by recommendation or private contacts and permission often depends on the political connections of the applicant. Moreover, the public cannot enforce access to information by legal means or through the courts.[142]

However, information on arms procurement decisions and expenditure is accessible in government publications and reports, legal journals and *Hansard*. The fact is that even where access to information is possible inside or outside Parliament the opportunities are little used. Question Time is poorly attended. Research on military organizations and national defence policies has received scant attention in studies sponsored by the government; fear of the Official Secrets Act may have discouraged research in this field.[143]

[140] Sharifah Munirah Alatas (draft, note 8), p. 25.

[141] Faridah Jalil and Noor Aziah Hj. Mohd. Awal (note 47), p. 8.

[142] Faridah Jalil drew the author's attention to this problem.

[143] Chandran Jeshurun (note 31), pp. 194–95; and Chandran Jeshurun, presentation at the UKM–SIPRI workshop, 18 Aug 1997.

Another possible reason for the indifference of the public in security issues is that the government is seen as the protector of the economic and physical well-being of the country in a paternalistic sense.[144] Security issues are left to the government while the citizens engage in the economic sector. Historically, the task of safeguarding the interests of the Malay polity was left to the ruler, and the people did not question this as long as it worked. This attitude, combined with the legacy of the colonial past, the race riots of May 1969 and the communist insurgency, worked against developing public awareness and oversight of defence policy. The Internal Security Act and Official Secrets Act played their part in building a culture of secrecy, which has still not been overcome. The major concerns of the public in arms acquisition processes are that the MAF are not fobbed off with inferior or outdated equipment and that delivery is on time.[145]

The Malaysian deference towards authority seems to impede the emergence of a questioning attitude in society. The benefits of strong leadership have corresponding drawbacks: strong government can become authoritarian government. That the present Prime Minister, Mahathir Mohammad, has realized these problems is indicated by his call for a mature democratic society, liberal, tolerant, self-confident and subservient to none—in short, a functioning civil society.[146]

Recommendations

Among the major recommendations made by the experts for improving transparency and accountability are: (*a*) an information policy which is as open as possible on the financial powers, decision-making methods and responsibilities of arms procurement; and (*b*) powers, methods and capacities for parliamentary committees and the Public Accounts Committee to examine arms procurement expenditure and decisions, to sanction expenditure, to censure and to institute action, especially where confidentiality is necessary.

1. A short-term measure to improve accountability would be to identify where responsibility for decisions lies at various levels in the arms procurement decision-making process, outlining the military, technical and administrative imperatives governing decisions.

2. The public could be generally informed about what types of armament are under consideration—for instance, whether the government will be buying combat aircraft or transport aircraft. Access to such information through official reports or elected representatives will improve the quality of debate in the press and among the public and allow corrections to be suggested. If the public are

[144] Sharifah Munirah Alatas (note 34), pp. 33–36.

[145] This happened with the F-2000 frigates ordered from the GEC-Yarrow shipyard in 1994. The delivery date of 1996 was not kept and has now been put back for the 4th time. *New Straits Times*, 28 Apr. 1998, p. 6.

[146] YAB Seri Mahathir Mohamad, 'Malaysia: the way forward', Centre for Economic Research and Services, Malaysian Business Council, Kuala Lumpur, 1991.

aware of the reasons for decisions, even if the decisions cannot be theirs, this would help to achieve a balance between socio-economic priorities and the military's requirements.[147] Transparency in this sense would produce better, more rational equipment choices. However, constitutional provisions that are deemed essential for regime control, like the Internal Security Act and the Official Secrets Act, are unlikely to be repealed in the near future.[148]

3. Accountability in decision-making processes would develop predictability, which in turn would facilitate forward planning for capability building. It would also help the armed forces to project their needs in the long and medium term.

4. Accountability in the arms procurement process could start with improved methods for budgeting and auditing, such as a modified PPBS which, although time-consuming and intricate, provides a stringent method of checking spending and programme evaluation. Arms procurement programmes should be judged according to comprehensive criteria based on rationality. These factors could be considered along with the overall foreign and socio-economic policies in deciding the national budget.

5. A White Paper as an implementation blueprint, as requested by the military, could also improve political accountability.

6. The influence of civil society in encouraging restraint in arms procurement is not definitely established.[149] Society might clamour for weapons rather than concern itself with the interests of peace. On the other hand, giving the military the final say in the acquisition process may not give due weight to other national priorities, which are best expressed through the society's elected representatives.

The research on which this chapter is based set out three objectives: to describe the decision-making process within the context of military, budgetary and economic constraints; to ascertain its efficiency in terms of stated goals and guidelines and/or identify obstacles to efficiency; and to discuss checks and balances in the system and whether they are working as intended.

The process of arms procurement as designed seems to be efficient, if time-consuming, in part because of strong centralization. Both civil and military executives have sometimes wished for more autonomy and blamed intervention by outside agencies for introducing inefficiencies, whether these are government departments not normally concerned with arms procurement, politicians taking decisions on arms procurement for reasons of economic or foreign policy, or private interests. The military has sometimes had to contend with decrees from outside or a higher level of authority which foisted unsuitable equipment on it. Here the role of an informed public was seen by the participants in this study as most crucial.

[147] Robless (note 54), para. 9.

[148] Mak (note 15), p. 19.

[149] One participant remarked during the UKM–SIPRI workshop that without a civil society the USA would probably produce and export even more arms, and that the Japanese pacifist stance was not a matter of choice but externally induced. However, this does not invalidate the basic argument.

While the formal arms procurement process is monitored by the relevant agencies in the executive branch, the principles underlying it are hardly discussed and criticized or put to public scrutiny even after decisions have been made, let alone before.

Could this system of arms procurement become more responsive to the objectives of public priorities and harmonize accountability with military confidentiality? In the Malaysian context, this is difficult. The agencies that play a role in this regard—the non-governmental organizations (NGOs), the press and the media who would have an interest—are seen as irritants and liable to censorship, or to censor themselves so that their potential to bring about change is limited.[150] The military establishment or the general public could influence the government in the direction of more, or less, or different arms procurement, but they are unlikely to do this in the near or medium-term future. Civil society will probably continue to take little interest in arms procurement or defence policy in general. Threat perceptions are muted and the country is considered stable and secure. The economic situation is currently having a restraining influence on arms procurement. The arms industry is in its infancy and not in a position to generate a powerful arms lobby. The voice of the military will only become decisive if the country becomes completely stable or if a definite security threat emerges. Until that happens the military will acquiesce in making do with what they are given, without questioning the process too much.

[150] Mak (note 15), p. 4.

Appendix 4A. Offset arrangements between Malaysia and the German Naval Group*

In September 1998 Malaysia concluded a contract with the German Naval Group (GNG) consortium for the purchase of new-generation patrol vessels (NGPVs). The arrangements are as follows:

1. The contractor should partly compensate his profits in Malaysia through investment, industry development projects, licensed production or counter-trade (for instance, through the promotion of Malaysian products overseas).
2. Malaysian industry participation should be at least 30 per cent (in the GNG contract it was 50 per cent).
3. Foreign currency compensation should be 70 per cent, that is, money that leaves the country as profit should be reinvested or otherwise compensated.

Before bidding for the NGPVs started, the potential bidders set up partnerships with local companies in order to be better positioned for the contract. As the Lumut Naval Dockyard (Naval Dockyard Sdn Bhd, NDSB) was the prime local contractor, the German bidders were in a comparatively strong position, since they had helped to set up the dockyard and had experience in working together with the company.[1] Other bidders made their bid through the heads of their operations in Malaysia—Ericsson, for instance, through a former Chief of the Royal Malaysian Navy.[2]

The NDSB deals in:

1. Design and licensing for the production of the vessels in the country by the GNG.
2. Major equipment and systems (MES). For these, tenders are issued directly from the Ministry of Defence (MOD), which subsequently makes a recommendation to the Treasury, from where a shortlist goes to the GNG and the NDSB. The latter two check for Malaysian industry participation and offset arrangements and then make their own recommendation, the GNG on the basis of risk calculation, the NDSB on the basis of profit. The Treasury then puts out the modified final list.
3. Non-MES items, for which no tender goes out. These are either products which have a small ratio of Malaysian participation or mass products which are already manufactured in-country, in which case a Vendor Development Programme (VDP) is applied to promote certain dual-use products made by industries participating in the project. The NDSB chooses the vendors under the VDP directly.[3]

[1] Gamal Fikry, 'Malaysia's NGPV programme enters decisive stage', *Asian Defence and Diplomacy*, Special Issue 1995, p. 75.

[2] *Asian Defence and Diplomacy*, Special Issue 1995, p. 77.

[3] *Asian Defence Journal,* no. 8 (1995), p. 18.

* The information in this appendix, except where otherwise indicated, was kindly provided by D. E. Dasberg, Senior Project Leader, GNG, 20 Oct. 1998. His patience in clarifying the GNG project for the author is gratefully acknowledged.

For the NGPV project the Treasury appointed the NDSB as the local main contractor for the whole platform. One intended side-effect of the project will be the upgrading of the shipyard and possible construction of vessels for a wider market.

The navy participated in the tender stage, giving the specifications, while the Treasury determined the permissible cost. The navy took part in the negotiations again at the stage when building and construction specifications were to be formulated.

5. Poland

Pawel Wieczorek and Katarzyna Zukrowska[*]

I. Introduction

The arms procurement procedures which currently apply in Poland have been created since 1994, within the framework of the transformation of the country's political and economic systems. The changes made in the arms procurement legislation had three goals: a more transparent decision-making procedure than that in use before 1989; civilian and democratic control in the planning and procurement of weapons and military equipment, bringing them closer to the standards applied in NATO member countries; and adaptation of the rules for the placing of orders by the Ministerstwo Obrony Narodowej (Ministry of National Defence, MoND) to the general rules applied in the member states of the European Union (EU), which Poland will join.

Arms procurement planning and implementation have not been studied in Poland in the past. Consequently there is very little literature available in this field, and what is available is fragmentary in nature. The main sources for research in this field have been interviews with people engaged in the arms procurement decision-making process working in different institutions.

This chapter aims, first, to present the main elements of the arms procurement decision-making process currently in operation. Second, it indicates in which directions this process should ideally go to improve public accountability, which would serve to increase transparency. This would involve the strengthening of civilian and democratic control over the military in this regard. Third, it identifies barriers and limitations in introducing the proposed changes.

II. The management of national security and defence planning

Defence management in Poland began to be brought fully under civilian control with the political changes that were launched in 1989. Since the 1990 elections the MoND has been headed by civilians.[1] The military staff is responsible for

[1] Article 26, point 2 of the Constitution of the Polish Republic of 6 Apr. 1997 states that 'Military forces keep political neutrality and are subject to civilian and democratic control'. The constitution was approved on 25 May 1997 in a national referendum and signed by the President on 16 July 1997.

* The authors gratefully acknowledge the help of Dr Janusz Reiter and the Centre for International Relations for assistance with the workshop, held at the Institute of International Affairs in Warsaw on 26 Nov. 1997, within the framework of the SIPRI Arms Procurement Decision Making Project. Twelve working papers prepared by Polish researchers as part of the project have been used in preparation of this chapter. They are not published but are deposited in the SIPRI Library. Abstracts appear in annexe B in this volume.

purely military and technical matters but not for the running of the ministry or for defence policy. At the beginning of the political changes, authority over the armed forces was not clearly divided between the parts of the executive branch—the President, Prime Minister, Defence Minister and Foreign Minister. However, with the passage of time there has been increasing clarity, even though some ambiguity remains as to roles and authority.

The Interim Constitution of 1992[2] stated that a will for cooperation and compromise should prevail among the main centres of power. The experience of the period of the Interim Constitution was not encouraging, mostly because the party system in Poland was not well developed and power was often seen in terms of personal position and influence rather than in terms of effective government. When the President and Prime Minister came from different parties they were not eager to cooperate. Moreover, members of Parliament played their own games according to their relations with the President and Prime Minister (although to some extent the fact that different political parties were involved helped to provide a check on the decision-making process, in a way which might not be possible with only one party represented).

These shortcomings have been removed by the 1997 Constitution, which has clarified the division of powers.[3] Article 134 states that the President is the supreme commander of the armed forces. During peacetime his power in this regard is exercised indirectly by the Minister of National Defence. The Chief of the General Staff and the commanders of the armed services are directly subordinated to the defence minister.

The President nominates the Chief of the General Staff and the Chief Commanders of the individual armed forces for a specified period of time. The method and conditions of their removal from these posts are set out in separate regulations, as is the Commander-in-Chief's subordination to the organs of state. In time of war the President appoints the Commander-in-Chief of the Polish Armed Forces, on the recommendation of the Prime Minister. All the powers of the President over the military are clearly stated in the legislation.

Poland does not publish a defence White Paper, although at this stage of development of defence policy-making methods it would be advisable. A publication of this kind should contain the main principles of the defence policy of the state: (*a*) threat assessment; (*b*) the military budget; and (*c*) preparation of the domestic defence industry to meet expected demand (and the share of imports). The MoND does prepare material on these issues for the Komisja Obrony Narodowej (National Defence Commission) of each of the two chambers of Parliament, but these documents do not give adequate details.

'Konstytucja Rzeczypospolitej Polskiej', *Dziennik Ustaw Rzeczypospolitej Polskiej* [Journal of legislation], no. 78 (6 Apr. 1997), poz. [item] 483. See also URL <http://www.sejm.gov.pl/english/konstytucja/kon1.htm>.

[2] 'Ustawa Konstytucyjna' [Constitution act, commonly known as the 'small' constitution], *Dziennik Ustaw Rzeczypospolitej Polskiej*, no. 75 (30 July 1992), poz. 367; and no. 84 (17 Oct. 1992), poz. 426, 23.

[3] See note 1.

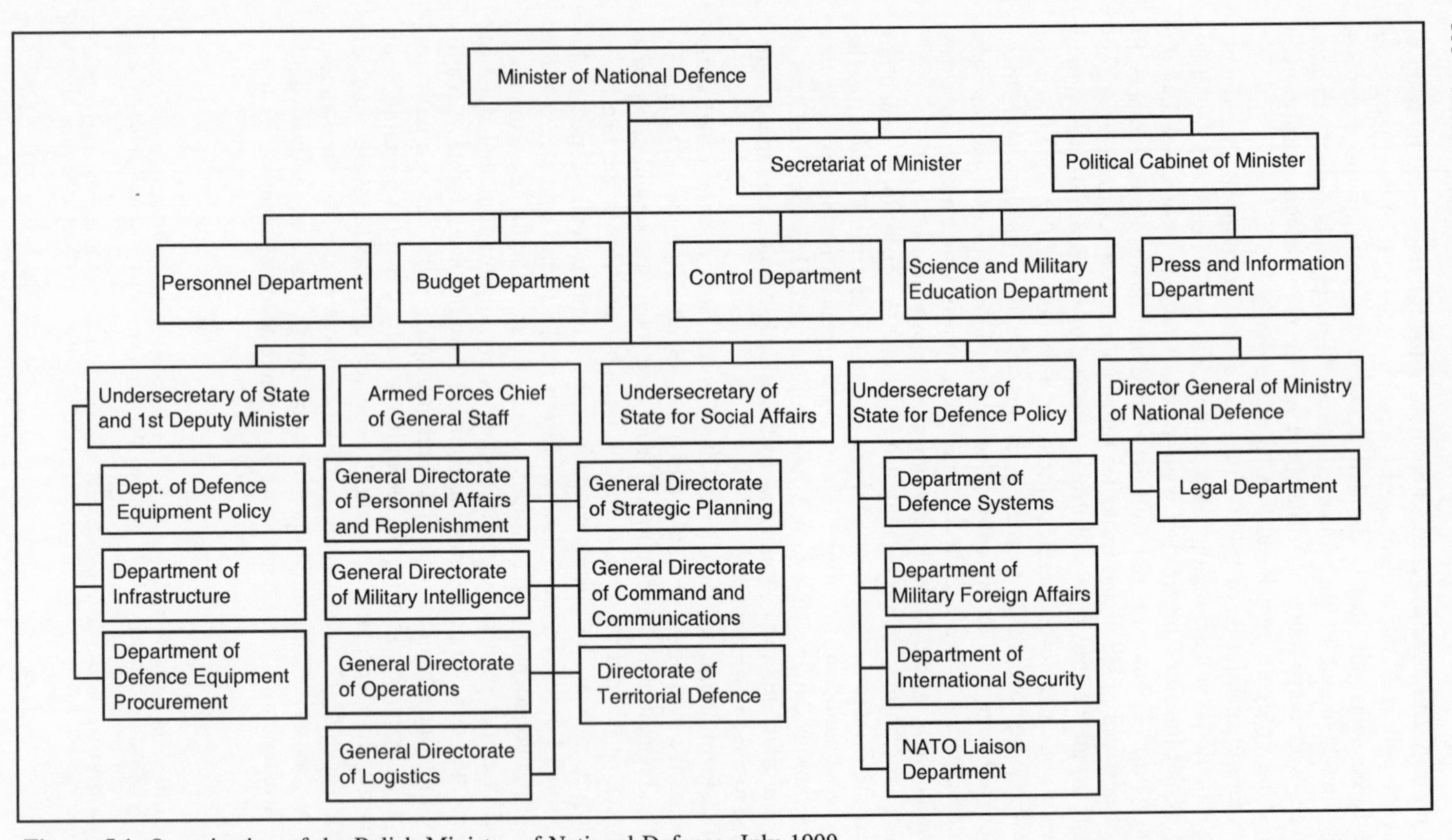

Figure 5.1. Organization of the Polish Ministry of National Defence, July 1999

Source: Polish Ministry of National Defence, 'Structure of Ministry of National Defence', URL <http://www.wp.mil.pl/orgresa.html>.

The actors in defence policy making and the arms procurement process

The actors in the executive branch include the MoND, the Urzad Zamówien Publicznych (Office of Public Procurement), the Komitet Spraw Obronnych Rady Ministrow (KSORM, Defence Committee of the Council of Ministers), the Ministry of Economy, the Biuro Bezpieczenstwa Narodowego (BBN, Bureau of National Security) and the Rada Bezpieczenstwa Narodowego (RBN, National Security Council). The number of actors is changing: until 1 January 2000 the Ministry of the Treasury was also involved. In certain instances the role of these bodies is limited to approval of documents; in other cases it includes evaluation, the formulation of opinions, expert advice and plans for execution. It also includes the drafting of the MoND budget requirements and the financial basis of the defence budget.

Other actors such as producers of military equipment, the trade unions and lobbies are also engaged in the process. In formal terms their role is limited, but in practice the political influence and intervention of the trade union organizations often makes the whole process highly politicized. The roles of these institutions are discussed in section VII of this chapter. The roles of Parliament and the Najwyzsza Izba Kontroli (NIK—Highest Chamber of Control or Polish Auditing Office, the main auditing and control body) are considered further in sections IV, V and VIII.

The Ministry of National Defence

The organizational structure of the MoND as revised in July 1999 is shown in figure 5.1. The MoND is responsible among other things for defence policy, armaments, infrastructure and the defence budget. The latter engages several departments besides the Budget Department: for instance, the General Staff works out the Chief of the General Staff Guidance for Material–Financial Planning in the Armed Forces. The departments of procurement, equipment policy and infrastructure are the responsibility of the same undersecretary of state. The Undersecretary for Defence Policy deals with foreign military affairs, international security and liaison with NATO. The Chief of the General Staff has six directorates whose roles and designation are similar to those of the corresponding US military organizations, and a seventh for territorial defence. Other functions are as follows:

(*a*) the central logistics organizations—the three services' Commands-in-Chief, and under the General Staff the Material Directorate, the Technical Directorate and weapon and equipment users. The commands of the different branches of service also participate in working out the Requirement Specifications for Weapon Systems and Military Equipment Procurement, Maintenance, Research Work and Implementation;

(*b*) the departments under the Undersecretary of State (who is usually the First Deputy Minister). They prepare the Weapon Systems and Military Equipment Procurement and Maintenance Annual Plan and the Research Work and Implementation Annual Plan;

(*c*) the Undersecretary of State/First Deputy Minister and the Chief of the General Staff, who both accept the above plans; and

(*d*) the Minister of National Defence, who gives final approval to the plans.

The plans approved by the Minister of National Defence are the basis for the MoND Department of Defence Equipment Procurement to start the executive procedures and for the military research centres and institutes to begin to implement the research projects approved.

Two other departments in the MoND are also involved in the procedure. First, the Control Department—the internal audit department—supervises and monitors ordering procedures in the individual departments of the MoND—the legal and formal correctness of the proceedings and compliance with the regulations concerning public procurement and the budget law. (External control is done by the NIK and Parliament.) Second, the Legal Department protects the activities of the MoND institutions which place orders.

There is also a special Komisja Bezpieczenstwa Narodowego (Commission on National Security) within the framework of the MoND, which was established to give Members of Parliament (MPs) who represent the interests of the Polish defence industry an insight in this area. Since 1997 representatives of the Parliamentary Defence Commissions have also participated in meetings of this commission, without having the right to vote. Hitherto there has been no formal component of the Parliamentary Defence Commissions to deal with these problems.

The Ministry of Economy

The ministry is involved: (*a*) when military technology is received from foreign sources; (*b*) when there is a need to issue licenses; (*c*) when military equipment has to be imported and exported; (*d*) when dual-use technology is transiting through Polish territory; and (*e*) when dual-use technology is exported from Poland.[4] The main role in this particular case is played by the Department of Export Control.

The Ministry of Economy was also recently involved in the industrial restructuring plan and drafting the 1999 rules for compensation in arms purchases, which provided the guidelines on offsets in arms purchases.[5] These are discussed further in section VI of this chapter. Since 1 January 2000 it has taken over from the Ministry of the Treasury[6] direct responsibilities in arms procure-

[4] Sliwowski, J., 'System kontroli eksportu w Polsce' [The system of export control in Poland], Paper prepared for the international conference on Cooperation of Enterprises with State Administration on Export Control, Warsaw, 13–14 May 1999.

[5] 'Ustawa o niektórych umowach kompensacyjnych zawieranych w zwiazku z umowami dostaw na potrzeby obronnosci i bezpieczenstwa panstwa' [Regulation on certain compensation agreements concluded as part of agreements concerning supplies for the defence and security needs of the state], 10 Sep. 1999, *Dziennik Ustaw Rzeczypospolitej Polskiej*, no. 80 (1999), poz. 903.

[6] 'Rozparzadzenie Rady Ministrow z dnia 17 listopada 1999 w sprawie wykazu spolek, przedsiebiorstw panstwowych i jednostek badawczo-rozwojowych, prowadzacych dzialalnosc gospodarcza na potrzeby

ment decision making in the case of enterprises that it owns—the 38 companies in which 25–100 per cent of shares is controlled by the state.[7] When the share it owns does not exceed 51 per cent, the control is limited to delegating representatives from the ministry to the management boards of these companies.

The Defence Committee of the Council of Ministers

All opinions concerning defence issues are evaluated by the KSORM. It comprises representatives of all the ministries that are important from the point of view of state security and often calls on the opinions of experts in developing its recommendations. It also has specialized working groups, including one which deals with arms procurement issues. Its meetings are not regular, and their frequency depends on the problems that arise.

The Presidency

Two more organizations come under the President. First, the BBN is the advisory body to the President, set up in 1991 to assess the security threats to the state.[8] Its duties are set out in Article 135 of the constitution. It includes representatives of all the state bodies that deal with security issues as well as independent experts. Second, the RBN replaced the former Komitet Obrony Kraju (KOK, Country Defence Committee) in 1998.[9] The division of responsibilities between the RNB, the BBN and the MoND may require further improvement. The members of the RBN are nominated by the President.

The Office of Public Procurement

The Office of Public Procurement monitors compliance with the regulations on public procurement. It guarantees that public funds are spent according to the requirements formulated in the law, which also includes competition. In the case of defence orders its role is rather limited: there are specific regulations for procedure here.[10]

Each ministry has a department to deal with its own procurement. The Department of Defence Equipment Procurement in the MoND is responsible for arms procurement in that ministry.

bezpieczenstwa i obronnosci panstwa' [Regulation of the Ministerial Council, 17 Nov. 1999, on the list of joint-stock companies, state enterprises and research units conducting economic activity for state security and defence purposes], *Dziennik Ustaw Rzeczypospolitej Polskiej*, no. 95 (1999), poz. 1102.

[7] See section VI in this chapter.

[8] Koziej, S., *Kierowanie Obrona Narodowa Rzeczypospolitej Polskiej/National Defence Management of the Republic of Poland* (Wydawnictwo Adam Marszalek: Warsaw, 1996), pp. 41, 42 (in English and Polish).

[9] The KOK, a body created under the communist system and not provided for in the new constitution, was responsible for drawing up policies related to national security and defence management and worked according to the decisions of the Council of Ministers. Koziej (note 8), pp. 39, 40. For background to the setting up of the RBN, see Stachura, J., 'Arms procurement decision making in Poland', SIPRI Arms Procurement Decision Making Project Working Paper no. 94 [1998], p. 7.

[10] See section IV below.

The Polish defence policy-making process

Figures 5.2 and 5.3 summarize the process. It involves the following stages.

1. All the government ministries concerned and the BBN participate in formulating the defence policy.
2. The MoND prepares the initial project with the participation of representatives of other ministries and institutions. This is drafted by a team headed by the First Deputy Defence Minister and approved by the Defence Minister.
3. The KSORM analyses the draft policy and makes recommendations, consulting the RBN. Contentious issues are resolved and the final draft is sent to the Council of Ministers for acceptance.
4. If a document has to be approved by the Sejm (the lower house of Parliament), the Council of Ministers sends it to the Defence Commission of the Sejm for discussion. It then goes to the Sejm as a whole.[11]

The participation of a number of bodies, in some cases with overlapping responsibilities, slows down the whole process by making it more complicated. On the other hand, it ensures that defence policy is formulated with the participation of all organizations responsible for its execution and in keeping with the requirements of legal and democratic procedures.

Transforming defence policy into defence programmes

Before defence planning comes defence forecasting—scientific forecasting of the future shape of the national defence and probabilistic estimates of the political, economic, social, military and other national and international factors which influence the shape of nation's defence, defining its needs. It creates the basis for decision and is done by scientific institutions, both state and private—the Akademia Obrony Narodowej (AON, National Defence Academy), the Rzadowe Centrum Studiów Strategicznych (RCSS, State Center for Strategic Studies), the Instytut Studiów Strategicznych Miedzynarodowego Centrum Rozwoju Demokracji (Institute of Strategic Studies of the International Centre for Democratic Development) in Cracow and the Centrum Stosunków Miedzynarodowych (Centre for International Relations) in Warsaw—and interdisciplinary research teams. A new centre, to assess threats to security, is also being set up in Warsaw under the auspices of the Warsaw School of Technology.

Defence planning consists of defining how to carry out the tasks identified by defence policy. It includes the creation of defence doctrine; war planning (including political and strategic defence planning); and long-term programming of the defence system and the armed forces. This leads to the formulation

[11] Koziej (note 8), pp. 42, 43. On the Parliamentary Defence Commissions, see section VIII in this chapter.

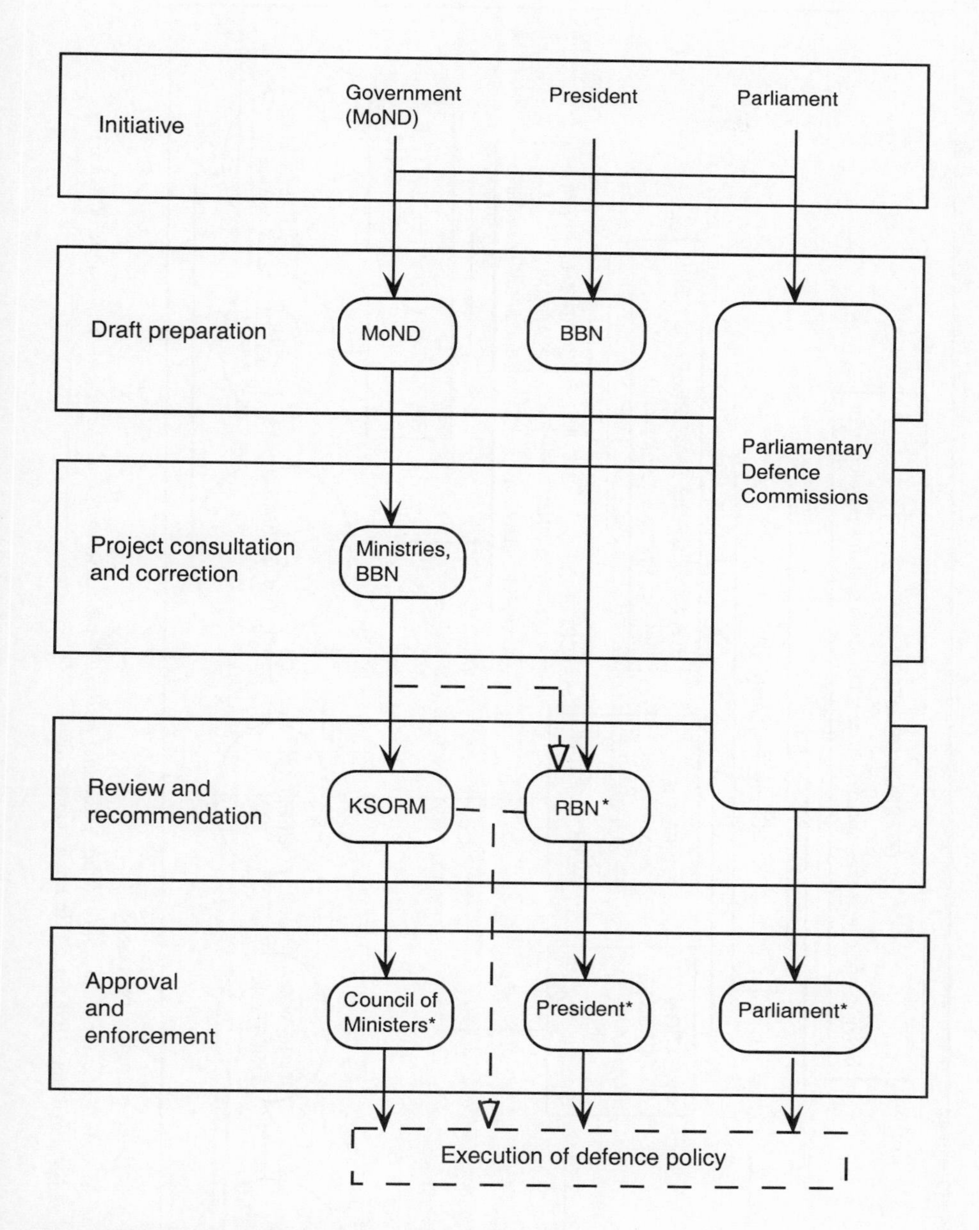

Figure 5.2. The Polish defence policy-making sequence

Notes: MoND = Ministry of National Defence; BBN = Biuro Bezpieczenstwa Narodowego; KSORM = Komitet Spraw Obronnych Rady Ministrow; RBN = Rada Bezpieczenstwa Narodowego. * = Bodies with decision-making authority.

Source: Koziej, S., *Bezpieczenstwo Narodowe i Obronnosc Rzeczypospolitej Polskiej/National Security and Defense of the Republic of Poland* (Wydawnictwo Adam Marszalek: Warsaw, 1996), p. 53 (in English and Polish).

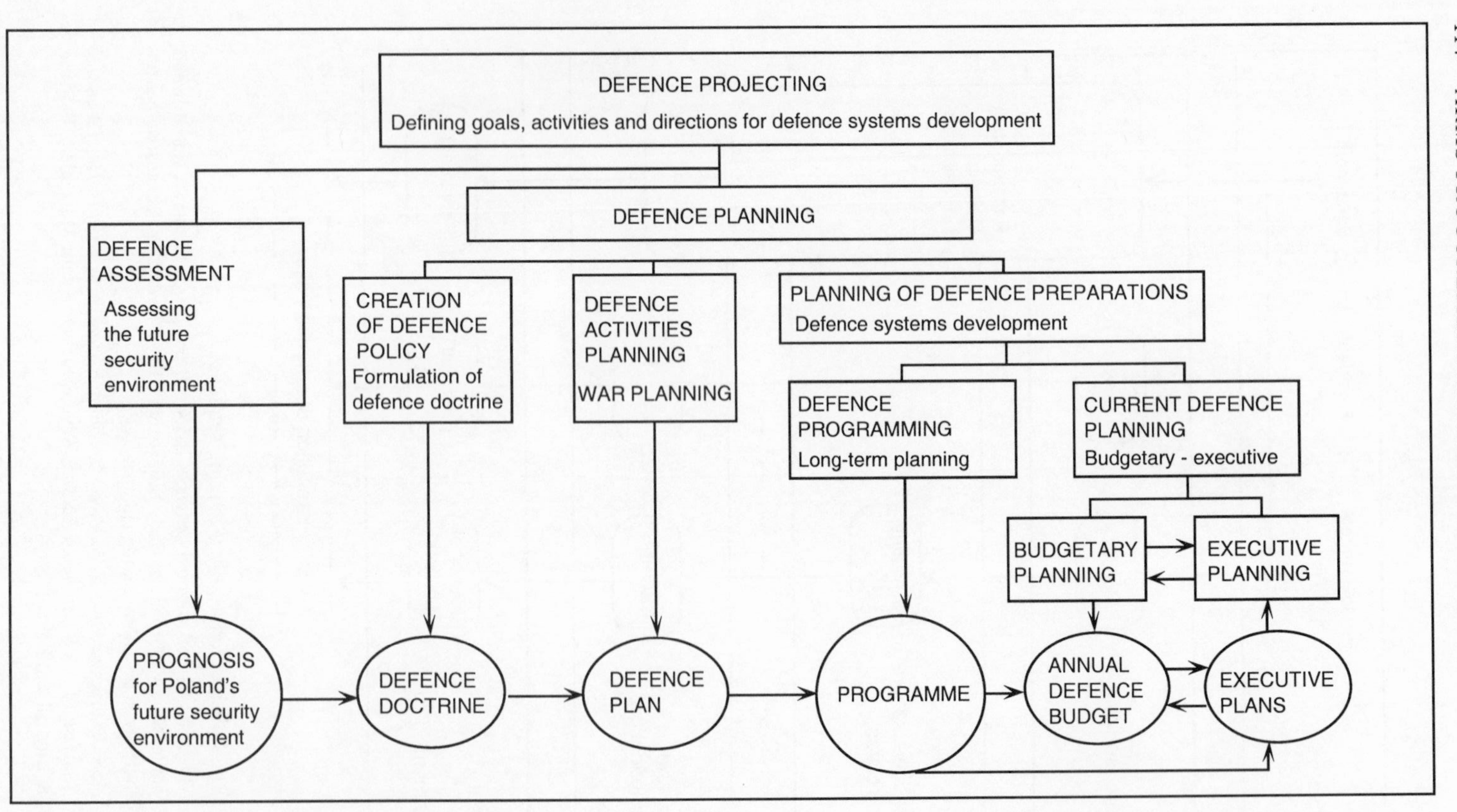

Figure 5.3. The Polish defence planning process

Source: Koziej, S., *Bezpieczenstwo Narodowe i Obronnosc Rzeczypospolitej Polskiej/National Security and Defense of the Republic of Poland* (Wydawnictwo Adam Marszalek: Warsaw, 1996), p. 54 (in English and Polish).

of the Political and Strategic Plan of Defence of the Polish Republic, which is developed by the MoND (including the General Staff) in close cooperation with the Ministry of Foreign Affairs, the Ministry of the Interior, the Central Planning Office and the Council of Ministers. The method of long-term planning in the Polish armed forces is still evolving. It was only in 1997 that the 15-year long-term plan for the modernization and restructuring of the armed forces was put into place.[12]

III. The arms procurement decision-making process

The arms procurement decision-making process consists of the preparation and planning stages, which are followed by execution. At the preparation stage the following guidelines are formulated: the Armed Forces Development Plan, the Ministerial Guidance and Budgeting Limitations, and requirement specifications. Technical analyses are conducted at the level of the General Staff and in the departments of the MoND that deal with procurement (the Department of Defence Equipment Procurement) and finance (Budget Department).[13] The planning stage involves the preparation of a Material–Financial Plan, the Weapon Systems and Military Equipment Procurement and Maintenance Annual Plan, and the Research Work and Implementation Annual Plan. At this stage the above plans are accepted by the Undersecretary of State and Chief of the General Staff, and finally approved by the Defence Minister.

There are two problems with the planning stage of the decision-making process. First, the institutions involved are still undergoing reform and their competences and responsibilities are still not finally established. The organizational structures of the MoND, the General Staff and the central military institutions underwent fundamental modifications in 1993, September 1996 and July 1999, reflecting changes required by Poland's membership in NATO.[14] Second, the organizational behaviour generally observed in this stage is characterized by passivity, play-safe decisions, and the avoidance of risk and responsibility. The planning process drags on during the initial stages, but accelerates rapidly towards the end, which reduces the time for decision making. This time limitation influences strongly the shape of the Weapons Systems and Military Equipment Procurement and Maintenance Annual Plan.

If a contract has a value of over 200 000 euros (about $200 000) then a member of the Parliamentary Defence Commissions and a representative from the MoND Department of Defence Equipment Procurement are invited to witness the work of the commission of experts that is set up to decide on each tender

[12] See section V in this chapter.

[13] Information kindly provided by Dr Andrzej Karkoszka, former Deputy Minister of Defence, Poland, Oct. 1999.

[14] The 2 latter implemented a bill on the Office of the MoND of 14 Dec. 1995. On 9 July 1996 and 19 Feb. 1999, new statutes for the MoND were adopted by the Council of Ministers. New regulations were promulgated by the Minister of National Defence on 3 Sep. 1996, whereby the General Staff was integrated into the MoND, and on 14 July 1999. This last stage brought the structure of the General Staff much closer to the models used in other NATO countries. Information kindly provided by Dr Andrzej Karkoszka, Oct. 1999.

operation,[15] but without a right to vote. However, working rules for this type of decision-making procedure have not been precisely defined so far. Computer-assisted methodologies to reduce subjectivity or preconceived judgements have not yet been implemented at all levels. It is impossible to avoid external influences or pressures entirely in any procedure of this type, but some attempt should be made to increase the objectivity of the decision-making process.

IV. Arms procurement procedure

In general two models of arms procurement can be distinguished. The first is followed for procurement within the budget of the MoND. It includes the MoND, arms producers, both houses of Parliament, the KSORM and the NIK. The second model concerns the procurement process outside the MoND budget (which covers what are called 'central projects') and is carried out with capital and technological cooperation between Polish and foreign companies. The second model applies when infrastructure investments are financed in cooperation with external sources. This model includes the institutions mentioned above and the Ministry of Economy.

The legal basis of the arms procurement procedure was formulated in the law on public procurement of June 1994.[16] The law on civilian and military public procurement is in principle the same and stated in the same regulation. Despite that, there are some major differences. As a temporary measure, orders for arms and related equipment aim to protect national producers.[17] This means that most tenders are not open to foreign suppliers. This is a temporary solution and is regulated by internal MoND regulations. The general rules for civilian and military procurement financed from public sources are regulated according to the solutions that are in use in the advanced democracies, as an effect of the harmonization of Polish law with the EU *acquis communautaire*.

Under this law, orders to be paid from public funds can be placed using the following procedures: (*a*) unlimited tender (treated as the basic method of ordering); (*b*) limited tender; (*c*) two-stage contracting; (*d*) competitive negotiations; (*e*) inquiry on price; and (*f*) 'free choice' procedure, meaning that purchase can be made in any shop.

Unlimited tender means a public tender open to all suppliers regardless of their location, size or organization. In the case of limited tender the invitation to participate is sent to a particular group of suppliers. In two-stage contracting the contract is granted to the winner of the second stage of a two-stage competition. Competitive negotiations mean that negotiations are conducted in parallel with different suppliers. Inquiry on price, sent out by the potential customer, helps to select a group of suppliers who can offer the most competitive conditions. The

[15] See section IV in this chapter.

[16] 'Ustawa z dnia 10 czerwca 1994, o zamówieniach publicznych' [Law on public procurement], 10 June 1994, *Dziennik Ustaw Rzeczypospolitej Polskiej*, no. 76 (1994), poz. 344, and subsequent amendments. See also URL <http://www.uzp.gov.pl/english/a_index.html>.

[17] See note 16.

free choice method is used in the case of minor purchases which can be made by the buyer without the requirement for competitive methods to be applied: the buyer can buy what he wants in any shop he chooses. The assumption is that competition on the market is sufficient to press prices down in the case of goods that are commonly used by both military and civilian customers (for instance, stationery and pens, although if large quantities of such goods are involved tender is often applied).

The value of an order is an important criterion in deciding procedure. If the order does not exceed 200 000 euros, the client decides alone, using the law on public procurement, if it is possible to depart from the general requirement of unlimited tender. If the order exceeds 200 000 euros any departure from the general rule requires the approval of the Head of the Office of Public Procurement. Up to 30 000 euros, the buyer can place the order according to the free choice procedure, and the regulations accept a simplified method, which means that some documentation is not required. If it is over 30 000 euros the order can only be placed in agreement with all elements of the procedure; in particular, full documentation linked with the contract has to be prepared.

In the case of arms procurement, a specific procedure which deviates from the general regulations can be applied in three cases—natural disaster; defence of the internal and external security of the state; and protection of state secrets. The regulations for these exceptions are formulated in a regulation of the Council of Ministers dated 20 August 1996[18] and concern such issues as the openness of the procedure, basic documentation, announcements about the procedure and its results, the time-limit for tenders, the supplier's right to cancel an order, and the requirement of approval from the Head of the Office of Public Procurement for any departure from the rule on unlimited tender.

If an order is placed abroad the regulations concerning preference for national supplies can be suspended. The general rules on preference for Polish suppliers require that at least 50 per cent of the value of goods and services offered by the supplier are produced with the use of Polish raw materials and products. If these conditions are met, a Polish supplier can expect that the order will be placed with him, even when the price of his products or services is 20 per cent higher than that offered by a foreign competitor.[19] In the case of orders for arms and military equipment national preference can be utilized when the input of raw materials and national products is lower than 50 per cent and the difference in price is more than 20 per cent.

If regulations linked with the protection of national security or state secrets apply and the order is to be placed abroad, the buyer can use the limited tender procedure instead of unlimited tender. When it is to be financed with public

[18] 'W sprawie okreslenia szczególnych zasad udzielania zamówien publicznych ze wzgledu na ochrone bezpieczenstwa narodowego, ochrone tajemnicy panstwowej, stan kleski zywiolowej lub inny wazny interes panstwa' [On defining special rules for the placing of public orders taking into account the protection of national security, the protection of state secrecy, natural disaster and other important state interests], *Dziennik Ustaw Rzeczypospolitej Polskiej*, no. 109 (1996).

[19] [Regulation of the Council of Ministers, 28 Dec. 1994, on utilization of national preference in planning public orders], *Dziennik Ustaw Rzeczypospolitej Polskiej*, no. 140 (1994).

money and foreign financing within the framework of an international agreement which provides for a different procedure from that defined by the law on public procurement, specific conditions may apply. This is so, for instance, when Poland is granted military aid credits by a state or international organization (NATO, the EU or the European Bank for Reconstruction and Development, EBRD) to assist adjustments towards international requirements, for example, development of military infrastructure: the suppliers of goods and services may, for instance, be provided by the supplier of the credit. This has not happened hitherto in Poland, but the regulations have to be flexible enough to foresee the possibility. Such regulations can also be used in cases when Poland finds a financial sponsor for activities planned within the framework of the armed forces modernization programme.

A regulation of December 1994 requires the ministers or the heads of the central organs of state to prepare, in consultation with the Head of the Office of Public Procurement, specific internal regulations to guide the discharge of their responsibilities. The procedure for placing orders for arms and military equipment follows the internal regulations of the MoND. These regulations cover the availability of documentation and the authority of different organizations in the process. According to a regulation approved in January 1997, the Director of the MoND Department of Defence Equipment Procurement has to present detailed regulations on the procedure.

Execution involves 10 separate steps: (*a*) analysis and review of the Weapon Systems and Military Equipment Procurement and Maintenance Annual Plan; (*b*) completion of 'procurement situation' estimates; (*c*) decision of the Office of Public Procurement and announcement of procedure in the *Official Journal of the European Communities*; (*d*) preparation of a list of suppliers; (*e*) announcement of the procedure for tender, followed by preparation of draft agreements and documents; (*f*) appointment of a commission consisting of at least five experts to check the tenders and negotiate, adjust technical requirements, identify selection criteria and establish voting principles; (*g*) selection of the offer to be accepted; (*h*) preparation of the final documents; (*i*) closing the selection procedure; and (*j*) signature.[20]

The process in the execution stage is still in a state of flux. Previous instructions of 14 September 1995 detailing the internal working procedures have been cancelled, while revised instructions are not yet ready. Because of the continuing organizational changes it is difficult to find any statistical indication of the advantages or disadvantages of the present procedure, which was introduced in 1996. The changes made then were the result of common sense and experience rather than a scientific approach. The absence of any obligation to use technology assessment (TA) methodologies is a distinct deficiency in the existing procedure. Methods are created in an ad hoc fashion and are largely dependent on the competence of the group assigned to carry out assessment.[21]

[20] Miszalski, W., 'Alternative procedures for military technology assessment and the selection of equipment', SIPRI Arms Procurement Decision Making Project, Working Paper no. 92 (1997), p. [17].
[21] Miszalski (note 20).

However, time is needed before the working of the procedure can be observed and evaluated on the basis of systematic evidence.

V. Financial planning and the defence budget

All decisions on procurement are taken strictly within the framework of the national budget, which is approved by Parliament and published, and within it the MoND budget.

The share of procurement of weapons and military equipment in total military expenditure is relatively low compared with that of other NATO members—9.7 per cent in 1999.[22] No increase in procurement expenditure is planned for 2000.[23] This is in spite of the plans for modernization of the armed forces in the context of Poland's joining NATO. It is expected that the share will increase when procurement and modernization go into the second stage, of active replacement of old systems by new ones, and when the share of personnel costs is reduced.[24] Most of the prognoses made by military specialists at the beginning of the systemic transformation (1990–93) expected the equipment the MoND had to be used by the armed forces for about 15 years, that is, until the budget had increased sufficiently to pay for renewal. The transition period for the armed forces should enable the producers to adjust to new demands and requirements and the MoND to prepare plans according to the requirements of NATO and EU membership in a changed security environment. It should also permit a progressive restructuring of the MoND budget and a gradual reduction of personnel costs.

The Programme Foundations for Modernization of the Defence Forces of the Polish Republic for the Years 1998–2012 were adopted in September 1997 in preparation for membership of NATO,[25] and the Programme of Integration with the North Atlantic Treaty Organization and Modernization of the Defence Forces of the Polish Republic in the Years 1998–2012 was published in 1998.[26] The total cost of this programme is estimated at 4.8 billion zlotys.[27] Expectations of an increase in the defence budget to match seem to be excessive, although an increase of MoND expenditure in real terms is possible as Poland is expected to return to high rates of growth in 2000.[28]

[22] Based on Poland's submission of statistics to NATO using the NATO standard definition of procurement. NATO, *Financial and Economic Data Relating to NATO Defence*, Press release M-DPC-2 (1999)152, 2 Dec. 1999, URL <http://www.nato.int/docu/pr/1999/p99-152e.htm>.

[23] Reply to SIPRI questionnaire by the Polish Ministry of Defence, 10 June 1999.

[24] A further fall in the number of troops is expected, to be followed by cuts in length of service and an increase in the ratio of professional soldiers to number of conscripts.

[25] Sköns, E. *et al.*, 'Military expenditure and arms production', *SIPRI Yearbook 1998: Armaments, Disarmament and International Security* (Oxford University Press: Oxford, 1998), p. 212.

[26] The foundations of the programme are presented in *Raport: Wojsko, Technika, Obronnosc* [Report: army, technology and defence], July 1998 and Mar. 1999.

[27] 'Ustawa budzetowa na rok 2000' [Budget law for year 2000], not published at the time of writing. Information provided to the author from the draft budget law, Dec. 1999.

[28] Annual growth of over 5% was expected by both the European Commission and the Organisation for Economic Co-operation and Development (OECD). *Rzeczpospolita*, 27 Dec. 1999.

In the 1999 budget personnel costs accounted for 49.9 per cent of expenditure, operating and training for 32.9 per cent, 'investments' 1.68 per cent and modernization programmes 15.5 per cent. The MoND share of the national budget is now increasing after falling for some years in succession.[29]

The national budget must be presented to the Sejm by the Council of Ministers three months before the beginning of the new fiscal year (although this strict timetable can be varied). The MoND budget, as part of the state budget, therefore has to be prepared, reviewed, presented, adjusted and accepted according to the same timetable.

Preparation of the budget consists of several stages.[30] The first is preparation of the guidelines for general social and economic policy. These are approved by the Council of Ministers and then the Sejm and Senate. In the second phase the Minister of Finance sends all interested departments and institutions a 'budget note' which defines the method, timetable and conditions of preparation of the budget.[31] The budget project is prepared in accordance with these, using the previous year's expenditure and revenue figures. The materials submitted by individual ministries and other units consist of detailed objectives as well as projects of individual parts of the state budget.[32]

The whole process of budget preparation is scheduled in detail from April to 15 November. In early April the various ministries are informed of new regulations that can influence their expenditure. The MoND establishes the broad outlines of its budget request at the end of April. In June all departments under the MoND and other ministries with their financial departments prepare their own budget projects, which are checked in the second part of the month by the Ministry of Finance against expected revenues. In July the preliminary budget requests are considered against the economic forecasts for the coming year—level of salaries, inflation, rate of growth in gross domestic product (GDP) and so on. The budget is discussed in August and the Ministry of Finance receives the MoND's budget proposal. In September the project is completed. October brings parliamentary debates and preparation of the detailed structure of spending. In November the budget bill is finalized and presented to Parliament for approval.

The approval of the budget lies in the competence of the Sejm and Senate. There are three readings. It is evaluated by appropriate commissions, the last one being a sitting of the Commission of Public Finances with the participation of standing committees concerned with sectors of the economy and other interested MPs. The sittings usually end with the preparation of a list of disagreements, which are then resolved by the Commission of Public Finances.

[29] Polish Ministry of National Defence, *Basic Information on the MoND Budget for 1999* [MoND: Warsaw, 1999] (in English; also available in Polish).

[30] Krasowska-Walczak, G., *Finanse Publiczne* [Public financing] (Wyzsza Szkola Bankowa: Poznan, 1997), pp. 98–99.

[31] Jaskiewicz, J., *Prawo Finansowe* [Financial law] (Wydawnictwo Uniwersytetu Gdanskiego: Gdansk, 1988), p. 77.

[32] Falkowski, A., *Pecunia Nervus Belli: Ksztaltowanie Budzetu Obronnego Polski* [Money is the sinews of war: shaping the Polish defence budget] (Bellona: Warsaw, 1998), p. 81.

After approval of the general framework of the budget there is a joint meeting of the Commission of Public Finances with the Commission of Legal Regulations. This is the second reading of the law and the deputies are entitled to introduce further remarks and corrections. If there are a great many changes the budget law is passed back to the two commissions before the third reading. The third reading leads to approval of the budget law by the Sejm.

After that step the budget is passed to the Senate, and can be returned to the Sejm. The Senate must approve the law within 20 days. The budget law is approved by an ordinary majority of those voting. The quorum is 50 per cent of MPs. If the law is not approved within three months of presentation of the first project, the President is authorized to dissolve the Sejm.

The President has seven days to approve the budget law from the moment when the Marshal of the Sejm presents it to him. He is authorized to turn to the Constitutional Tribunal to ask if the budget law is correct from the constitutional point of view[33] and has no right to reject it so long as the Constitutional Tribunal considers that it is.[34] After approval by the President the budget law is published in the *Dziennik Ustaw Rzeczypospolitej Polskiej* (Journal of legislation of the Polish Republic).

Poland attaches great importance to transparency in military budgeting. Before the budget law is passed, general information about the level of the defence budget and its structure is published in the specialized military press and the daily newspapers—*Rzeczpospolita* and *Gazeta Wyborcza*. Parliamentary proceedings are transmitted by one of the state television programmes and in some cases they are available on the Internet.[35]

However, it is not sufficiently clear what the defence budget includes. The breakdown of the published budget is fairly detailed and covers all military expenditure, but the expenditure heads sometimes overlap. The list of budget expenditure in the MoND covers all those who have budget money at their disposal, in all 21 positions (such as the Commander of Land Troops, the Commander of the Navy, the Commander of Air Defence, the Director of the Department of Defence Equipment Procurement and so on). Individual units' expenditure is divided into planned and actual expenditure. Equipment and weaponry are divided into 32 groups which define in detail all types of weaponry, systems and spare parts. Finally, there is a document which sets out the 28 budget heads of the MoND—salaries and money owed to soldiers, goods supplied to soldiers, social expenses, ammunition and explosives, maintenance, weapons and military equipment, research and development (R&D), integration with NATO, interoperability, international obligations and so on.

[33] This happened with the budget law of 1995, when the Sejm decision to order the Minister of National Defence to buy defined type of weapon systems from domestic producers was questioned. 'Poslowie bronia decyzji o Irydzie' [Deputies defend decision on Iryda], *Rzeczpospolita*, no. 46 (1995).

[34] A similar procedure applies to other regulations. 'Konstytucja Rzeczypospolitej Polskiej' (note 1), art. 122, 224.

[35] URL <http//:www.sejm.gov.pl> (in Polish).

Since 1997 the MoND Budget Department has published an annual booklet on the current budget.[36] It contains information on the share of the MoND budget in GDP and overall state expenditure, the structure of military expenditure, the costs of reaching goals defined in the programme for modernization of the armed forces, and so on. It is clear and understandable, and reflects all defence expenditures.

Monitoring of the budget is done primarily by the Sejm, as is clearly stipulated by the regulation on the budget law. In practice the monitoring is carried out by means of a report on the implementation of the budget after six months. Later the Council of Ministers presents to the Sejm and the NIK reports on the implementation of the budget law, supported by accounts of revenues and expenditures on the central and local levels. The budget law defines the contents of these reports, which are examined as background for preparation of a vote on acceptance of the accounts for the government. If the accounts are not accepted in this vote, the Council of Ministers must resign.

VI. The Polish defence industry[37]

The Polish defence industry goes back mainly to the 1930s. During the cold war Poland maintained a substantial defence industry potential which was subordinated to the needs of the Warsaw Treaty Organization (WTO). In 1989, military production (or 'special production') was delivered by 128 companies, 39 of which manufactured final products and 89 of which produced dual-use goods or were engaged in repairs and maintenance. Many were also producing civilian goods. The range of products was rather limited. The Polish defence industry did not develop advanced types of military production because of technological and economic barriers.

The Polish armed forces were to a great extent reliant on deliveries of military materials from foreign sources. Demand from the Polish armed forces was and still is relatively small—the size of the defence industry considerably exceeds the country's needs—and production of a diverse range of armaments in small quantities was not economic. Opportunities to export were limited. The production of some types of weapon on the home market therefore became uneconomic. In numerous cases Poland faced the dilemma whether to keep production capacities in defined types of equipment or to reduce the range of national production and cover part of the armed forces' needs by imports.

The arms market for all WTO members was limited to the Soviet sphere of influence. This geographic concentration of arms imports was driven by

[36] *Basic Information on the MoND Budget for 1999* (note 29).

[37] This section is based on Wieczorek, P. and Zukrowska, K., 'The influence of equipment modernization, building a national arms industry, arms export intentions and capabilities on national arms procurement policies and procedures', SIPRI Arms Procurement Decision Making Project, Working Paper no. 98 (1998); and Mesjasz, C., 'Restructuring of defence industrial, technological and economic bases in Poland, 1990–97', SIPRI Arms Procurement Decision Making Project, Working Paper no. 91 (1998). See also Kiss, J., 'Poland', ed. J. Kiss, SIPRI, *The Defence Industry in East–Central Europe: Restructuring and Conversion* (Oxford University Press: Oxford, 1997).

economic, technological and political reasons. Cooperation within the socialist bloc was reinforced by apprehension on the part of Western states that cooperation with the socialist bloc in defence technology would have negative consequences for their national security.[38] Poland's relations with the USSR and other WTO states were guided by the avoidance of anything that would lead the ruling Soviet elite to doubt its reliability as an ally and a firm WTO member. This could be observed especially clearly in the 1970s. Furthermore, the R&D potential of the Soviet Union was greater than that of all the other WTO countries together. As a result, none of the WTO East–Central European countries, Poland included, had the R&D capacity to produce a wide range of military equipment on its own.

The end of the cold war in 1989 and the collapse of the WTO in 1991 put military industrial capacities under pressure to adjust to tougher competition and a shrinking market. It also brought the end of the command economy in Poland and opened up markets. Military production in Poland fell dramatically, reaching its lowest level in 1993. It has been increasing since, but in 1997 was at only 55 per cent of its 1991 level. Employment in the defence industry over that period fell from 100 000 to 66 000. Deliveries for the civilian market have, however, been growing faster than the production of military goods.[39]

Since 1990, 38 companies in which between 25 and 100 per cent of shares is controlled by the state have been considered as the core of the Polish defence industry. Since the majority of companies in the defence sector are single-owner joint-stock companies, where the state holds a controlling part of the shares, the Ministry of Economy participates in the procedure of tendering for supply of arms and military equipment.

The legal and institutional aspects of the defence industry and 'special' production are regulated by:

(*a*) a regulation of November 1967 on the Common Obligation to Defend the Polish Republic. This regulation is based on the assumption that a country must keep a national industrial potential that is capable of supplying necessary military equipment, regardless of cost. Some elements in this regulation are no longer appropriate: it was prepared for a command–distributive economy in which costs were not the most important factor in decision making;

(*b*) regulations of December 1988 concerning economic activities in Poland. These foresee that some types of activity require concessions, such as the production of arms and military equipment;

(*c*) regulations concerning the control of transfers of arms and related technology introduced since 1989. The principal document is the law of 2 December 1993.[40] This was supplemented by detailed regulations and instructions, for

[38] This was evidenced by barriers such as the Coordinating Committee on Export Controls (COCOM).

[39] Wieczorek and Zukrowska (note 37), p. 3.

[40] 'Ustawa o zasadach szczególnej kontroli obrotu z zagranica towarami i technologiami w zwiazku z porozumieniami i zobowiazaniami miedzynarodowymi' [Law on special control of trade with other countries in goods and technologies in relation with international agreements and obligations], *Dziennik Ustaw Rzeczypospolitej Polskiej*, no. 129 (1993), poz. 598, with later amendments.

instance, on the registration of companies which can participate in arms transfers and the documents needed for such transactions;[41] and

(*d*) a regulation of December 1997 on the handling of arms exports and imports by Polish companies and the transit of arms through Polish territory.[42]

The regulations create rules for the control of the export, import, re-export and transit of goods and technologies which are on the international control lists of the Wassenaar Arrangement.[43] The Ministry of Economy coordinates this type of control. Poland is in the process of preparing companies to apply the same 'catch-all' principle as the USA, which creates the conditions for control of technologies that are not on control lists but can be used in the production of military goods.[44] The regulations governing the defence industry are published in the *Dziennik Ustaw Rzeczypospolitej Polskiej*.

The survival of Polish arms-producing companies will depend on an adjustment strategy, which involves restructuring, consolidation, privatization, conversion, cooperation and internationalization, and an inflow of foreign direct investment (FDI). This has created opportunities as well as challenges.

Restructuring

The companies of the defence industry are gradually adjusting to new requirements prepared according to guidelines set by international organizations and countries that have relevant experience. The message of the regulations is clear: current conditions create a new relationship between the companies and the government. Companies have to become self-reliant by utilizing all available sources of information and knowledge. All, both in the defence sector and in civilian industry, are facing difficulties in coming to terms with the new functioning. There are several requirements if these difficulties are to be overcome. Companies' passivity, which is based on past experience, has to be replaced by active initiative, which includes seeking new partners, maintaining financial liquidity, adjusting to new conditions, and preparing and promoting new products to meet market requirements. This will require a process of learning what types of information are needed and where to find it. Companies have to adjust to new relations with the Ministry of Economy and to be more oriented to developing cooperation with international partners, international sources of financing, markets, technology, the organization of production and know-how,

[41] 'W sprawie ustalenia wzoru rejestru osob prowadzacych obrot specjalny z zagranica, sposobu jego prowadzenia, a takze wzoru wniosku o dokonanie wpisu do rejestru oraz okreslenia niezbednych dokumentow i informacji, które nalezy dolaczyc do wniosku' [On definition of the pattern of register of legal persons engaged in special trade with foreign countries, methods of conducting it, and the application procedure for registration and definition of the required documents and information, which should be submitted with the application, Order of the Minister of Economy, 19 Jan. 1998], *Dziennik Ustaw Rzeczypospolitej Polskiej*, no. 12 (1998), poz. 47.

[42] *Rzeczpospolita,* 31 Dec. 1997–1 Jan. 1998.

[43] The Wassenaar Arrangement on Export Controls for Conventional Arms and Dual-Use Goods and Technology, an informal grouping of states established in 1996. In 1999 there were 33 members.

[44] Sliwowski (note 4).

international regulations and so on. They will have to face competition and take part in international exhibitions.

This also requires certain adjustments on the part of government. The government has to collect necessary information, and process and disseminate it. It is playing a new role in preparing legislation to meet international obligations, harmonize Polish regulations with NATO and EU requirements, and represent the national interest (although according to the constitution international obligations are binding and superior to national arrangements).

A new concept of the defence sector was born in the Ministry of Economy in early 1999 and resulted in the preparation of a Programme of Restructuring of the Arms Industry.[45] An earlier programme, of April 1996, the Programme of Restructuring the Defence and Aviation Industries in Years 1996–2010, has not been implemented because of the lack of funds outside the state budget to cover the costs, and it has been largely overtaken by other developments.[46] The new programme is intended to stimulate investment and strengthen companies through specialization. These changes are expected to improve the negotiating position of the arms producers in their talks with foreign investors.

Privatization

In the case of the defence industry the main route of privatization has been commercialization by turning state companies into joint-stock companies owned by the State Treasury. This was an intermediate phase leading to capital privatization. It was also a condition of beginning conciliatory proceedings with the banks to negotiate the return of credits.

The first ownership changes in the defence industry took place in 1991. Two enterprises, Stalowa Wola Steel Mill and WSK PZL-Swidnik, were registered as limited liability companies. In 1992–93 three other companies were transformed. The privatization process intensified in 1994, when 22 companies became joint-stock companies. Also in 1994, with the approval of the Council of Ministers, shares of two companies were transferred to creditors of those companies. At least some military enterprises followed the pattern of privatization of civilian industry.

This kind of adjustment to market 'rules of the game' is referred to in the literature as bank-led restructuring. In other countries of the former socialist bloc a more paternalistic approach was taken, with greater involvement of the government in the process through funding from the state budget. In the case of Poland the involvement of the state central institutions was limited and generally less than in any other country of East–Central Europe.

[45] 'Program restrukturyzacji przemyslu obronnego i wsparcia w zakresie modernizacji technicznej sil zbrojnych' [Programme of restructuring of the defence industry and supporting the technical modernization of the armed forces], Ministry of Economy, Warsaw, 8 Feb. 1999.

[46] Since 1980 the Polish authorities supervising arms-producing enterprises have launched 7 programmes of restructuring of the defence industry. None of them has been implemented because of lack of finance.

All the 38 principal enterprises in the defence industry were transformed into joint-stock companies by the end of 1996. Of these, 25 were exclusively owned by the State Treasury. (Until 1996 the Ministry of Industry and Trade had this function. In 1997 it was taken over by the State Treasury, directly after that body was established, and in 1999 by the Ministry of Economy).

The privatization of each company of the defence industry requires the approval of the Council of Ministers. It decides on the privatization of defence companies on the basis of the following criteria: (*a*) the companies must continue to be able to meet the goals of national defence policy; (*b*) government control of company activities must be preserved in the field of special production[47] (attempts are currently being made to find a new formula, to enable the state to control companies by means other than control of stocks); (*c*) the companies must be strengthened through capital input and technological advancement; and (*d*) the current level of employment must be maintained.

Privatization of defence companies is progressing relatively slowly for a number of reasons. Potential investors (except the creditors) show rather limited interest in the defence sector, treating it as a high-risk investment because its future is uncertain and the chances of any meaningful increase in demand for its products are small. Moreover, the location of defence companies in the past was typically subordinated to military considerations, not to intrinsic factors. It is also important that production entities are economically viable, as they include vast territories and expensive capital investment. Meanwhile the companies in the defence industry do not always manifest sufficient will to privatize as they are afraid of losing state support.

Privatization is proceeding faster in companies that produce small arms than in the case of suppliers of more complex weapon systems.

Cooperation and internationalization

Examples of cooperation in defence production are still limited and they do not involve R&D, which would reduce the costs of prototypes produced in Western companies considerably. According to the Ministry of Economy, cooperation will be advantageous if it: (*a*) produces more orders for the Polish defence industry through offsets; (*b*) introduces new technologies and efficient organizational arrangements; (*c*) stabilizes the economic and financial situation of the defence industry; (*d*) increases export opportunities; (*e*) reduces the costs of technical modernization of the Polish armed forces, including the programme aimed at reaching NATO standards; and (*f*) preserves the defence potential of the country in the sense of mobilization readiness in the event of war or danger of war.

The Polish defence industry lags behind the standards achieved by the developed NATO countries in its organizational structure, its financing mechanism, the level of technological development of its machine tools and its methods of

[47] Continuation of special production does not necessarily mean that the company engages 100% of its production potential in supplies for the military market.

production. Internationalization and concentration will enable Poland to use its military budgets and their R&D components better by eliminating the inefficiencies of a small industry, which cannot achieve efficiencies of scale because the country's defence expenditure is low. In other words, it will enable more economic production of sophisticated weapons and thus in turn reinforce security. The assumption that market-driven international cooperation can improve Poland's competitiveness is based on the following arguments: (*a*) the costs of production of certain components are lower in less developed partner countries, as in civilian production; (*b*) enlargement of the market will offer economies of scale; and (*c*) cooperation in R&D will be possible.

As is evident in the 38 leading companies, restructuring, which requires diversification of production, adjustment to the changes in the market, the introduction of new technologies and modernization of machinery, is also being hindered by the economic and financial situation of the defence industry. As a result, most of these companies are not able to undertake major investment. They are therefore searching for foreign partners in order to enhance their production capacities through advanced technologies and attract export orders. Western companies are showing increasing interest in the Polish defence industry, as is indicated by the number of business inquiries and promotion missions sent by different companies to specialized exhibitions in Poland.[48]

It seems that the internationalization of arms production will be accelerated when decisions on political integration in the EU are made.

These comments relate mainly to the production of major items of conventional military equipment such as ships, tanks, armoured vehicles and aircraft. They do not concern so much small arms, which most probably will continue to be built on a national scale and standardized internationally. The cost of R&D in the case of small arms makes it possible to retain the current national production, while the increasing costs of R&D for major equipment require international cooperation, sharing costs and enabling the production of more sophisticated weapons, especially when cooperation is linked to the creation of a cooperative security system based on NATO, the EU, the Western European Union (WEU), the Organization for Security and Co-operation in Europe (OSCE) and other regional organizations.

The NATO countries can be expected to develop cooperation with Poland because their defence industries are tending to look beyond national boundaries to the development of international structures in the defence sector, and shrinking demand for arms and military equipment on the national and international markets is forcing them to seek new markets. Cooperation in such conditions can be considered a precondition of increased sales. Three factors may encourage production cooperation between NATO and EU countries and the East–Central European countries. Most of the countries in transition lack advanced technologies; they possess skilled labour forces; and they need to re-equip their

[48] About 40 foreign companies seeking joint ventures, including the most important and well-known producers in the world, participated in the Fourth International Industry Exhibition, organized in Kielce in Sep. 1998.

armies according to the requirements of the emerging security system. In terms of common interests, these three things can be the basis for closer ties between the military industries of the two groups of countries.

Joint ventures

The most common approach to cooperation with foreign partners is the establishment of joint ventures. International cooperation in arms production is decided at the level of enterprises, not at the level of government, which is a new approach compared to the period before 1989. The first agreements of this type were concluded in late 1993 and early 1994.[49]

All the current joint-venture programmes concentrate on cooperation in production. Cooperation in large R&D projects has not been established, although some studies have been conducted with foreign partners on a limited scale.

It should be stressed that all industries are treated on similar terms, which means that there is no preferential treatment from the side of government. The procedure for establishing a joint venture involves several steps similar to those found in most market economies which try to attract foreign investors. Foreign capital can be involved in a company without permission being needed from a state body. (The only exception to this is the banking system). Foreign investors can establish two types of companies in Poland, a limited liability company or a joint-stock company, the legal requirements being different.[50] Information on action to be taken by investors is provided by the Ministry of Economy or the Panstwowa Agencja Inwestycji Zagranicznych (PAIZ, Polish Agency for Foreign Investment).

Poland has not prepared a technological programme based on joint ventures. Decisions on technologies to be used are taken by the companies.

Collaboration between the Polish defence industry and civilian industry is also limited, despite the absence of any legal barriers between the two and the fact that after 1989 military R&D centres became more dependent on the civilian market to promote their products.[51]

From the perspective of the Polish defence industry, joint ventures provide the most important forms of technology transfer. In addition to the capital flows into Polish companies and expansion of orders, joint ventures lead to longer-term commitments.

[49] The creation of a joint venture between the RADWAR company in Warsaw and the French Thomson, over a 'friend or foe' identification kit, is a good example. Another is cooperation between WSK PZL-Kalisz and PZL-Rzeszów with the Canadian company Pratt and Whitney, which forms the framework for production of parts for aircraft engines. Recently McDonnell Douglas presented a letter of intent to produce the F-16 aircraft in Mielec.

[50] Generally, the minimum share in the case of a joint-stock company is 1 zloty, the minimum value of company capital is 100 000 zlotys and the minimum number of founders 3, while in the case of a limited-liability company the corresponding figures are: share—50 zlotys, value—4000 zlotys, and number of founders—1. PAIZ data sheets, Warsaw, Jan. 1999, pp. 2–12.

[51] Mesjasz (note 37), pp. 18–20.

Offsets

Offsets seem to be the guiding idea in shaping the defence industrial policy as indicated by the Programme of Restructuring of the Arms Industry. Offsets have become a precondition of arms imports. A regulation of September 1999 introduces two conditions in the case of larger arms import transactions by the MoND: the foreign supplier which receives an order has to place an order of equivalent value on the Polish market (this can be for all industries,[52] not only military production[53]) and it has to participate in privatization of the Polish defence industries.[54] The regulation defines such notions as an offset agreement, offset obligation, foreign supplier, Polish enterprise, direct and indirect offset obligation, and an offset multiplier. The multiplier falls between 0.5 and 2.0, which means that offsets should amount to 50–200 per cent of the value of the orders placed by the Polish MoND abroad.

The offset regulations are very important in the context of the approaching second phase of adjustment towards NATO interoperability requirements and realization of the programme for modernizing and restructuring the Polish armed forces in the years 1999–2012. The procurement of combat aircraft, which has been suspended for a long time, is entering the realization phase. The introduction of the offset regulation also brings the prospect of orders being placed with the Polish aircraft producers (mainly PZL-Mielec).

It is as yet difficult to identify priorities in the Polish offset policy which might indicate a view of the desired future shape of the defence industry. However, some fields of specialization can be identified indirectly on the basis of competitiveness, quality and volume of sales. They are in small arms, radio detection systems and aircraft production. Capacities for the production of armoured vehicles, tanks and aircraft should be reduced.[55]

The scope of public information

Despite improvements since 1990, less information is publicly available about the defence industry and defence R&D than about other branches of industry. This is the case in the advanced democracies and Poland seems to be following the same path.

Under current regulations, all joint-stock companies, including producers of military goods, publish information on their financial situation. Defence industry issues are thoroughly studied by the Defence Commissions in the Sejm and Senate and in the Defence Department of the NIK. In addition, the appropriate ministries (the Ministry of Economy, the Ministry of the Treasury and the MoND) provide the media with a wide range of information on the defence industry and its problems. Information in an aggregated form is usually given at

[52] Referred to in the regulation as indirect offset.
[53] Referred to in the regulation as direct offset.
[54] 'Ustawa o niektórych umowach kompensacyjnych . . . ' (note 5).
[55] This is the authors' opinion. No official statement on this subject has been presented by the government or officials.

specific events related to the defence industry, such as the launching of the Programme of Restructuring of the Arms Industry. Information about the sector and its enterprises can also be obtained from the Glowny Urzad Statystycany (GUS, Central Statistics Office) in its publications on industry and trade—*Rocznik Statystyczny Przemyslu* (Yearbook of industrial statistics)—or general statistics—*Rocznik Statystyczny* (Statistical yearbook).

The Ministry of Economy releases specialized information sheets and publishes periodical reviews of the economy showing the state of all branches, including military and aircraft production. These publications are irregular but there are some indications that they will continue. The most recent were published in Polish and English.[56] Recently, the ministry published a set of books reviewing the branches of Polish industry in which the defence industry was also discussed.[57] They include a set of six charts illustrating the economic and financial results of the military and aircraft sector. Less complex, more general information is given in the reviews published by the RCSS.

The next channel of information about the defence sector is interviews with journalists or scholars. Both produce short items published in the media.

Many defence companies run promotion and advertising campaigns about themselves and their products. They can also be considered important sources of information. Information on military producers can be found in exhibition catalogues[58] and a growing number of military and other periodicals, such as *Polska Zbrojna*, *Mysl Zbrojna*, *Wojsko i Wychowanie*, the monthly *Raport*: *Wojsko*, *Technika*, *Obronnosc,* or *Wprost*, *Polityka*, *Zycie Gospodarcze* and *Nowe Zycie Gospodarcze* (published weekly).

The availability of information on the defence industry generally can be illustrated by the stages in which the Programme for Restructuring of the Arms Industry was released to the public in 1999. In the first stage the public was informed by the mass media that a programme was being prepared. In the second stage it was said that the programme was being discussed and approved by the KSORM. In the third stage the general outline of the programme was released in the newspapers (*Rzeczpospolita*, *Gazeta Wyborcza* and the specialist press dealing with military issues, such as *Raport*). Detailed information on the sector and on military expenditure can also be found in the annual budget approved by Parliament. A second volume of the budget law also provides detailed statistical data on the defence sector.

Information about the defence sector is available if the user knows where to look for it. Only a limited number of issues are covered by state secrecy. They include the current level of production, sales of armaments, reserves of capacity and conditions of deliveries (prices). The decision to stamp items 'state secret'

[56] Polish Ministry of Economy, *Poland's Report: Industry in 1998* ([Ministry of Economy]: Warsaw, 1999); *Poland's Report: Economy in 1998* ([Ministry of Economy]: Warsaw, 1999); *Poland's Report: Domestic Trade in 1998* ([Ministry of Economy]: Warsaw, 1999); and *Poland's Report: Foreign Trade 1998* ([Ministry of Economy]: Warsaw, 1999).

[57] 'Defense industry and aviation sector', *Poland's Report: Industry in 1998* (note 56), pp. 217–20.

[58] E.g., 'Katalog Polskiej Izby na rzecz obronnosci kraju', Swietokrzyska Agencja Rozwoju Regionalnego, Kielce 1999.

are taken in central government institutions according to regulations embodied in legislation. The regulations governing those decisions are harmonized with NATO and EU requirements.

A special public system of information control is in fact being built up around the defence industry in Poland. Information is no less readily available to the public than in most EU member states. This is a fairly new phenomenon: information on military production was formerly top secret. Nevertheless, there is still a need to disseminate information and make clear what is released and what needs to be made public.

VII. Factors influencing arms procurement

The implications of Polish membership of NATO

In January 1994 Poland signed a Framework Document of the Partnership for Peace (PFP). The Individual Dialogue between NATO and Poland, which opened a qualitatively new stage of cooperation between them, began and Poland formally joined NATO in March 1999.

Formerly, the main goal of the armed forces was to defend the country from invasion. Membership of NATO will mean new tasks, such as peacekeeping, and more selective and specialized arms procurement. National armies from different countries will form international specialized units. There will be opportunity for some specific transitional arrangements to be made before the new international security system comes into being, to cover two gaps—between supply and demand, and between technical capabilities and NATO standards. Those arrangements could involve temporary leasing of major equipment from one or more NATO members. Above all, membership of NATO will mean modernization of the armed forces and interoperability of defence equipment. The 1997 Programme Foundations for Modernization of the Defence Forces for 1998–2012 set out the directions of change needed according to the resources available.

For all the NATO countries, Poland is an interesting market. To that extent, Polish membership of NATO allows pressure from Western arms suppliers to influence the decision makers directly. Poland would prefer to choose equipment from the NATO countries for reasons of interoperability. The NATO countries are also counting on the 'imitation effect' of the Polish example on other candidates for NATO membership from East–Central Europe.

One important issue for Poland, as a member of NATO, will be keeping the balance in its arms purchasing between the USA and European members of NATO. This, however, will be resolved by the progressive internationalization of the defence industry.

The implications of future membership of the European Union

Its future membership of the EU has profound implications for Poland's future arms procurement, arms procurement procedures and arms export policies. Further internationalization will be encouraged.

There are also implications for procurement procedures. In the restructuring of the economy and the legal system, the principles guiding official procurement were changed in keeping with EU requirements, including procurement by the MoND. EU procedures are, however, still not fully applied in everyday practice, and some solutions are applied which protect Polish producers to allow the country to build up its arms production potential before fully facing the competition from the EU and NATO countries. Certain suspensions of the usual regulations on public tender have been introduced. Polish arms procurement will, however, shortly be fully subordinated to regulations imposed by the EU law on public tendering. In a longer perspective, protection could result in the defence industry becoming uncompetitive.

In the EU member countries, the abolition of economic borders and shrinking budgets will enforce a more stringent approach to defence budgets. The EU is to include defence matters in its external policies and develop a capability for conflict prevention and crisis management missions. The WEU is to be incorporated into the EU structures. The Council of the European Union will be empowered to take decisions that affect the modernization and integration of the armed forces and defence industries of the member states.

Poland's adaptation to the EU is not complete. It has introduced most of the legal arrangements required of all future members, but full application of the new laws and democratic mechanisms will depend on other, intangible factors—the knowledge of those engaged in the process, access to information and the ability to discriminate between correct information and false. If these are not developed the democratic mechanism cannot work effectively even when proper institutions and the legal mechanisms which work in advanced democracies are in place.

The influence of interest groups on the arms procurement process

Suppliers of weapons and equipment

The suppliers for the military market include domestic producers, foreign manufacturers and foreign trading companies. Companies play a role in the arms procurement process insofar as they respond to invitation to tender, market their goods and demonstrate their potential to produce the weapons needed, but it is a passive role and their ability to force arms procurement decisions on the government is limited. The number of companies that are not state-owned is increasing. Their obligations and the rules of behaviour of the market are well defined in Polish law.

Foreign suppliers also respond to invitations to tender, but their influence on the Polish arms procurement decision-making process does not go beyond the rules applied in other democratic states. Arms producers and other companies which supply the MoND participate actively at exhibitions of military equipment and services, demonstrate weaponry and equipment, and advertise and market their goods in the usual way.

The defence industry trade unions

The defence industry workers in Poland are organized in two unions.[59] One, the Union of Military Industry Workers, is linked with the Sojusz Lewicy Demokratycznej (SLD, Democratic Left Alliance). The other is a branch of the Solidarity Union. In most cases they cooperate, representing the workforce in their relations with enterprises owned by the State Treasury. Since the problems of the Polish defence industry are politically sensitive and because the industry employs thousands of workers,[60] politicians try to win their support. Apart from the SLD and Solidarity, the defence industry can count on strong support from the Polskie Stronnictwo Ludowe (PSL, Polish Peasants' Party) and the Ruch Odbudowy Polski (ROP, Movement for Reconstruction of Poland), both of nationalist orientation.

The defence industry trade unions are very visible in the mass media, where they try to present their needs and the views of the industry or factory they represent. They exert constant pressure on government institutions at the national level. In some cases they have even tried to bring pressure to bear for an order to be placed in Poland by organizing demonstrations in front of government buildings. They have their representatives in Parliament and most of the members of the Parliamentary Defence Commissions act as lobbyists, analysing the defence budget from the point of view of industrial potential and looking for opportunities for 'their' enterprises to win contracts.

Lobbying

Lobbyists engage parliamentarians who represent regions in which arms production is concentrated. A specific role is also played here by lobbying organizations such as the Polish Industrial Lobby (PLP, Polskie Lobby Przemyslowe) in which corporations, industrial unions and industrial research organizations aim to strengthen Polish industry through a combination of research reports, strikes and demonstrations.[61] The Polish Chamber of Defence Producers organizes trade shows, exhibitions and presentations of military equipment and weapons. It also organizes seminars, conferences, courses and publications to

[59] Stachura (note 9), pp. 16–17.

[60] In 1998 the sector (defence industry with aviation) employed 66 010 people—6.9% fewer than in 1996. This was *c.* 2.5% of total employment in industry. The sector produced 1.3% of total goods and services produced.

[61] Tarkowski, M., 'Arms procurement decision making: process, pressure groups, inter-elite controversies and choices', SIPRI Arms Procurement Decision Making Project, Working Paper no. 95 (1998), p. 5.

disseminate information on changes both within and outside the defence industrial sector that have an impact on arms production.

A similar role can be ascribed to the regional development agencies, which participate in organizing exhibitions and conferences where they act as intermediaries between companies located in their regions and government organizations, EU and NATO bodies, and other producers, including foreign companies. The regional agency from Kielce, for instance, organizes an annual international arms trade fair accompanied by a conference on new trends in arms production and sales, defence industrial policy, and the experiences of downsizing and restructuring of the arms industries in the USA and Europe.

Economic and financial conditions

The cost of procurement projected in the programme of integration and modernization of the armed forces up to the year 2012 will be gradually linked with changes in the structure of the MoND budget. While the demands of the Polish armed forces for modern equipment are huge, the long-term prognoses of the growth of GNP and state budget are optimistic.[62]

Poland (like the two other new NATO members, the Czech Republic and Hungary) has been planning to buy modern combat aircraft but has been prevented by the lack of funds from the state budget. It is currently hoping to lease 60 aircraft from major defence companies.[63] If it buys aircraft from abroad those producers who are ready to participate in an offset agreement placed with Polish producers (military or civilian) will have the best prospects.

Diversification of imports

Poland's former dependence on military technology, products and components from the USSR and other WTO countries had far-reaching military and political as well as economic and financial consequences, even after the end of the cold war. In the post-war period, and especially in 1980–81, Poland was under pressure from the USSR, as was evidenced by restrictions on supplies of military equipment and spare parts. This included supplies for civilian markets: merely postponing some deliveries was enough to produce tension in a 'deficit economy' that did not have the necessary reserves. Still, in the years 1995–97, when Polish membership of NATO was pending, Russia, possibly as part of a political campaign against Polish membership, highlighted the fact that the Polish Army was equipped with large quantities of weaponry produced by the former USSR or in Poland under Soviet licence. In 1992 Russia linked withdrawal of its objections to Polish membership of NATO to the maintenance of contracts for deliveries of military technology. Poland rejected this as being

[62] See note 28.

[63] The Jas-39 Gripen (produced by a consortium formed by Swedish Saab and British Aerospace), the Mirage 2000-5 produced by Dassault of France, and the F-16 produced by Lockheed Martin in the USA. Information provided by the SIPRI Military Expenditure Project, Nov. 1999.

totally opposed to the new directions it was taking and to basic concepts of its foreign policy.

The list of suppliers to Poland, including Western suppliers, has grown considerably since 1989. However, no major contracts have been concluded since 1989, with some exceptions, such as the order in 1999 for six 155-mm artillery turrets from the UK, prior to licensed production.

In the years 1995–98 the share of imports in Poland's total arms procurement was about 5 per cent.[64] Resources were used to buy spare parts and cover costs of repairs, mainly on the former USSR and WTO markets, although some components and assemblies needed for the upgrading of equipment produced in Poland were also imported from Western markets.

In the past Poland (like other countries) sought to limit one-sided dependence by diversifying suppliers or establishing supplies within one alliance. Poland now has freedom of choice, restrained by military and economic considerations and legal provisions. The strategy is now to create relations of interdependence. This is based on the patterns that have developed in civilian production: internationalization of production followed naturally on increasing trade and international capital flow.

The negative experience of dependence on a dominant arms supplier (although it has not been studied deeply by Polish scholars, civilian or military) belongs mainly to the period of the cold war. In NATO, an alliance of a quite different kind from the WTO, the political and military repercussions of cooperation within a narrow group of suppliers should be different. This is illustrated by the experience of NATO members, which only to a limited degree, if at all, try to diversify their patterns of cooperation within the framework of the defence industry and arms production. The main division between them is in transatlantic relations and rivalry between the countries that form the European and North American pillars of the Alliance (also reflected in the pattern of competition of the main weapon producers and suppliers). Nevertheless, this is changing as shrinking arms markets and revision of Article 223 of the 1957 Treaty of Rome force the companies to cooperate in order to enter foreign markets within the alliance.

The diversification policy in the contemporary period has grown out of the activities of the MoND and the Ministry of Economy. The MoND evaluates equipment and shows how it matches NATO requirements; the Ministry of Economy examines the requirements of the regulations on control of advanced technology transfers. To a lesser extent there are pressures from the enterprises. One example is the discussion that accompanied the choice between upgrading national production by imported electronics and broader cooperation between Polish companies and foreign partners, resulting in incorporation of the most advanced technologies into the process of production.

[64] Polish National Statistical Office, *Statistical Yearbook of Foreign Trade* (annual), various issues 1995–97.

VIII. Democratic oversight

While Poland was a member of the WTO, its arms procurement decisions were based on purely administrative considerations. They were taken in the MoND within the framework of burden-sharing decisions of the WTO (although they were formally approved by the Sejm). The planning and implementation of arms procurement programmes were not subordinated to democratic control by Parliament or coordinated with other state or public institutions. Nor did this issue attract the interest of the scientific community or of the media.

The process of achieving transparency in defence planning and strengthening democratic control over the armed forces in Poland has been as much part of building a democratic state and a modern defence system as of meeting the requirements of NATO membership. The process has developed in three main stages. In the first phase, from 1989 to 1991, elements of Communist Party control over defence and national security were eliminated. Political indoctrination of the armed forces was forbidden and the officer corps was depoliticized. In the second phase, 1991–92, a civilian defence ministry was created along with a mechanism for parliamentary control over the armed forces. In the third phase, 1992–99, an integrated MoND was created and the Chief of the General Staff was subordinated to it as an integral part. The legal foundations of the defence system were completed.[65]

There is a clear link between Poland's political transformation and its adjustments towards membership of NATO and the EU. The two processes should not be treated as separate. The political transformation was accelerated by the membership negotiations and by the guidelines and requirements of NATO and the EU. This resulted in far-reaching harmonization of legislation and procedural and institutional changes in the field of public procurement, including the arms procurement process. Poland has fully supported the Alliance's point of view on democratic control and transparency in decision making as formulated in the *Study on NATO Enlargement* in September 1995.[66] Its plans in this field were presented in an Individual Discussion Paper on NATO Enlargement, which was presented to NATO Headquarters in 1996.

The methods and processes applied in advanced democracies for democratic control over the armed forces often differ. Nevertheless, it is possible to define some universal principles, which are being incorporated into the Polish processes. The armed forces should not be given any autonomous authority in making security policy for the state: this would alienate them from the society and could result in their interests being promoted at the expense of society's. They should not be given special political or social privileges. They should not have strong or extensive relations with the economic sector, in particular with the defence industry. The processes for civilian and democratic control over the

[65] Polish Ministry of National Defence, 'Report on Poland's integration with NATO', Feb. 1998, p. 19 (in English).

[66] *Study on NATO Enlargement* (NATO: Brussels, Sep. 1995).

armed forces and the defence budget have to be institutionalized. This control should be carried out by transparent structures within the MoND. The state authorities should have access to independent expertise on defence matters, including those related to defence expenditure. Finally, information dissemination is a vital element of democratic control over the armed forces, arms procurement and the defence budget.

These principles of democratic control can be found in constitutional regulations and in the laws and resolutions approved by the Sejm.[67] Nevertheless, there is still a need for skills and expertise in the processes and procedures relating to the armed forces' activities. This includes an understanding of the arms procurement decision-making mechanism.

Parliament

Parliament's main role is in advising on the defence budget and in the participation of its two Defence Commissions in defence policy making. The Committee for Foreign Affairs also oversees defence policy related to treaty obligations and integration with NATO.[68]

The two chambers of Parliament, the Senate and the Sejm, each have a Defence Commission. The Sejm consists of 460 MPs and the Senate has 100 members, all elected in direct, secret and general elections. The parliamentary term is four years. The powers of the two commissions are determined for each parliamentary session by internal rules of the Senate and the Sejm. This means that the work of the commissions reflects the ideas of the majority parties during the life of a parliament.

Neither commission has a specific structure. Like other committees of the Senate and the Sejm, they are headed by a chairman chosen from one of the parties of the coalition in power. The Deputy Chairman usually represents one of the opposition parties. The Sejm Commission consists of about 20 members, that of the Senate of 6–8. Outside experts can be invited to attend their meetings. The Commission in the Sejm, which is the more important of the two, has been building up parliamentary oversight from scratch.

The Commissions play a consulting and advisory role. They can give their opinion on the defence budget and on the arms procurement plan but do not have powers to force government agencies to execute their decisions. The Sejm Defence Commission, as mentioned above, also receives reports on the implementation of the MoND budget, which includes orders placed for arms and military equipment. In 1996 it suggested setting up a mechanism for monitoring the defence budget and in 1997 a joint working group was set up with the Committee on Economic Policy, Budget and Finance.

The inexactness of the regulations creates vast possibilities for rather free interpretation, which can be seen especially in the executive instructions. In the last Parliament, as mentioned above, the Commissions delegated a represent-

[67] See note 1.

[68] 'Report on Poland's integration with NATO' (note 65).

ative to participate in the procedure for placing orders worth above 200 000 ECU. One case brought before the Constitutional Tribunal in 1994 concerned the Sejm's influence on the defence budget in the case of the Iryda jet trainer aircraft. The Sejm decided to spend 300 billion zlotys ($75 billion) on buying this aircraft, which is made by the PZL-Mielec company. The Tribunal ruled that the Sejm was abusing its powers and infringing the principle of the separation of powers. The verdict clarified that the Sejm can only make general changes in the budget and can only make recommendations when it comes to detailed allocations to projects.[69] Some questions of a similar character involving the competence of the Sejm in setting detailed policy are still unresolved in the current parliament.

Three difficulties expressed by the Sejm in its scrutiny of the budget are the lack of a detailed breakdown, the limited time available (since the budget is submitted to it on 30 November and has to be passed by the end of the year) and lack of expertise.[70] In their work the commissions use reports prepared by experts in the Sejm Biuro Studiów i Ekspertyz (Bureau of Research, BSE). They also receive help from the BSE Budget Analysis Department and from outside experts working on strategic, economic or political matters from research departments of different ministries, the academic world and independent consultative organizations.[71] However, serious attempts are being made to address the problems of lack of expertise: Warsaw University has started a graduate programme in security studies and legislators are offered courses on security and military issues by the National Defence Academy.[72]

Parliament may order special investigation committees, but responsibility for audit functions belongs to the NIK.

The national audit authority

The NIK is the highest office set up to monitor and audit the government's and other state agencies' compliance with the law and efficient use of resources. It can initiate inspections when asked by the President, the Prime Minister or the Sejm or on its own initiative. Its reports are accessible to the public to a limited extent[73] and its conclusions are widely reported in the media. The NIK is headed by a politically independent chairman and his deputies, whose term of office lasts six years and can be prolonged only once. It works as a collegial body and consists of specialized departments which reflect the structure of

[69] Stachura (note 9), pp. 11–12.

[70] Stachura (note 9), pp. 10–11.

[71] The following institutes are those most often engaged in expert work: the Rzadowe Centrum Studiów Strategicznych (RCSS, State Centre of Strategic Studies), the Instytut Rozwoju i Studiów Strategicznych (Institute of Development and Strategic Studies in Warsaw), the Instytut Studiów Miedzynarodowych (Institute of International Studies) in the Warsaw School of Economics, the Instytut Studiów Politycznych (Institute of Political Studies) of the Polish Academy of Sciences in Warsaw, the Instytut Stosunków Miedzynarodowych (Institute of International Relations) of Warsaw University and the Miedzynarodowe Centrum Rozwoju Demokracji (International Centre for Democratic Development) in Cracow.

[72] 'Civilian interference redefined by Poland', *Congressional Quarterly*, 7 Feb. 1998, p. 279.

[73] Stachura (note 9), p. 13.

government activity as well as the structure of the economy. The NIK monitors all activities of state bodies in the executive branch which use public funds. It analyses the consistency of actual expenditure with: (*a*) the budget law; (*b*) the obligatory decision-making procedures; (*c*) the principles of cost-effectiveness; and (*d*) the justifications for the expenditure. The NIK also carries out post-procurement performance audit, by continuing monitoring by an auditor present in the MoND or by annual check-ups to verify that the guidelines of state policy and the budget law are observed.

The NIK reports on its work to the Sejm by presenting its analyses of implementation of the budget law and principles of financial policy; opinions on the subject of the accounts for the Council of Ministers; information on the results of audit; and recommendations and presentations, which are defined in separate regulations. The focus of its monitoring is mainly on the formal and legal aspects of implementation of the budget.

According to currently accepted practice, the NIK audit is considered to be effective, although it has no executive powers. If it finds evidence of incorrectness or fraud, it can only direct the case to court.

Others

The media are considered to be a major element of public control. Sometimes they report very detailed information concerning major orders for weapons and military equipment. Recently the press even released information that protocols on secret sittings of the two Parliamentary Defence Commissions can be found on the Internet.

All the elements enumerated here form a fairly tight web through which arms planning and procurement are monitored publicly. The effectiveness of this system depends to a great extent on the knowledge of the people engaged in the decision-making process and in monitoring and on their ability to find, interpret and use the available information.

IX. Conclusions: an 'ideal type' of arms procurement decision-making process for Poland

The arms procurement decision-making process in Poland is based on general principles introduced by the law on public orders and recommendations of the Council of Ministers identifying the procedures to be followed. Public control of the arms procurement process is developing gradually.

This does not mean that the decision-making process will not need improvement in the future. Changes if any should be guided by the principle of democratic control over the whole arms procurement decision-making process, in order to: (*a*) increase the influence of representative institutions on the process; (*b*) increase the effectiveness of the use of public funds spent for defence purposes: this is especially important in the light of the financial constraints on the MoND, reflected among other things in the limited resources available for arms procurement; (*c*) eliminate corruption; (*d*) eliminate personal linkages between representatives of the armed forces and the managements of defence companies; and (*e*) educate the officials who participate in arms procurement decision making about democratic control over the procedure.

To improve civilian and democratic control, the following steps are the most desirable. First, the status of the Senate and Sejm Defence Commissions should be enhanced. There is a need to formulate a legal basis for their functioning by a permanent law (not an internal regulation of Parliament, which changes according to the balance of power in each Parliament). Second, better-quality professional advice on defence matters is needed for the members of the two Parliamentary Defence Commissions and the NIK. This would help them to participate more actively in the discussions on the structure of the defence forces, the programme of restructuring of the defence industry and individual defence programmes. Third, stronger financial and tax control of arms producers should be introduced along the lines indicated by the EU.[74] Fourth, a more active information policy is needed on the part of the MoND.

The role of the armed forces and their position in the Polish political system seemed to be clearly defined from the 1990s, but the concept of civilian control of the armed forces has not been easy to accept and has been realized only gradually. A big step towards clarification of the system was achieved by approval of the new constitution.

The introduction of the changes to build accountability in arms procurement decision making has encountered numerous difficulties in Poland. These can be categorized into political, psychological, economic, technical and industrial factors.

First, the concept of democratic oversight of the armed forces, including arms procurement, is still new and not understood among the military to the extent it is in the developed democracies. This implies that, despite the introduction of new institutions and legal regulations to match the methods in Western Europe, much still has to be done to educate the officials engaged in the decision-making process and in control of the system.

The second problem is linked with the limited budget of the MoND, including its arms procurement budget. This means that procurement plans are often changed when procurement priorities are re-assessed (often under pressure from producers), which makes it very difficult to have a consistent policy.

The third problem is the need to maintain confidentiality and avoid information leaks. There are circumstances when information should not be made publicly known, for instance, concerning decisions on Poland's adjustments to the standards applied in NATO member states.[75]

[74] These concern among other things more effective tax collection from companies. Poor performance in this field is considered by the EU as invisible state support for enterprises.

[75] 'Ustawa o ochronie informacji niejawnych' [Law on protection of non-public information], 22 Jan. 1999, *Dziennik Ustaw Rzeczypospolitej Polskiej*, no. 11 (1999), poz. 95.

The fourth problem derives from the restructuring of the defence industry. This includes ownership changes, establishing a large number of companies based on trade regulations. It has involved various kinds of misuse and departure from legal procedures, for instance, in selecting arms suppliers or subcontracting companies, although this has not occurred on a large scale.

The last and not the least important problem is the monopolistic or quasi-monopolistic position of the national arms producers. The regulations that have been introduced limit open tenders procedure in order to protect the national defence industry.

Following the guidelines defined by the EU and NATO, the arms procurement process and its democratic control have been constantly improved and are evolving in step with the changes taking place in Poland's institutions. With the advance of reforms and democratization, appropriate changes will be incorporated in keeping with the policy formulated by the member countries of the EU and NATO. This includes decision-making mechanisms, institutional and legal regulations, procedures for information dissemination, rules on payment, and mechanisms of democratic control and education of the officials engaged in the process. Arms procurement procedure can be expected to change in the direction of more intense competition as a natural consequence of changes in the defence industries in the EU and NATO countries, downsizing and internationalization of production. Adjustments in arms procurement decision making and democratic control of this process will be accelerated by the continuation of the reforms, generation changes among the decision makers and the continuing education of officials.

Arms procurement is an inseparable part of public procurement generally, where despite resistance the requirement of competition has been introduced. This will gradually extend to military production and procurement. Barriers that are still in place are being eliminated and the ground for new mechanisms of democratic control in arms procurement decision making is being prepared.

6. South Africa

*Gavin Cawthra**

I. Introduction

One of the few countries in the world to have had a mandatory United Nations arms embargo imposed on it,[1] South Africa developed a unique system of arms procurement[2] for which the Armaments Corporation of South Africa (Armscor) was created. During the apartheid era—especially after the mandatory UN arms embargo was imposed—arms procurement was necessarily a secretive, often covert, affair carried out with minimal democratic accountability and driven almost entirely by Armscor and the South African Defence Force (SADF).

During the transition from apartheid to democracy—between 1990 and the first non-racial national elections in April 1994—substantial changes occurred in the arms procurement process, which was increasingly subjected to multi-party political scrutiny. Many large defence procurement and development projects were scrapped or put on hold and the defence budget went into sharp decline, a trend which continued after the inauguration in May 1994 of the Government of National Unity, which was dominated by the African National Congress (ANC).[3]

The ANC came to power on a platform which promised a democratically accountable and transparent government that would concentrate on social and economic advance rather than military security. Insofar as the ANC's security polices were concerned the movement argued that 'National security and personal security shall be sought primarily through efforts to meet the social, eco-

[1] The embargo imposed through UN Security Council Resolution 418 on 4 Nov. 1977.

[2] It should be noted that Armscor makes a distinction between 'procurement', which is defined as 'the process required to obtain goods and services from outside the organisation [Armscor]' and 'acquisition', which is transforming 'an operational capacity into a commissioned system'. Sparrius, A., 'Quality in armaments procurement', SIPRI Arms Procurement Decision Making Project, Working Paper no. 110 (1997), pp. 3–4. In other words, procurement is 'off the shelf' purchase while acquisition entails project development. This chapter does not make this distinction except when referring to Armscor's internal processes.

[3] Cawthra, G., *Securing South Africa's Democracy: Defence, Development and Security in Transition* (Macmillan: London, 1997), pp. 27–60.

* This chapter draws extensively on 14 papers commissioned for the SIPRI Arms Procurement Decision Making Project which were written during the first half of 1997 by South African academics, senior Armscor employees, Department of Defence officials and representatives of the defence industry. Under the aegis of the Institute for Democracy in South Africa (IDASA)—SIPRI's South African partner organization in this project—and in cooperation with Armscor, the papers were presented at a workshop in Pretoria on 6 May 1997. Many parts of this chapter are drawn directly from these papers, which represent a cross-section of informed views about the acquisition process.They are not published but are deposited in the SIPRI Library. Abstracts appear in annexe B in this volume.

nomic and cultural needs of the people'.[4] This, combined with cuts in the defence budget, made it difficult for the new government to embark on any major arms procurement projects, despite the fact that it had inherited a defence force which faced obsolescence in many areas as a result of the years of isolation under apartheid and the UN arms embargo. Although a considerable domestic defence industrial capacity had been developed during the 1980s, the apartheid regime had been unable to develop or acquire some major weapon platforms, notably combat aircraft and naval vessels.

Procurement decisions were also made dependent on a coherent mission and force design for the defence force, which was only agreed on by the government in 1997 as the result of a protracted process of drawing up a White Paper on National Defence and a Defence Review. Major procurement decisions thus became the subject of heated political debate over national priorities, exemplified by the policy vacillations between 1994 and 1997 over a proposal to equip the navy with four corvettes.

The armaments policy debate also involved related arms export control issues and the role of the South African defence industry as an arms exporter. Two major policy initiatives were taken by the Cabinet in this regard: the appointment in late 1994 of the Cameron Commission of Inquiry into some South African arms transactions, and the establishment in August 1995 of a new system of arms controls under the Cabinet-level National Conventional Arms Control Committee (NCACC). During 1997 the NCACC began the process of developing a White Paper on the Defence Industry, the remit of which included acquisition procedures and processes.

There was thus some uncertainty about South Africa's arms procurement processes in 1997–98, when this chapter was written, although the broad outlines of processes were becoming apparent and the Defence Review had established a force design which provided a basis for acquisition planning.

II. Arms procurement under apartheid and during the transition to democracy

This chapter concentrates on arms procurement during the post-apartheid period. However, since many of the institutional arrangements and processes were inherited from the apartheid system or were established during the negotiations to end apartheid, it is necessary briefly to examine the history of South African arms procurement decision making.[5] Before it left what was then the British Commonwealth in 1960 and declared the Republic in 1961, the country had been closely integrated into the UK and Commonwealth defence systems, and its acquisition policies reflected its international alliance commitments.

[4] African National Congress, *Ready to Govern: ANC Policy Guidelines for a Democratic South Africa* (ANC: Johannesburg, 1992).

[5] This section is based largely on Batchelor, P., 'Balancing arms procurement with national socio-economic imperatives', SIPRI Arms Procurement Decision Making Project, Working Paper no. 100 (1997), pp. 3–10.

Table 6.1. Arms procurement decision making in South Africa, 1961–94

	Procurement options	Determinants of procurement	Procurement institution[a]	Government policy on procurement
1961–68	Primary: • direct imports Secondary: • licensed production • indigenous production	Primary: • economic • industrial Secondary: • strategic • military	Munitions Production Office (1951) Armaments Production Board (1964)	White Papers on Defence, 1964/65, 1965–67
1968–77	Primary: • direct imports • licensed production Secondary: • indigenous production	Primary: • strategic • economic Secondary: • military • industrial	Armaments Board (1968)	White Papers on Defence and Armament Production, 1969, 1973, 1975
1977–89	Primary: • indigenous production Secondary: • illegal imports	Primary: • strategic • military Secondary: • economic • industrial	Armscor (1977)	White Paper on Defence, 1977 White Papers on Defence and Armaments Supply, 1979, 1982, 1984, 1986 Briefing on the organization and functions of the SADF and the Armaments Corporation of South Africa, 1987 White Paper on the Planning Process of the SADF, 1989
1989–94	Primary: • indigenous production Secondary: • imports with offsets • illegal imports	Primary: • strategic • economic Secondary: • industrial • military	Armscor (1992)	Draft national policy for the defence industry, Transitional Executive Council, Apr. 1994

[a] Dates in brackets are year of establishment.

Source: Batchelor, P., 'Balancing arms procurement with national socio-economic imperatives', SIPRI Arms Procurement Workshop, Working Paper no. 100 (1997), p. 4.

Although a defence industry had been built up during World War II, it was dismantled after the war and South Africa imported the completed weapon systems it required principally from Britain.[6]

A schematic outline of arms procurement decision making in 1961–94 is shown in table 6.1.

[6] Cawthra, G., *Brutal Force: The Apartheid War Machine* (International Defence and Aid Fund: London, 1986), pp. 9–13.

The early 1960s saw the rapid isolation of South Africa and the imposition in 1963 of a non-mandatory UN arms embargo, which narrowed its procurement options, although many countries—notably France and Italy—were still willing to supply it.[7]

In 1964, partly in response to the embargo, the Armaments Production Board was established as an autonomous body within the Department of Defence (DOD) with the aim of handling all procurement as well as re-establishing a domestic defence industry, largely by supporting private-sector activities.[8] Domestic procurement constituted only around 10 per cent of arms procurement in the 1960s, but after 1968 the government shifted towards licensed and indigenous production, with the aim of achieving strategic self-sufficiency in an increasingly hostile world, and to support the policy of import-substitution industrialization. In that year Armscor (until 1977 called the Armaments Development and Production Corporation) was established and domestic arms production was accelerated. After 1977 Armscor assumed sole authority for arms acquisition and military research and development (R&D). It also carried out around 80 per cent of domestic production, which increased rapidly after the imposition of the mandatory UN arms embargo in November that year. The 1977 embargo had profound effects on procurement policies and processes. South Africa, already severely constrained, now had either to develop domestic production capabilities (and even then it had to covertly import key technologies, components and machinery) or to establish covert supply channels. It often had to accept what it could get and pay a considerable premium to middlemen and others. Elaborate schemes were developed involving the establishment of front companies, deals with other 'pariah' states and smuggling networks. While the development of a domestic industry had some economic benefits, it was primarily driven by strategic concerns and often involved establishing production facilities with high set-up costs and short production runs.[9] Evidence also suggests that the domestic arms industry 'crowded out' civilian R&D and had a negative effect on economic growth.[10]

In the 1970s and 1980s procurement decisions were based largely on a perceived need to build up defences against a possible attack from communist countries, possibly Cuba acting as the Soviet Union's proxy, and from African countries to the north. This led to a relative neglect of the navy: by the end of the 1980s, 90 per cent of the defence budget was allocated to the army and air force, with the result that the navy was left in a position where it was arguably unable to carry out its assigned roles.[11] Throughout this period there was very

[7] Cawthra (note 6), p. 91.

[8] Buys, A., 'The influence of equipment modernization, building a national arms industry, arms export intentions and capabilities on South Africa's arms procurement policies and procedures', SIPRI Arms Procurement Decision Making Project, Working Paper no. 101 (1997), p. 1.

[9] For a detailed account of the effects of the embargo see, Landgren, S., SIPRI, *Embargo Disimplemented: South Africa's Military Industry* (Oxford University Press: Oxford, 1989).

[10] Batchelor (note 5), p. 8.

[11] Mills, G. and Edmonds, M., 'New purchases for the South African military: the case of corvettes and aircraft', SIPRI Arms Procurement Decision Making Project, Working Paper no. 108 (1997).

little room for public debate on defence posture, force design and acquisition. The defence force had promoted the concept of a 'communist total onslaught' against the country which demanded a 'total strategy' in response. This required the centralized coordination of state activities under the State Security Council, which was dominated by military and police officers and officials.

During the 1980s procurement was driven largely by pressing operational requirements, and especially by the war in Angola, and there was little linkage between the arms procurement process and technology development. The armed services would identify a requirement in general terms, which would be translated by Armscor into equipment specifications, which would then be covertly procured from abroad or developed and then industrialized in South Africa.[12] Many of these initiatives involved Armscor and other officials in illegal activities and drew them into contact with pariah states such as Iraq and the Chile of General Augusto Pinochet.[13]

Arms procurement during the transition from apartheid

The accession to the presidency of F. W. de Klerk led to a break with the militarization of the 1980s. In February 1990 de Klerk lifted the long-standing prohibitions on free political activity and announced his intention to free Nelson Mandela and other political prisoners and to negotiate an end to apartheid. This ushered in a period of negotiations which lasted until the first non-racial national elections in April 1994. During this period the defence budget went into free fall, a reassessment of the threat environment took place and multi-party (effectively ANC–National Party) negotiations and consultation began to take the place of the formerly monolithic decision-making process.

Change in the defence arena was relatively slow, however, as both the ANC and the National Party saw little advantage in politicizing defence issues. It was only in April 1993 that the first face-to-face meetings between members of the ANC's armed wing, Umkhonto we Sizwe (MK), and the SADF took place. Later that year, when a multi-party transitional authority became a reality, the two forces together with the 'independent' homeland armies became part of the Joint Military Co-ordinating Council (JMCC). The JMCC was composed of senior military officers from the Transkei, Bophuthatswana, Venda and Ciskei (TBVC) states (homelands), the SADF and the MK. It was charged, *inter alia*, with drawing up a threat analysis and a force design for the immediate post-apartheid period and thus providing a basis for arms procurement decisions. In practice, the JMCC—which had only four months to complete this task and thus lacked preparation as well as resources—drew substantially on the SADF's

[12] Truscott, E., van der Merwe, W. and Wessels, G., 'Alternative procedures for technology assessment and equipment selection', SIPRI Arms Procurement Decision Making Project, Working Paper no. 111 (1997), pp. 12–14.

[13] Crawford-Browne, T., 'Arms procurement decision making during the transition from authoritarian to democratic modes of government', SIPRI Arms Procurement Decision Making Project, Working Paper no. 104 (1997), p. 3.

1994 Strategic Planning Process (SPP), a system which had been implemented in late 1992 in all the armed services in an attempt to develop an integrated force planning process. This meant that the force design which emerged was little changed from the previous one.[14]

Nevertheless, the reality of a rapidly declining budget, the end of conflicts with neighbouring countries to the north, South Africa's new international acceptability and the impending introduction of democratic accountability and transparency in governance had a significant impact on the arms procurement decision-making process. Some major procurement projects were terminated or put on hold while nuclear, chemical and biological warfare and putative space programmes were abandoned during this period. However, other procurement projects continued, despite the rapidly declining defence budget—for example the replacement of air force trainers, the upgrading of the Mirage combat aircraft and the acquisition process for corvettes for the navy. Although the UN arms embargo remained in place until after the inauguration of the new government (it was rescinded by the Security Council on 20 May 1994), the expectation that it would be lifted led to a new approach to procurement whereby competitive international tendering could take place. In the case of the new air force trainers, this resulted in a decision to purchase the Swiss Pilatus, even though a prototype trainer (known as Ovid) had been developed by the South African defence aviation industry in the expectation of a domestic contract.[15]

While Armscor remained responsible for procurement during the transition period, in April 1992 it lost its production functions, which were transferred to a new state-owned company, Denel, which came under the Ministry of Public Enterprises. Denel inherited most of Armscor's production and research facilities and over 15 000 employees were transferred to the new structure. Both Armscor and Denel made efforts to become more transparent, representative and accountable, publishing annual reports for the first time in 1993 and appointing new board members. The separation of production and procurement functions allowed Armscor to introduce a more flexible and competitive procurement process, emphasizing competition for contracts and value for money, and introducing fixed-price rather than cost-plus contracts. A new policy for counter-trade (offsets) was introduced: all contracts worth over 5 million rand would need to include at least 50 per cent counter-trade.[16] Having lost its privileged and protected position and facing a dramatic drop in demand owing to the new strategic situation, the domestic industry was forced to shed more than half its jobs: employment fell from 150 000 in 1989 to just over 70 000 in 1993, while the share of defence R&D as a proportion of the country's total R&D fell from 48 per cent to 18 per cent.[17]

[14] Williams, R., 'South African force planning', SIPRI Arms Procurement Decision Making Project, Working Paper no. 113 (1997), pp. 9–15.

[15] Mills and Edmonds (note 11).

[16] Batchelor (note 5), p. 9. See also section III in this chapter.

[17] Batchelor, P. and Willett, S., SIPRI, *Disarmament and Defence Industrial Adjustment in South Africa* (Oxford University Press: Oxford, 1998), p. 74.

III. The current arms procurement process

The reintegration of South Africa into the international community and the lifting of the UN arms embargo naturally had a profound effect on Armscor's activities. South Africa's good international standing means that it can now trawl the international market as it wishes as well as drawing on existing domestic capabilities. Previously, many market sectors were dominated or entirely controlled by one domestic company, whereas now procurement takes place in a multi-source environment within a much freer market. This has resulted in a number of policy changes which are examined later in this chapter.

The political and military context of the decision-making processes regarding arms procurement has changed substantially since the transition period and the establishment of the Government of National Unity, and the processes are still being revised.

Politico-military questions

In apartheid South Africa the military played a powerful role in politics and in the coordination of the overall strategy of the state: it was therefore a priority for the new government to stabilize civil–military relations and to ensure effective civilian control and oversight over defence policy, including procurement. The new government thus moved swiftly to draw up a defence White Paper, pointedly entitled *Defence in a Democracy*.[18] The White Paper paid little attention to the size and shape (and hence procurement requirements) of the new defence force: instead, it went back to first principles, establishing a framework for civil–military relations and establishing the legal and normative context of defence. It also reiterated the government's position that socio-economic issues were a greater challenge than defence and that resources needed to be allocated accordingly.

The White Paper included only a brief threat analysis, in effect concluding that there was no conceivable conventional military threat to the Republic of South Africa for the foreseeable future and that force planning therefore needed to take place in a 'threat-independent' manner—in other words, to prepare for generic rather than specific contingencies. This marked a fundamental departure from previous threat analyses, which were predicated on the concept of the 'total onslaught' which was deemed to be orchestrated by the communist bloc and to be manifest in violence emanating from other African countries.[19] As there was no perceivable threat, the White Paper argued that the South African National Defence Force (SANDF) could be scaled down to a 'core force' which could be expanded to a 'war force' should this become necessary. In other

[18] South African Department of Defence, *White Paper on National Defence: Defence in a Democracy* (Government Printer: Pretoria, 1996).

[19] Although the threat analysis developed by the JMCC did not mention a communist threat, force planning was still predicated on the concept of a conventional attack from Africa, although implicitly from a non-African power which had established a base there.

words, it would seek to retain all its key capacities, which would be 'balanced', but not at sufficient levels to fight a war. It did not spell out what this might entail: this task was delegated to a Defence Review process which took place in 1996–97.[20] The White Paper also stated that South Africa should have 'a primarily defensive orientation and posture' and would be committed to 'the international goals of arms control and disarmament'[21]—a fundamental departure from past policies.

Arms procurement decisions were effectively put on hold until the force design component of the Defence Review was completed in the first half of 1997 and approved by Parliament in July of that year.[22] In any case, further cuts to the defence budget made any major procurement initiatives impossible as the SANDF, faced with a growing personnel bill as a result of the incorporation of former guerrilla and 'homeland' forces, was obliged to cut back severely on capital expenditure. By 1998 expenditure on weapon acquisition had fallen to less than 10 per cent of the defence budget.[23]

The Defence Review was a remarkable consultative process, involving extensive discussions with non-governmental organizations (NGOs), other government departments and the general public: three large conferences were held where the document was discussed in public. The parliamentary Joint Standing Committee on Defence (JSCD), a multi-party committee involving members from both the National Assembly and the Council of Provinces (formerly the Senate), played a crucial role in the outcome of the process, as it did with the White Paper.[24] The Defence Review drew up four options for force designs and spelt out to the level of items of main equipment. The first of these was described as the DOD's 'long-term vision'; the second as the 'growth-core force design' (effectively a scaled-down version of option 1); the third as a 'demonstration option' which would be the result of further budget cuts, obliging the SANDF to concentrate on secondary rather than conventional roles; and the fourth as a 'defensive operational concept' design, drawn up to illustrate how a commitment in the White Paper to a 'primarily defensive' posture might be operationalized.[25]

All these force designs reflected the 'threat-independent approach', although the Defence Review did conclude that defence needed to be predicated on defence against possible attack from a middle-level power or another African country with support of a major or middle-level power. The models also drew on the 'core force approach' to varying degrees in that they sought to retain a balanced force which could be expanded through additional acquisition pro-

[20] Cawthra (note 3).

[21] *White Paper on National Defence* (note 18), p. 6.

[22] South African Department of Defence, 'South African defence review 1998', chapter 8, URL <http://www.mil.za/Secretariat/Defence%20Review/Table%20of%20Contents.htm>.

[23] South African Department of Defence, 'Accelerating transformation, Address by the Minister of Defence, Mr J. Modise MP, on the occasion of the defence budget vote, National Assembly, 26 May 1998', URL <http://www.mil.za/DoD/Secretariat/address.htm>.

[24] Williams (note 14), pp. 16–20.

[25] See note 22.

grammes if a threat emerged. The models were generated in part through discussion between the various armed services, the SANDF, the Ministry of Defence (MOD) and other key stakeholders, and in part through a computer modelling process known as Project Optimum which, as its name suggests, was meant to optimize force design in the light of calculated risk (defined as a combination of probability and impact) and cost. It was much criticized by those who were not involved in its creation on the grounds that it was too technical and that its value-inputs were questionable, in that could have reflected a set of assumptions about the validity, for example, of offensive defence.[26] It was claimed, however, that Project Optimum led to savings of 22 per cent, radically improved the cycle time for strategic planning and clearly showed what tasks could be prepared for within a given budget.[27]

Force Design 2, as it was then known, was recommended and duly approved by the JSCD and eventually by the Cabinet in May 1997. With its detailed breakdown of main equipment, it should have provided a firm basis for acquisition. However, it immediately became evident that the force design would be hostage to political decisions regarding the size of the defence budget. At around the same time as the force design was approved, the Cabinet also demanded an unexpected additional cut of 500 million rand in the 1997/98 defence budget. The force design itself was predicated on the assumption that by a transformation and rationalization process (involving downsizing from around 100 000 full-time personnel to around 70 000) the SANDF would be able to achieve a target of a 30 per cent share of defence expenditure for capital and equipment renewal. It was evident, therefore, that major procurement decisions would remain politically charged regardless of the apparent consensus over force design and the specifications for equipment in the Defence Review: trade-offs would need to take place, as it was unlikely that there would be enough money to pay for what was specified.

The nature of these changes is also reflected in the ongoing debate on an attempt by the South African Navy to procure corvettes. The navy viewed the transition from apartheid as an opportunity to redress the historic imbalance in spending between the arms of service and put forward a request for four corvettes. A frigate/corvette requirement had in fact been identified in the early 1970s but had fallen victim to the international embargoes, budget constraints and inter-service rivalry. A Naval Staff Requirement for four patrol corvettes was approved by the Defence Command Council in May 1993 and, after deciding to seek international bids, Armscor sent out a Request for Information (RFI) to ascertain what was available on the international market—despite the fact that the UN embargo was still in place. Fourteen proposals were selected, narrowed down by Armscor to five by the end of 1993, and then to two. By the end of March 1995 Armscor was ready to recommend one proposal—from Bazan in Spain—to the new ANC-dominated Cabinet.

[26] Williams (note 14), pp. 16–18.
[27] Truscott *et al.* (note 12), p. 47.

The corvette decision-making process had been conducted in virtual secrecy and little attempt had been made to consult with other government departments. It was only in August 1994 that the public became aware, through media leaks, of the decision to purchase corvettes, and not until February 1995 did the new Parliament, through the JSCD and the Joint Finance Committee, become actively involved in the issue. There was a widespread, and accurate, perception that the deal had been cooked up behind closed doors in the 'old way' and there was strong press and public opposition to the deal in the light of the new government's commitment to social and economic issues. In this climate, the Minister of Defence withdrew the item from the Cabinet agenda on 17 May 1995, when the project should have been approved.[28] However, the first report of the Defence Review, which was approved by all parties in Parliament in 1997, specified that the navy should be provided with four corvettes. This was approved by the Cabinet during the course of 1997, but it was still not clear where the money would come from.

The structure of arms procurement decision making

The current relationship between the MOD, the DOD, the SANDF and Armscor is set out in figure 6.1. The MOD includes the minister and his staff, the Defence Secretariat and the office of the Chairman of Armscor and the DOD; the DOD includes the offices of the Secretary for Defence, the Chief of the Defence Staff and the SANDF but excludes Armscor. The creation of the post of Secretary for Defence and the institution of an integrated civilian–military MOD operating through 18 divisions has limited the powers of Armscor and the SANDF in acquisition and institutionalized a system of civilian checks and balances.

The arms procurement function was investigated during the course of 1994 and 1995 by a specialized MOD Acquisition Project Team (MODAC) and a Ministerial Steering Committee. As a result, three reports were published, dealing with technology and armament acquisition management, defence industry policy and organizational structure of the defence acquisition programme management. These reports were incorporated into a wider study carried out into acquisition during the Defence Review process, which led to the publication of a final draft chapter on the acquisition management process in May 1997.

These reports established the institutional and other arrangements for acquisition decision making. The division of responsibility is broadly as follows:

1. The Minister of Defence is the highest authority and bears ultimate political responsibility for the acquisition function. He or she is accountable to the Cabinet, the President and Parliament.
2. The SANDF defines and prioritizes its acquisition needs, and is also responsible for management of the user system, including personnel and facilities.

[28] Mills and Edmonds (note 11), pp. 5–20.

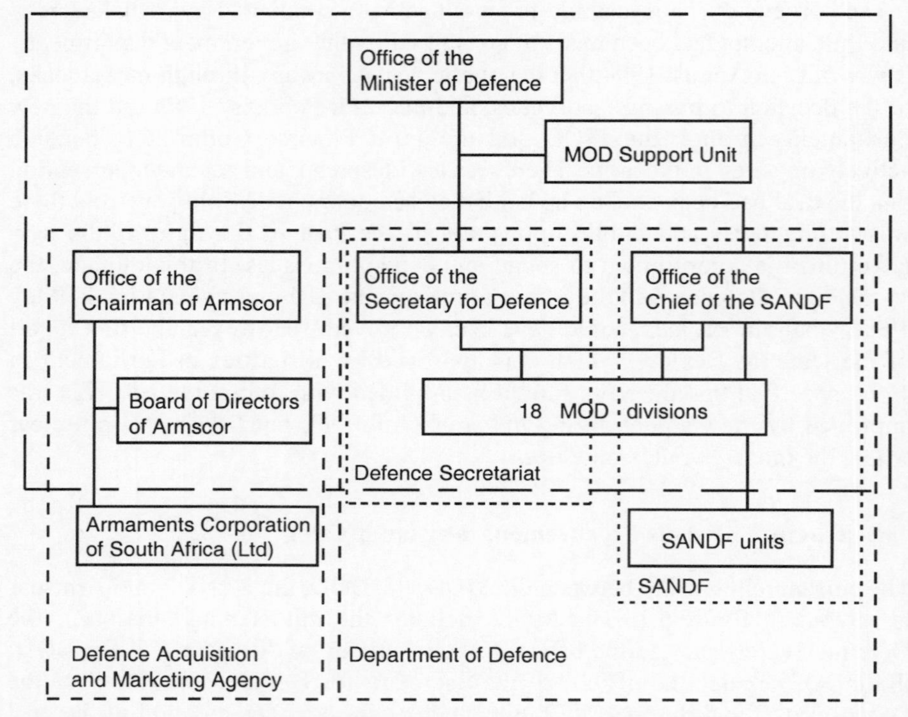

Figure 6.1. The structure of the South African Ministry of Defence

Source: Buys, A., 'The influence of equipment modernization, building a national arms industry, arms export intentions and capabilities on South Africa's arms procurement policies and procedures', SIPRI Arms Procurement Decision Making Project, Working Paper no. 101 (1997), p. 12.

3. Armscor is responsible for programme management and contracting of industry during the execution of acquisition programmes, for ensuring the technical, financial and legal integrity of the process during contracting, and, with the DOD, for overseeing industrial development in support of acquisition programmes.

4. As Accounting Officer of the DOD, the Secretary for Defence is responsible for ensuring that all acquisition activities are executed within the framework of national objectives, policies and constraints.[29]

The approval structure for project submissions is shown in figure 6.2.

Projects are classified as cardinal or non-cardinal in order to decide the level of top management involvement. Criteria used for classification include political profile, national strategic interest, inherent risk, cost profile, urgent operational need and influence on existing capability and size.[30]

[29] South African Department of Defence, 'Defence Review chapter on the defence industry: the acquisition management process (sixth draft)', 7 May 1997; and Buys (note 8), p. 12.

[30] Buys (note 8), p. 11.

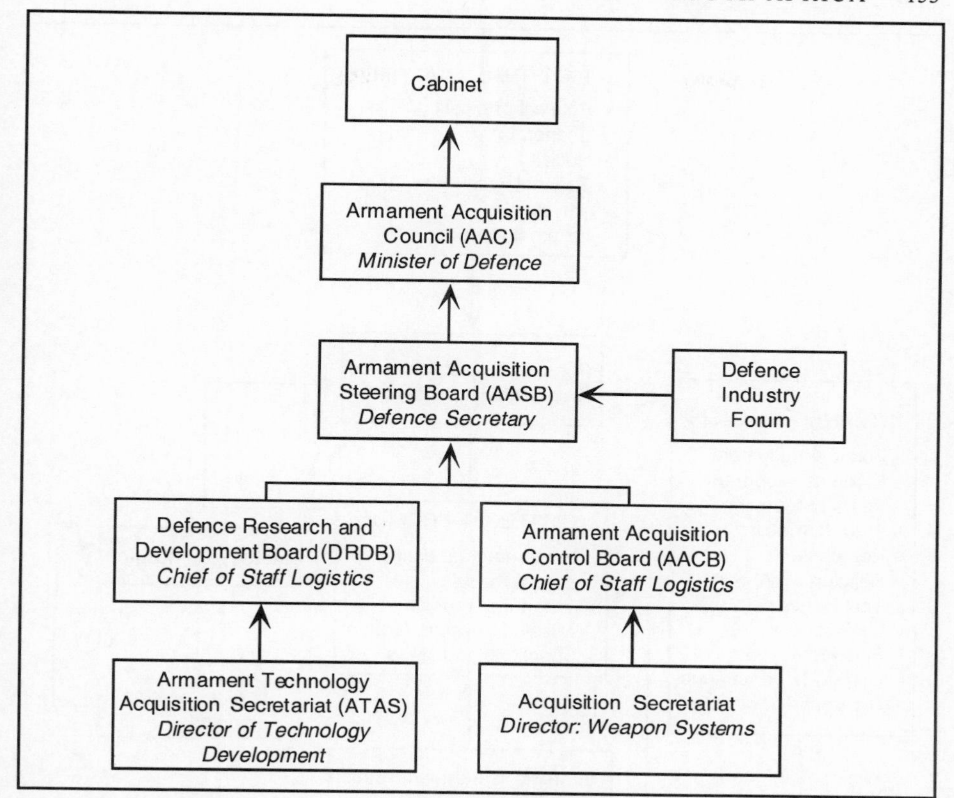

Figure 6.2. The arms procurement decision-making process in South Africa
Source: Batchelor, P., 'Balancing arms procurement with national socio-economic imperatives', SIPRI Arms Procurement Decision Making Project, Working Paper no. 100 (1997), p. 12.

Cardinal projects need to be approved by the Armament Acquisition Council (AAC). It consists of members of the Council of Defence, which consists of the Minister of Defence, the Chief of the SANDF, the Secretary for Defence and the Executive Chairman of Armscor. Such projects also have to be presented to Parliament for approval. The AAC also makes the final decisions regarding the selection of successful contractors and available finances.

Non-cardinal projects are approved at the level of the Armament Acquisition Steering Board (AASB). Chaired by the Secretary for Defence, the AASB consists of senior SANDF, Defence Secretariat and Armscor officials and also screens cardinal projects. The third level of approval is the Armament Acquisition Control Board (AACB) which screens all projects and other routine programmes.[31]

Once projects are approved, contracts are placed with the industry for project execution. Tender adjudication is the responsibility of the Armscor Board of

[31] Buys (note 8), p. 13; and South African Department of Defence (note 29), p. 15.

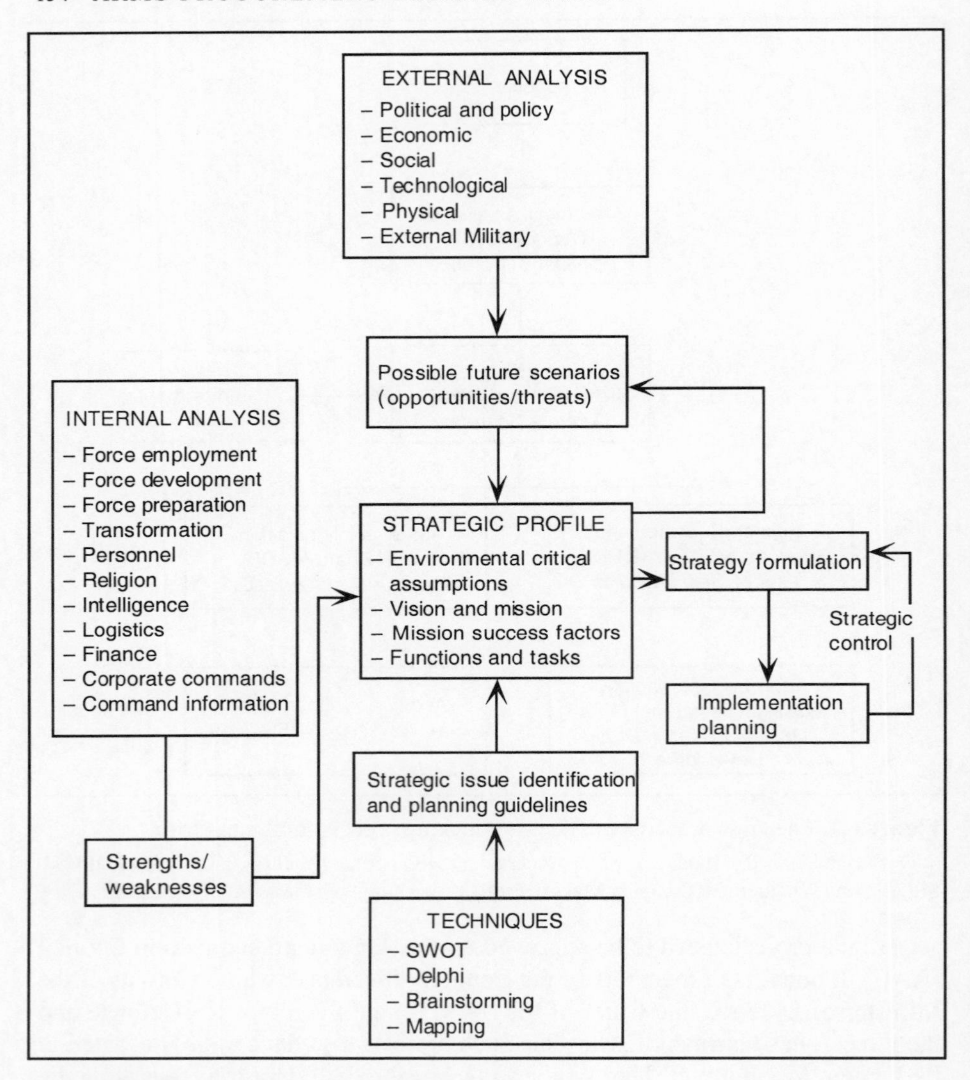

Figure 6.3. Strategic planning in South Africa

Note: SWOT = strength–weakness opportunity analysis.

Source: Griffiths, B., 'Arms procurement decision making', SIPRI Arms Procurement Decision Making Project, Working Paper no. 105 (1997), p. 5.

Directors, which acts as the tender board for this purpose, although this may change as a new decentralized state tendering policy is developed. The Board is appointed by the Minister of Defence (usually from the business sector but more recently also from the professional sector) and includes the Secretary for Defence and the Chief of the SANDF as *ex officio* members.[32]

[32] Buys (note 8), p. 13.

While the Minister of Defence bears ultimate responsibility for all acquisition, the process basically involves the SANDF determining requirements. These are translated into more specific requirements by Armscor, which is then responsible for the programme management and contracting processes. The DOD is responsible for strategic planning, high-level programming, budgeting, and control and auditing of expenditure.[33]

Figure 6.3 summarizes the strategic planning process. It is initiated by the Secretary for Defence and performed by a Joint Strategic Workshop of DOD, SANDF and Armscor representatives. (Input from the defence industry can also be incorporated.) All three organizations are involved through the various committees and are linked on a project and committee basis. A first cut of required resources in terms of functional allocations across arms of service may be estimated at this point.

The responsibility for conducting threat assessments, divided broadly into an external and an internal strategic analysis, leading to the development of a strategic profile, is driven by the SANDF and coordinated by the DOD and the MOD. The White Paper, the Defence Review, the constitution, other legislation and defence policies provide the policy context for this process. Key challenges are broken down into problems with various elements. A strategy is formulated which assesses objectives, ways of achieving the objectives, strategic gaps, contingencies and risks. This leads to an implementation planning phase resulting in plans, programmes and broad financial estimates. These outputs are then integrated and plans for resources, structures, capital, human resources, finances and so on are established.[34]

The stages of the procurement process

While there is some variation in the acquisition process depending on the size and nature of the project, implementation, which is managed by Armscor, generally follows three generic phases: project study, acquisition and contract. Each is divided into a number of different stages.[35]

The project study phase consists of the following:

1. The client (usually the SANDF, although Armscor also carries out acquisition for the South African Police Services) defines a user staff requirement within the framework established by the policy, budget and strategy.
2. A programme manager is appointed by Armscor and a project officer by the MOD and, after a programme plan has been drawn up, a project team consisting of DOD and Armscor officials is appointed. This consists of pro-

[33] Hatty, P., 'The South African defence industry', SIPRI Arms Procurement Decision Making Project, Working Paper no. 106 (1997), p. 17.

[34] Griffiths, B. N., 'Arms procurement decision making', SIPRI Arms Procurement Decision Making Project, Working Paper no. 105 (1997), pp. 4–6.

[35] This section is based on van Dyk, J. J., 'The influence of foreign and security policies on arms procurement and decision making', SIPRI Arms Procurement Decision Making Project, Working Paper no. 112 (1997), pp. 29–35.

fessionals qualified in the technical, financial and legal fields as well as in quality assurance, programme management and industrial participation (IP—offset or counter-trade). For cardinal programmes a formally constituted project steering committee will also be established to oversee the process.

3. The project team draws up a list of all possible contenders, locally and globally. An RFI is sent to the contenders, listing basic requirements.

4. The RFI responses are screened in terms of a Level 1 Value System which seeks to assess each contender's experience, capacity and IP proposals.

5. The contenders selected for the next phase are sent a Request for Proposal (RFP) which contains further details of the project and more detailed IP requirements. The RFP requires contenders to respond with: (*a*) a comprehensive specification; (*b*) a certification and integration test plan; (*c*) a quality assurance plan; (*d*) a configuration management plan; (*e*) an integrated logistic support plan; (*f*) an acceptance test procedure; and (*g*) confirmation of IP compliance.

6. The RFP respondents are screened in terms of a Level 2 Value System which addresses issues such as risk, manufacturing and integration capability, facilities, quality assurance and configuration management expertise, logistic support capability, management experience, financial stability, compliance with the user staff requirement, cost, timescales and IP. Respondents may be visited by the project team for a technical inspection.

7. A project study report is then generated and a tenderer is nominated.

In the acquisition phase, the following steps take place:

1. Negotiations are held with the prospective tenderer to generate the main agreement, which should include a Release to Service Plan. Depending on the size of the project this plan is approved by the relevant level of authority in Armscor, the Industrial Participation Control Committee and where applicable by the various arms acquisition committees and boards and the JSCD.

2. An acquisition plan is generated by the client (usually the SANDF) and Armscor. The acquisition plan is phased, involving sequential and parallel processes, a systems engineering approach and management by project teams. Armscor's acquisition services include the following: (*a*) feasibility studies to identify alternative system concepts and determine which will best implement the required operational capability; (*b*) specification of the selected system; (*c*) design, development, testing and evaluation of the selected system; (*e*) recruiting and training the operators and maintenance personnel for the system; (*f*) producing and commissioning the system in the required quantities; and (*g*) deploying and commissioning the system.

As part of this process, Armscor may buy in, usually from private companies: (*a*) services to perform concept and feasibility studies; (*b*) system engineering services to specify the system; (*c*) design, development, and test and evaluation services; and (*d*) manufacturing services to produce and deliver the system.[36]

[36] Sparrius (note 2), p. 4.

The third phase, the contract phase, involves a comprehensive process of programme management, including technical and administrative liaison, cooperation, monitoring and reporting to ensure fulfilment of contractual obligations by both seller and buyer—delivery, training, integration, support, local manufacture, import and export, invoicing, payment and so on.

Armscor

MODAC confirmed the role of Armscor as the acquisition agency for the DOD. Armscor is now a shadow of its former self: in the late 1980s it employed 23 000 people, now (1998) reduced to fewer than 1000. Most of its assets and staff were transferred to the newly created Denel in 1992 when it relinquished its manufacturing functions, but Armscor has since been further downsized and has lost some of its other functions—notably arms control, which was transferred to the Secretary for Defence in September 1995.

The Minister of Defence retains ultimate responsibility for Armscor, but control is exercised principally by a Board of Directors while day-to-day management is the responsibility of a Management Board (see figure 6.4).

Armscor's principal function is the acquisition of arms, which is funded through the Special Defence Account (2990 million rand in 1996). Armscor has very wide powers in relation to the acquisition function. In terms of the Armaments Development and Production Act (no. 57 of 1968) it is authorized to: (*a*) promote, coordinate and exercise control over the development, manufacture, acquisition or supply of arms; (*b*) sell or export arms or promote sales—in other words, act as a marketing agency for the South African defence industry (for example, through coordinating participation in international defence shows); (*c*) promote industrial development relating to armaments; and (*d*) render services to any agency which requires them as determined by the Minister of Defence (this has allowed Armscor to carry out limited non-military procurement functions for other government departments).[37]

Armscor also carries out technology development functions in relation to acquisition and controls the Elandsfontein Vehicle Test Facility and the Alkantpan Test Range for ballistics. In 1996 its assets amounted to 354 million rand and it received an allocation of 190 million rand for its activities from the state in financial year (FY) 1995/96. It carried out acquisition to the value of 3653 million rand, 94 per cent of which was for the SANDF and 5 per cent for the South African Police Service.[38]

As a creation of the apartheid government, which functioned in virtually complete secrecy (the Special Defence Account was not publicly audited during the apartheid era), Armscor was treated with considerable suspicion by the incoming government. As a result, it has made an effort to transform itself, implementing an affirmative action programme to promote the advancement of the

[37] *Armscor Annual Report 1995/96* (Armscor: Pretoria, 1996).
[38] *Armscor Annual Report 1995/96* (note 37), p. 38.

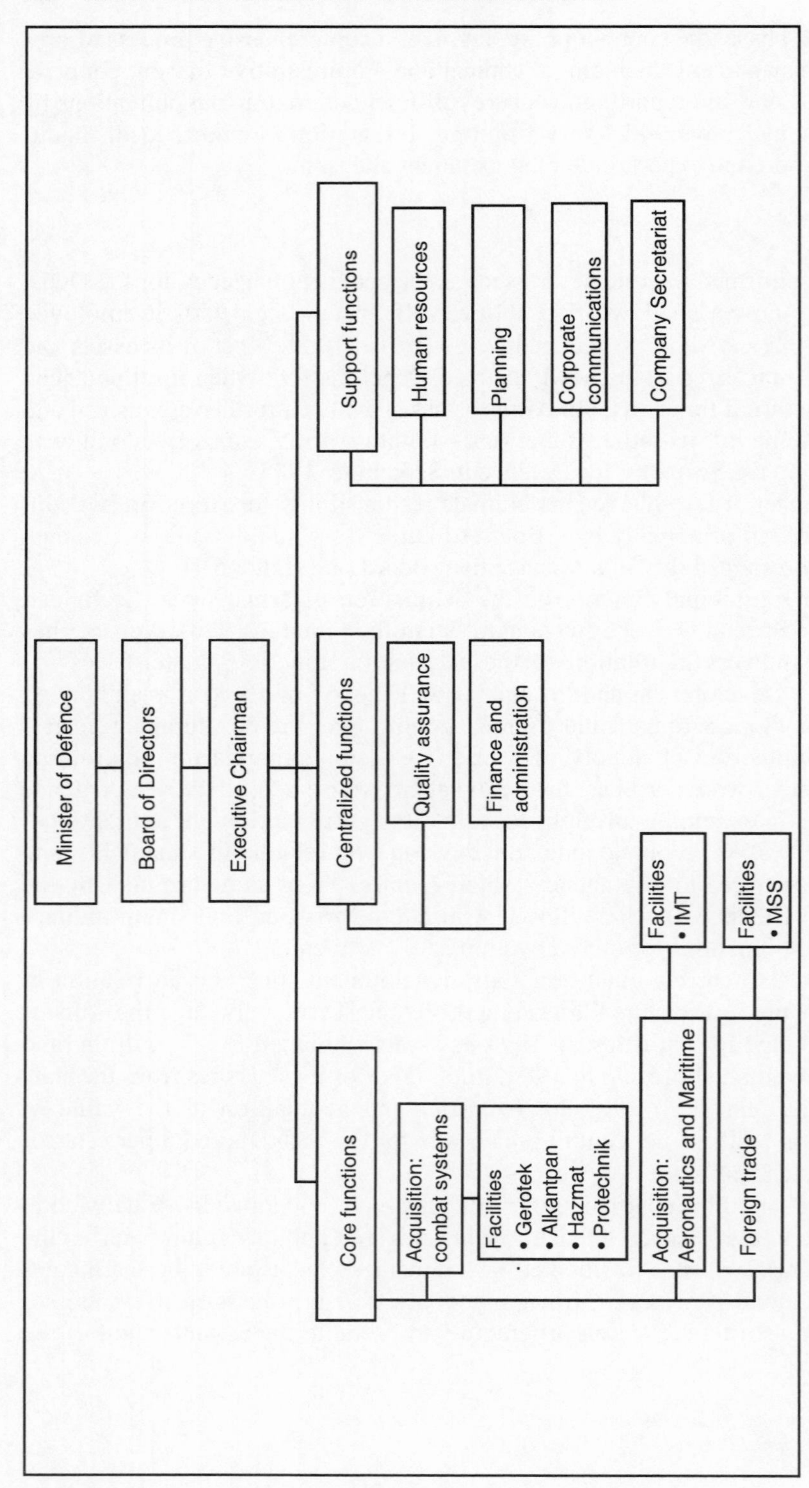

Figure 6.4. The organizational structure of Armscor

Source: van Dyk, J., 'Influence of foreign and security policies on arms procurement and decision making', SIPRI Arms Procurement Decision Making Project, Working Paper no. 112 (1997), p. 20.

black population, appointing new black members to its controlling board, and carrying out environmental and community projects to improve its public image. In February 1994 it adopted a new policy 'of transparency and accountability aimed at empowering our ultimate client, the South Africa public, to assess our acquisition and marketing decisions as well as our human resources and technology development policies . . . '.[39] The role and status of Armscor remain the subject of considerable debate, with some in government and the defence industry questioning whether it should continue to carry out such a diversity of roles related to acquisition. These issues are likely to be further discussed in the drafting of the White Paper on the Defence Industry.[40]

The Department of Defence

While Armscor is the DOD's acquisition agency, the department itself works alongside Armscor in ensuring that projects are implemented. The Secretary for Defence is the chief accounting officer for acquisition while the minister retains overall political responsibility. A secretariat was established in 1994 to support the Secretary and to carry out a number of tasks previously the responsibility of the defence force itself. The secretariat's functions were to include policy formulation, parliamentary liaison, financial control and budgeting and some aspects of personnel and public relations.

It took some time to establish an effective secretariat as there were few civilians trained and experienced in defence management, resources were inadequate and there was resistance from some sections of the SANDF. In an effort to overcome some of these problems, an integrated head office structure was devised in 1997, with 18 functional divisions (including one for acquisition) controlled by both the Secretary for Defence and the Chief of the SANDF and employing both civilian and military personnel. This essentially brought the DOD into line with the British system. The division of responsibility between the DOD and Armscor for acquisition was set out in the MODAC 1 report[41] and, as discussed above, in practice projects are managed by joint teams ensuring fairly efficient coordination between Armscor and the DOD.

Foreign policy

There has been considerable public debate in South Africa since 1994 on the role of arms exports in foreign policy, but virtually nothing has been said about the foreign policy implications of importing arms—whether it is appropriate for South Africa to procure from certain states or not, for example, or what the

[39] Omar, Y. A., 'Different perspectives on the relationship between national security, military security and military capability objectives', SIPRI Arms Procurement Decision Making Project, Working Paper no. 109 (1997), pp. 15–16.

[40] The White Paper was published in Dec. 1999. South African Government, *White Paper on the South African Defence-Related Industries*, Dec. 1999, URL <http://www.polity.org.za/govdocs/white_papers/defence/defenceprocure1.htm>.

[41] South African Ministry of Defence Acquisition Project Team, 'Technology and armament acquisition in the Department of Defence', [Pretoria], 8 Aug. 1996.

long-term implications of seeking a strategic arms transfer relationship with one or more of the major powers might be.

The ANC-led government proclaimed a strongly normative foreign policy framework, built around principles of human rights and democracy, a commitment to international law and the redressing of global inequalities. However, the practice of South African foreign policy in the first three years of the Government of National Unity reflected a more flexible approach in which these principles were modified by realpolitik and the government's perception of South Africa's national interests. In relation to the arms trade, while South Africa remained publicly committed to global disarmament and to a policy in which arms should not be exported to countries with poor human rights records or which were involved in conflicts, its decisions sometimes seemed out of kilter with these criteria.[42]

Coordination between foreign policy, domestic security policy (policing and justice) and defence policy takes place at a number of levels. Ad hoc coordination is common. At the highest level this happens through the relevant Cabinet committee. For example, the Cabinet Committee for Security and Intelligence, which acts as the nodal point for security decisions in the Cabinet, includes ministers and deputy ministers of defence, foreign affairs, home affairs, security and intelligence. Police and military cooperation occurs around a number of issues, especially crime control, and security policy integration is reflected in the 1994 White Paper on Reconstruction and Development[43] and the inter-departmental National Crime Prevention Strategy, but these are not sufficiently developed to result in an integrated or coordinated policy regarding arms acquisition.

In the absence of any foreign policy guidelines specifically aimed at arms procurement, it could be assumed that South Africa would be prepared to buy arms from those countries to which it would be willing to sell. However, no 'blacklists' are kept, each proposed sale is considered on an individual basis and there are few clear guidelines. South Africa enjoys a remarkable freedom in foreign policy terms, with good or potentially good relations with virtually all states. It is unlikely, however, that it would be prepared to procure from countries which were subject to international sanctions.

South Africa will have to consider the long-term foreign policy implications of the major procurement decisions it will soon need to make. In particular, the government will need to decide if it prefers to procure weapons from its main trading partners or whether it will seek to use arms purchases as means of making new alliances, for example in the Far East. It will also need to decide whether arms imports should be driven primarily by strategic and political considerations or by trade and industry-related issues. These decisions are likely to brought to a head over the 'package deals' which began to be offered after the adoption of the Defence Review. By August 1997 Germany and the UK had

[42] Crawford-Browne (note 13), p. 13.

[43] [South African Government], *White Paper on Reconstruction and Development*, Sep. 1994, URL <http://www.polity.org.za/govdocs/white_papers/rdpwhite.html>.

both put together packages built around South Africa's equipment requirements as specified in the Defence Review, and other countries seemed likely to follow. In the case of the UK this consisted of corvettes, Upholder Class submarines, Gripen combat aircraft, Hawk jet trainers and possibly anti-aircraft missiles, while Germany was also reportedly linking a submarine deal to the sale of corvettes and possibly helicopters and jet trainers. In the event South Africa chose to diversity its sources.

Arms control

Arms control has been a major concern of the Government of National Unity, which has taken a number of steps to ensure political control and to bring South Africa, once an 'outlaw' in arms control terms, into line with international norms. While South Africa's arms control system, like those of most countries, is mainly concerned with controlling the export or domestic production of weapons, it also affects procurement in two ways. First, the international conventions, treaties and regimes to which the new South Africa is party place restrictions on research into and the manufacture of certain types of weapon, notably those of mass destruction (although these provisions also affect a variety of potential dual-use equipment). Second, the manufacturing, acquisition and domestic sales are controlled by a complex set of permits administered by various government bodies. Commercially available arms are largely the responsibility of the South African Police Service.

The major international control mechanisms to which South Africa is now signatory are the 1968 Non-Proliferation Treaty (NPT), the 1972 Biological and Toxin Weapons Convention (BTWC), the 1981 Convention on the Prohibition and Restriction on the Use of Certain Conventional Weapons (CCW), the 1987 Missile Technology Control Regime (MTCR) and the 1993 Chemical Weapons Convention (CWC). South Africa is also a member of the Zangger Committee and Nuclear Suppliers Group but has not yet decided whether to participate in the Wassenaar Arrangement.[44] These agreements are enforced in South Africa through a number of bodies, notably the Non-Proliferation Council.

Domestic procurement, sales and production of various categories of weapon are all controlled through a variety of permit systems. Conventional weapons, whether produced domestically or externally, may not be exported, imported or marketed within or outside South Africa without a permit. While Armscor was previously responsible for issuing permits, these are now considered through a four-level process involving the departments of defence, foreign affairs, and trade and industry and, at the highest level, the NCACC, which was set up on 30 August 1995 in an effort to gain firmer political control over the process. Most of the processing work, however, is carried out by the Directorate for Conventional Arms Control (DCAC), which was set up in September 1995

[44] On these organizations see, e.g., *SIPRI Yearbook 2000: Armaments, Disarmament and International Security* (Oxford University Press: Oxford, 2000), pp. xxxi, xxxv and xl–xlii.

under the Secretary for Defence.[45] It is possible that the process will be streamlined as a result of the White Paper on the Defence Industry.[46]

IV. Financial and budget questions

Arms procurement budgeting

Budgeting occurs within a well-established defence budget cycle, each phase of which lasts for two and a half years and which establishes a rolling operational budget for a five-year period and a capital budget of 10–20 years, a requirement necessitated by the long lead times associated with the military acquisition process. The strategic implementation guidelines developed in the strategic planning process described above are used to fashion a medium-term environmental analysis (usually understood as up to five years) which is then used by the armed services to develop requirement guidelines.

Each arm of service—in South Africa, as well as the air force, army and navy, the medical service constitutes a separate arm—determines its main activities and requirements in a bottom-up process. At the operational level budgeting is divided into three categories: personnel and administration, operational requirements and capital replacement. The latter is controlled by Force Development Steering Committees which consist at present mostly of military personnel but are due to be civilianized. Capital funds are allocated from the Special Defence Account, which is kept separate from the rest of the defence budget. The budget is approved by the highest defence staff councils, the Department of State Expenditure and the Cabinet: at the latter level final decisions regarding the allocation of funds are made on the basis of SANDF programme requirements, rather than the specific requirements of the various armed services.[47]

Cost assessment

Armscor evaluates bids on the basis of value for money, not merely the lowest bid. This is assessed in terms of performance/cost and risk and of life-cycle costing (LCC). The evaluation model thus includes assessments of the bidding company's qualification requirements, a critical performance analysis, and analyses of cost, risk and discriminating performance analysis.[48] Factors such as affirmative procurement and industrial participation are also taken into account. Once a programme has been recommended, a political and economic impact analysis is carried out before a final choice is made (it is at this point that a domestic bid may be given preference over a foreign bid for reasons other than

[45] South African Directorate for Conventional Arms Control, *Guide to the Terms of Reference of Conventional Arms Control in South Africa* (Office of the Secretary for Defence: Pretoria, 1996), pp. 1–35.
[46] South African Government (note 40).
[47] Griffiths (note 34), pp. 6–7.
[48] Griffiths (note 34), p. 9.

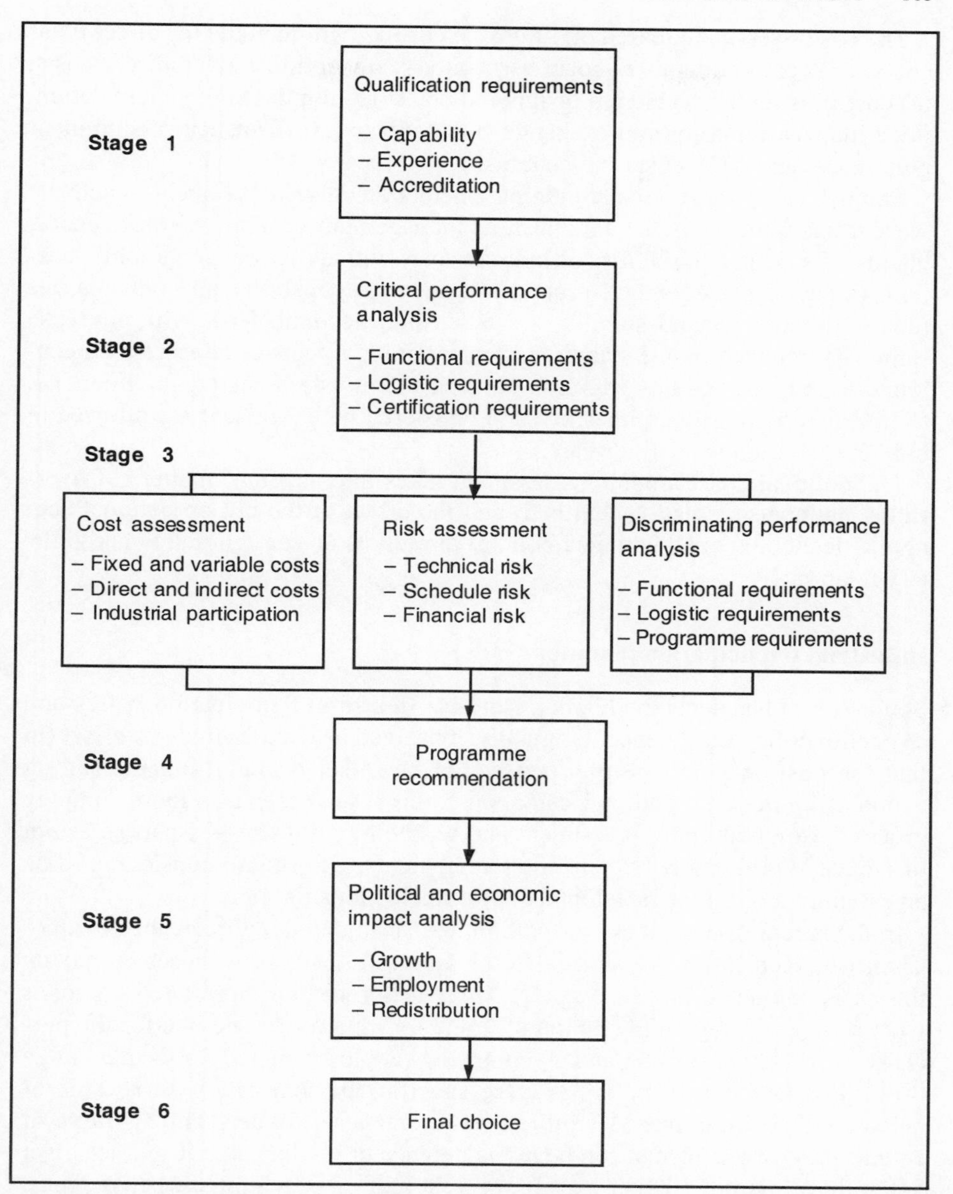

Figure 6.5. Evaluation of arms systems in South Africa

Source: Griffiths, B., 'Arms procurement decision making', SIPRI Arms Procurement Decision Making Project, Working Paper no. 105 (1997), p. 9.

the competitiveness of the bid itself). This process is primarily driven by Armscor, also using independent consultants, and is shown in figure 6.5.

The cost assessment, based on life-cycle costing, will include: (*a*) direct fixed costs; (*b*) cost ceilings; (*c*) costs associated with options; (*d*) indirect costs; (*e*) cost-plus items; (*f*) hidden or ignored costs; (*g*) duplication; (*h*) escalation, price increases and foreign exchange fluctuations; (*i*) affirmative procurement preference; and (*j*) IP costs and benefits.

The risk assessment aims at judging whether a contractor is likely to achieve what it has quoted for. Factors such as management capability, track record, quality of systems, possibility of indirect costs, infrastructure, geographic location, financial management, project management capability and other issues related to capacity and stability are taken into account. Risk is in any case somewhat reduced by the principle of maintaining a register of accredited contractors: only companies which have been previously assessed for financial, technical and security competence and registered with Armscor are allowed to bid. (Registration is for five years.)

The political and economic assessment takes into account 'matters of overriding national interest' (undefined) and the effect of the bid on national economic, technology and strategic policies as well as on foreign policy and military relations.[49]

Industrial participation (counter-trade)

South Africa has developed both a National Industrial Participation Policy and a specific policy for defence. Originally conceived as counter-trade or offset (in that the cost of a purchase abroad would be offset by requiring the selling country to purchase South African goods), this is now seen as a more complex process of mutual trade, investment and technology transfer—in part so as not to violate World Trade Organization agreements. Offsets are considered to be an essential tool for the development of a stable industrial base.

In the recent re-equipment of the South African defence forces, the Armscor Chairman, Ron Haywood, stated that the successful bids were chosen largely on the basis of their offset packages.[50] Three different but interrelated contracts were set up; one regulating civilian IP, one for military IP and one for the purchase of the actual system, and all contracts were to be signed for the deal to go through.[51] Defence IP policy is quite specific: all contracts with a value of between $2 million and $10 million require at least 50 per cent by value of counter-trade or IP, which can be in the defence or civilian area. Contracts over $10 million require 100 per cent IP, at least half of which must be in defence, with the aim of supporting the defence technology base and the export of value-added defence goods.

Detailed provisions for the management of IP—a complicated and often protracted process—have been drawn up. The Department of Trade and Industry has set up a list of 22 areas which are prioritized in the offset policy, for

[49] Griffiths (note 34), pp. 10–12.
[50] URL <http://area51.upsu.plym.ac.uk/dgdd/offsets/ofrsafr.htm#South africa>.
[51] Campbell, K., 'Pretoria's choices', *Military Technology*, Dec. 1998, p. 8.

example, transport systems, software and solar power technology.[52] A penalty system, of 5 per cent of the deal, has also been devised for cases of agreed IP failing to materialize, although it remains to be seen if this will work.[53] As the process evolves, it is bound to prove even more difficult in implementation than in planning and to have some unforeseen consequences: this is particularly the case with major arms transactions with rich industrial countries (for example, the British or German package deals).

V. Techno-industrial questions

The domestic defence industry plays a crucial role in acquisition by providing about 70 per cent of all military *matériel* acquired by the SANDF (a total of 2685 million rand in FY 1995/96).[54] While they are not involved directly in the decision-making process, the role South African defence companies play in R&D, manufacture, testing, maintenance, support and import of armaments makes them key actors in the acquisition process.

It is difficult to define exactly what constitutes the defence industry, given the extent of diversification and the overlaps between civilian and defence production. However, the companies which are members of the South African Aerospace, Maritime and Defence Industries Association (AMD) provide 94 per cent of the local defence equipment purchases of Armscor (and 76 per cent of turnover in these companies is defence equipment).[55]

Four major groups of companies supplied 67 per cent of Armscor's defence purchases during 1996, although hundreds of other companies and subcontractors are also involved. By far the biggest is Denel, Armscor's former manufacturing arm, which consists of 18 major divisions and subsidiaries and accounts for about 80 per cent of Armscor's defence acquisitions. Denel carries out a wide range of management, R&D, engineering and manufacturing activities: 74 per cent of its output is defence equipment or services, including missiles, armoured vehicles, aircraft and information technologies. Other large defence companies are Altech, Grintek and Reunert. All three are public companies with only a minor part of their turnover in the defence sector.[56]

The post-apartheid government has shifted decisively away from the inward-looking import-substitution economics of the apartheid era towards an outward-oriented approach focused on the achievement of national competitiveness, encompassed in the national macroeconomic strategy for Growth, Employment and Redistribution (GEAR). Key issues which influence acquisition policy, particularly in relation to the domestic defence industry, include: (*a*) a commitment to fiscal and monetary discipline as well as reducing the budget deficit; (*b*) liberalization of the capital account of the balance of payments and possible

[52] See note 51.
[53] Griffiths (note 34), pp. 14–15.
[54] Hatty (note 33), p. 2.
[55] Hatty (note 33), pp. 1–3.
[56] Hatty (note 33), pp. 2–3.

incremental abolition of exchange controls; (*c*) tariff reductions to facilitate industrial restructuring; (*d*) support for small and medium-sized enterprises; (*e*) strengthening of competition policy and the development of 'cluster support programmes'; and (*f*) the restructuring of state assets (privatization) and the introduction of schemes to allow the wider population to become owners of these assets—an issue that could potentially affect Denel.

As part of its restructuring and its efforts to cut costs the government also intends to reform national procurement policy, with potentially important effects on defence procurement. In April 1997 the Ministry of Finance and the Ministry of Public Works issued a Green Paper on Public Sector Procurement Reform in South Africa,[57] aiming to free up the tendering process and give easier access to the public sector for small, medium and 'micro' enterprises. The Green Paper proposed the abolition of existing state and provincial tender boards and their replacement by procurement centres at the departmental and provincial levels. If this were applied to defence, it would empower the Secretary for Defence, as Accounting Officer in the DOD, to carry out all procurement—a power presently invested in Armscor in its capacity as the State Tender Board for capital procurement.

Domestic arms production

One of the major issues in procurement decision making for any country with a domestic arms industry is whether and to what extent it should favour domestic procurement over imported equipment and what role, if any, the domestic industry should be allowed to play in the arms procurement decision-making process.

Defence equipment remains one of South Africa's most significant manufacturing outputs, although production has declined rapidly since the collapse of domestic demand following the end of the Angolan and Namibian wars in 1989. The opening up of South Africa to international trade has provided the South African arms industry with considerable opportunities, but it has also meant more open competition for domestic contracts: as a result the industry has been forced to downsize and diversify into civilian production. The estimated number of employees involved in one way or another in the industry has fallen from 160 000 in 1989 to less than 50 000 in 1997. Direct employment in the industry, in the sense of employees of companies which are members of the AMD and employed on defence work, is even lower, at around 17 000. Defence sales of companies which are members of the AMD increased between 1992 and 1995, from 3452 million to 3638 million rand, but this increase disappears if inflation is taken into account and is entirely attributable to improved exports.[58]

[57] South African Ministry of Finance and Ministry of Public Works, 'Green Paper on public sector procurement reform in South Africa', Apr. 1997, URL <http://www.polity.org.za/govdocs/green_papers/procgp.html>.

[58] Hatty (note 33), pp. 1–12.

The South African defence industry has often argued that it is 'world class' and represents a source of considerable scientific and technological skill and innovation. In part this is true. During the 1970s and 1980s the government ploughed R&D and other funds into defence, with the result that the industry developed a cutting-edge technological advantage over other sectors. Since 1990, however, there has been a rapid decline in defence R&D funding (see table 6.2). With a declining domestic market the industry is often unable to achieve the economies of scale needed to make it competitive on the international market. As time passes, without injections of further R&D, the South African industry is likely to atrophy further. It is possible that whole sub-sectors will collapse. The key to survival for industries unable to export will be to seek commercialization and diversification opportunities, or even full-scale conversion to civilian production, although this would probably require government assistance.

The local defence industry argues that there are many advantages in domestic procurement, including the following:

1. Maintenance, modifications and performance enhancement can be carried out locally, saving costs and making it possible to keep systems in service for longer.
2. Surprise in battle can be achieved, as capability is not known to the enemy.
3. The defence industry provides technological support to the defence force, which, due to high staff turnover, cannot develop the same capacities.
4. The existence of a domestic industry makes it easier to gear up the 'core force' to deal with a potential threat.
5. Equipment can be provided which has been designed for local conditions and needs.
6. The industry is a national asset that generates taxes for the state and saves foreign exchange (3500 million rand in 1994/95).[59]

Some of these points are disputed. Few commentators doubt the military and strategic benefits of domestic procurement, but economists are not in agreement about the economic effects. It has been argued, for example, that when domestic industries are small, domestic procurement is costly as economies of scale cannot be made. Furthermore, when developing countries like South Africa seek to maintain domestic defence industries, the effect on economic growth tends to be negative and the economy becomes skewed.[60] The evidence points to this having been the case in apartheid South Africa: it would be far better to reallocate resources to the development of more internationally competitive industries, especially those which are more labour-intensive. There are also important opportunity costs.[61]

[59] Hatty (note 33), pp. 10–12.
[60] See, e.g., Batchelor and Willett (note 17), pp. 9–19.
[61] Batchelor (note 5), p. 2.

A broad policy framework for domestic acquisition was set out in the Defence Review, although often in vague terms. The government was committed to achieving 'limited self-sufficiency in key areas' in which technology development would be concentrated. The Defence Review also stated that 'preference should be given to the procurement of defence products and services from local suppliers, providing such procurement represents good value for money', although 'fair and open competition will be used as far as is practicable . . . this will include the invitation of foreign tenders'.[62] It also stated that adjudication of tenders would not necessarily be based on the lowest price but on 'value for money and industrial development goals' and that 'life-cycle costs, DOD requirements, local industrial development goals, social responsibility (economic empowerment of previously disadvantaged persons), and subcontracting' will be taken into consideration. It is not clear, however, how these various factors will be weighted. In the case of single-source offers (previously the norm), 'bench-marking' against comparable foreign systems or products should be employed to ensure value for money.[63]

Another factor which needs to be taken into account is the pressure placed on Armscor to procure local products which have already been developed, even if the SANDF does not really require them, on the grounds that if they are not purchased locally they may be impossible to sell abroad and the development costs may be squandered. Some commentators believed that this was the rationale behind the DOD's decision in mid-1996 to purchase 12 indigenously developed Rooivalk attack helicopters at a cost of 876 million rand, although the Chief of the Air Force had opposed the project and the purchase was not approved or even discussed by the JSCD.[64] Armscor was at the time trying to sell the Rooivalk to the UK (the bid was rejected) and Malaysia (the contract was apparently still being negotiated a year later) and subsequently attempted to sell it to Turkey. (The sale was vetoed by the NCACC on the grounds of Turkey's human rights record).

The government is also committed to creating a more predictable environment for the domestic defence industry, for example by setting out medium- and long-term acquisition requirements and by introducing a more stable budgeting system. The DOD has undertaken to publish an annual acquisition master plan to indicate all projects required for political approval from the Minister, as well as a medium- to long-term Defence Requirements Statement to guide technology development and industrial planning, although it is not clear exactly how much detail these documents will contain.

Affirmative procurement

A peculiar aspect of South African procurement policy is 'affirmative procurement', introduced by the government in an effort to address the reality that the

[62] South African Department of Defence (note 29), p. 17.
[63] South African Department of Defence (note 29), p. 18.
[64] Crawford-Browne (note 13), p. 13.

domestic defence industry is overwhelmingly dominated by whites and thus unrepresentative of the demographics of the country. This is seen as a medium-term (10-year) requirement to assist in 'levelling the playing field'. The policy provides for up to 22 per cent preference for bidding companies split between equity (10 per cent) and added value (12 per cent) if this will benefit previously disadvantaged groups of people. However, there are many difficulties in the application of this policy, not least of which is deciding whether bidding companies genuinely represent the interests of previously disadvantaged groups—both equity ownership and employee profile need to be taken into account.[65]

Armscor has adopted the following principles with regard to affirmative procurement: (*a*) the defence industry should be committed to redressing previous imbalances; (*b*) the industry should support government initiatives to encourage previously disadvantaged people to become entrepreneurs and owners of productive wealth; (*c*) an organizational climate conducive to the management of diversity should be established within the defence industry; and (*d*) affirmative procurement will be guided, monitored and controlled by the Secretary for Defence.

The national technology base and R&D[66]

A key consideration in arms procurement is the extent to which procurement and R&D decisions will affect the national technology base. In South Africa this consideration is sharpened by the fact that, as a result of apartheid security priorities, defence R&D was far more advanced than civilian R&D. In addition, the concept of a 'core force' entails the retention of capacities to develop and manufacture arms, rather than the retention of a full complement of major weapon systems. Furthermore, even if armaments are mostly procured abroad, some domestic technological capacity is required in order to evaluate such procurements.

Before 1994 defence R&D decisions were taken largely on military and strategic grounds and with little attempt to integrate with civil R&D. Funds were allocated in isolation from the national R&D account through the Special Defence Account and were administered by Armscor. While Armscor still coordinates the DOD's R&D, the White Paper on Science and Technology published in 1996 recommended that defence R&D spending should also be reflected in the national R&D budget to allow 'government and the public to evaluate total R&D spending in an unfragmented way' and to provide for the integration of defence R&D with national R&D and the proposed National

[65] Griffiths (note 34), p. 14; and van Dyk (note 35), pp. 42–46.

[66] Armscor defines defence-related R&D very broadly as 'all scientific and engineering effort that precedes the production phase of any new item, i.e., operations research, basic research, applied research, experimental development, full-scale development, industrialisation and prototype manufacture'. South African Department of Defence, 'Input of the Department of Defence into the Green Paper on technology', [Pretoria], no date, p. 3, quoted by Cilliers, J., 'Defence research and development in South Africa', SIPRI Arms Procurement Decision Making Project, Working Paper no. 103 (1997), p. 3.

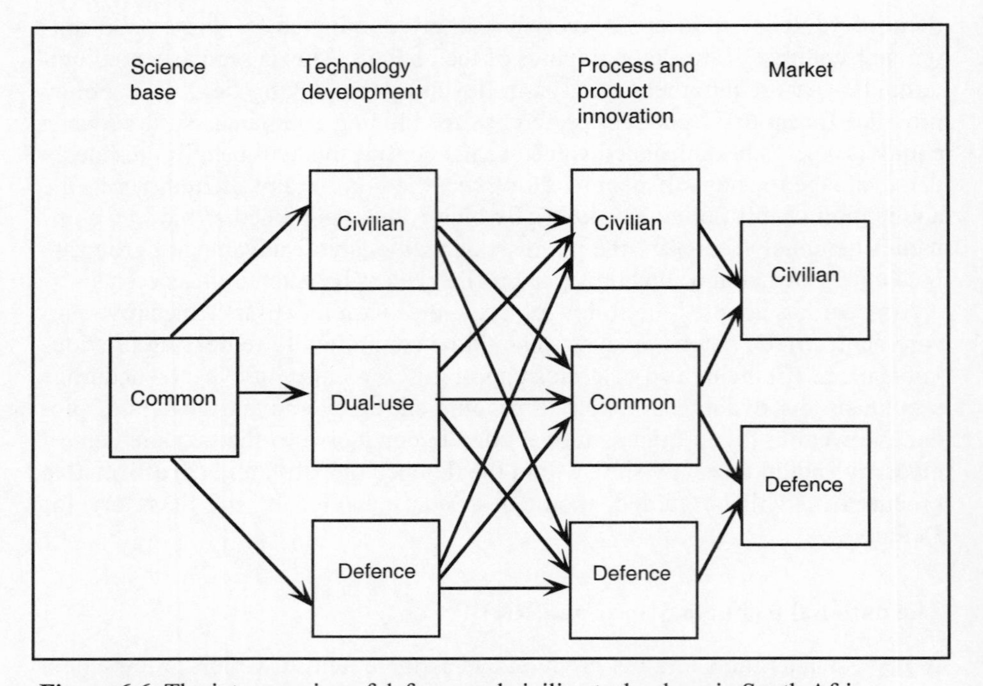

Figure 6.6. The interweaving of defence and civilian technology in South Africa

Source: Buys, A., 'The influence of equipment modernization, building a national arms industry, arms export intentions and capabilities on South Africa's arms procurement policies and procedures', SIPRI Arms Procurement Decision Making Project, Working Paper no. 101 (1997), p. 14.

System of Innovation.[67] This may lead to a further leaching of defence R&D as civilian priorities take precedence—a process likely to be accelerated by the increasing global tendency for civil R&D to lead defence R&D, rather than vice versa (the international norm during the cold war). The DOD is likely, as a result, to make far more use of off-the-shelf civilian technology.[68] The defence industry will continue to seek to leverage spin-offs in the civilian sector and to develop relationships with civilian institutions in the National System of Innovation to promote spin-offs. There are many civilian–military counterpart technologies, most notably computers, but also medical equipment, surveillance and intelligence systems, navigation systems, and clothing and food technologies. The relationship between defence and civilian technology has been conceptualized by Armscor as seen in figure 6.6.

There has been a rapid and significant decline in defence R&D spending in the 1990s (see table 6.2). As a result, only approximately 15 per cent of the acquisition budget of the SANDF—572 million rand in 1996—is spent on R&D

[67] South African Department of Arts, Culture, Science and Technology, *White Paper on Science and Technology: Preparing for the 21st Century* (Government Printer: Pretoria, Sep. 1996).

[68] Batchelor (note 5), p. 15; and Buys (note 8), p. 14.

Table 6.2. Defence R&D expenditure in South Africa, 1987–96
Figures are in million rand and current prices.

Year	Technology development	Full-scale development projects	Total
1987	249	1 546	1 795
1988	295	1 459	1 754
1989	292	1 311	1 603
1990	258	1 000	1 258
1991	534	376	910
1992	493	222	715
1993	317	202	519
1994	292	238	530
1995	300	225	525
1996	312	260	572

Source: Cilliers, J., 'Defence research and development in South Africa', SIPRI Arms Procurement Decision Making Project, Working Paper no. 103 (1997), p. 6.

and this is unlikely to improve significantly. Additional small amounts—perhaps 5–10 per cent of this figure—are spent on R&D by private-sector defence companies and by Denel.[69] Despite this, the White Paper on Science and Technology noted that the defence sector 'is a repository of considerable skills in instrumentation, controls and advanced materials handling. Extending or converting these skills to civil use could broaden our industrial skills base considerably'.[70]

The White Paper also argued that the core force concept adopted by the SANDF required a greater reliance on technology to increase the flexibility and responsiveness of a smaller military establishment. It set out a broad policy framework for defence technology retention and R&D.

The maintenance of a strong technology base is therefore a prerequisite of the new SANDF strategy and must serve a number of purposes: (*a*) maintaining the capability to detect threats; (*b*) creating an awareness of trends in military technology and their implications for the SANDF; (*c*) maintaining the ability to produce technology demonstrators that can be turned into military technology quickly; (*d*) maintaining the ability to provide expert advice for procurement purposes; (*e*) providing test and evaluation services; and (*f*) supporting upgrade and maintenance activities.[71]

In the procurement context this entails the retention of technology to support, upgrade and evaluate systems, to monitor technology trends and to produce technology demonstrators—the development and upgrading of prototypes of new weapon systems, without going into full-scale development or production (current or previous programmes include tanks, helicopters, artillery, advanced

[69] Batchelor (note 5), p. 15.
[70] *White Paper on Science and Technology* (note 67), p. 34.
[71] *White Paper on Science and Technology* (note 67), p. 34.

avionics, stealth technology, fighting vehicles and multi-purpose stand-off weapon demonstrators).[72] The coordination of technology development within the defence sector takes place through the Defence Research and Development Board, supported by an Armament Technology Acquisition Secretariat.[73]

Research institutions, both public and private, also play an important role in military R&D, the Council for Scientific and Industrial Research (CSIR) being the most important. One of its divisions, Aerotek, is extensively involved in defence research (in 1994/95, 74 per cent of its work was defence-related) and has an agreement with the SANDF to carry out research on a variety of technical issues related to military air capacity, many of which have an impact on procurement decisions.[74]

While the government is clearly committed to trying to retain some defence technological and R&D capability, it has recognized that independent local development of major weapons is no longer possible in a climate of budget constraints and reduced demand. Although relative latecomers to the international trend for partnerships and joint development, South African defence companies have moved swiftly into this arena. By the end of 1996, 12 companies were reported to have entered into a total of 93 ventures with companies in 20 other countries, notably in France, Germany, Malaysia and the UK. While this has potential advantages, such as economies of scale and utilization of synergies, in most cases the source of the technology in co-development ventures appeared to be South African, giving rise to concerns about technology outflows.[75] The DOD has also expressed its concern about the possibility of international companies buying out not merely South African technology but also South African companies, especially if Denel were to be privatized in line with government restructuring initiatives. The DOD has indicated its intention of protecting immaterial property rights to state-funded technology.[76] Beyond this, however, it is unclear what steps could be taken to prevent technology outflows or foreign ownership given the growing internationalization of defence industries and the fact that South Africa is such a small player on the world arms market, accounting for less than one-half of 1 per cent of global sales.[77]

VI. Organizational and behavioural issues

The division of responsibilities for acquisition within and between government departments and agencies is outlined above. This section examines the composition and roles of some of the institutions involved as well as the role played by non-government actors.

[72] Buys (note 8), pp. 21–22.
[73] Truscott *et al.* (note 12), pp. 38–39.
[74] Cilliers (note 66), pp. 11–12.
[75] Cilliers (note 66), p. 7.
[76] Buys (note 8), p. 21.
[77] See, e.g., Hagelin, B., Wezeman, P. D. and Wezeman, S. T., 'Transfers of major conventional weapons', *SIPRI Yearbook 2000* (note 44), p. 372.

Parliament and the executive

Control by elected civilian representatives is ensured by the fact that Parliament and the Cabinet have to approve cardinal projects. However, is it is not always clear exactly how this takes place, and the division of responsibility between the executive and the legislature remains a contested issue. South Africa inherited a Westminster-type system with a powerful executive and a fusion between executive and legislative functions (with the exception of the President, all Cabinet ministers sit in the legislature). Under the Government of National Unity, parliamentary committees have gained greater power.

Under the interim (1993) constitution the JSCD was vested with considerable powers, including the competence to investigate and make recommendations regarding the defence budget and armaments. In practice, the extent to which the JSCD has asserted itself in regard to procurement issues has depended on the interest and strength of the personalities of its members. Its functioning has also been hampered by a lack of expertise among its members on procurement issues. It has no secretariat or research support and verbatim records of its proceedings are not kept. Many decisions can be taken in the Cabinet or cabinet committees (such as the NCACC or the Cabinet Committee on Security) without reference to the JSCD. After early enthusiasm over the White Paper process, attendance at JSCD meetings dropped and some parliamentary commentators expressed disappointment at its performance.

Other parliamentary committees are also entitled to deal with defence acquisition issues. The National Assembly includes a Portfolio Committee on Defence with powers to consider legislation and make recommendations, while the National Council of Provinces has established a Defence Committee with advisory and legislative roles. The powerful Finance Committee can have and has had a say in acquisition. The distinction in roles, powers and mandates between the three defence committees is not always clear, although in practice the JSCD is the main locus of parliamentary oversight on defence.[78]

Historically, civilian organizations played little role in arms procurement decision making, while the media were severely constrained by the Armscor Act and other legislation, so that reporting was based largely if not entirely on official DOD information. There was virtually no public scrutiny or discussion around armaments acquisition. While the situation has changed dramatically since 1994, disclosure is still not complete and the capacity of the media is limited (see below). There has been considerable public debate over major issues, notably the corvette proposal, and a number of civil society organizations, especially peace or pacifist organizations, questioned the wisdom of spending such large sums of money on defence acquisitions or openly opposed the proposal. Many of these were small NGOs, such as the Ceasefire Campaign, but others

[78] Calland, R., 'An examination of the institutionalization of decision-making processes based on principles of good governance', SIPRI Arms Procurement Decision Making Project, Working Paper no. 102 (1997).

were mass-membership organizations such as the Congress of South African Trade Unions and the Anglican Church.[79]

Academics and experts outside the relevant government departments have played a minor role in influencing arms procurement decisions. A small group of defence policy analysts and a few NGOs specifically interested in defence matters have written articles in the press and contributed to policy development through White Papers and advisory roles to cabinet ministers. A number of academics and advisers also act as consultants to Armscor in a technical capacity.[80]

Accountability and transparency

The move towards greater accountability and transparency in arms procurement since 1990, and particularly since 1994, has been uneven and fraught with disagreements and difficulties. Principles of transparency are outlined in the Open Democracy Bill which was tabled for parliamentary and public debate during 1997. However, the bill aimed to introduce important limitations which would have an impact on arms procurement transparency. Information could be restricted on the grounds of protecting third-party commercial information, the defence and security of the Republic, South Africa's ability to conduct international relations, its economic interests and the commercial activities of government bodies. In particular, the bill states that information may be restricted if it jeopardizes the effectiveness of arms and equipment (including communication and cartographic equipment) used, intended to be used or being developed by disclosing its capabilities, quantity or deployment.[81]

In June 1997 the Cabinet approved a policy on transparency in defence issues which sought to integrate the letter and spirit of the Open Democracy Bill with the peculiarities of the arms trade. In particular it specified that transparency with regard to procurement was important because public funds were involved. However, commercial confidentiality clauses would need to be respected and technical specifications could remain secret. Major (cardinal) procurement programmes, it reiterated, would have to be approved by the Cabinet while the JSCD retained an oversight function which included guidance to the DOD regarding timing of tenders, submission of RFPs, IP obligations and so on. The policy also noted that international espionage on defence industrial and technology issues needed to be taken into account. One of the most visible steps towards transparency with regard to procurement has been the establishment of a monthly *Tender Bulletin* by Armscor, in which all tenders it adjudicates are listed and which is published both electronically and through the printed media.

[79] Crawford-Browne (note 13), pp. 10–15.

[80] The main NGOs involved in debates over procurement are the Institute for Security Studies, the Centre for Conflict Resolution, Ceasefire, and the Group for Environmental Monitoring.

[81] Draft Open Democracy Bill, 18 Oct. 1997, General Notice 1514/1997, *Government Gazette* no. 18381 (1997), URL <http://www.parliament.gov.za/bills/1997/opendemo.html>. The bill had not been passed at the time of writing (1997–98).

In July 1997, Kader Asmal, Chairman of the NCACC, clarified the government's policy on the arms trade in general, although his statements were made specifically in response to a public dispute over the disclosure by some South African newspapers of the name of a country with which Denel was negotiating a multi-billion rand arms export deal subject to a commercial confidentiality agreement. Asmal claimed that the government would disclose 'an unprecedented amount of information on arms transfers' that would be 'unique internationally' (this was disputed by some analysts) but that there would nevertheless be limitations, particularly with regard to commercial confidentiality.[82]

VII. Towards an 'ideal type' of arms procurement decision-making process for South Africa

This chapter identifies some of the key concerns and issues in the arms procurement process in South Africa on the basis of the research papers commissioned by SIPRI. In some of these, specific recommendations were made as to how procurement decision making could be improved. While it is impossible to reconcile or incorporate all of these views, this section draws on some of these proposals as well as the discussion above.

First, it must be stated that the process of governance never is and never can be 'ideal'. It is always the product of political compromise, historical inheritance, institutional and cultural character, and a host of other social, political and economic determinants. South Africa has undergone a remarkable transition and is self-consciously seeking 'best practice' in managing a democracy in a developing world context. At the same time, the legacy of the traumatic recent history of the country remains a heavy burden, while the institutional inadequacies and inequities which flowed from the distortions of apartheid remain. A centralized, secretive decision-making process was the norm both for the apartheid regime and for its opponents, who were hounded and driven underground or into exile.[83] In general terms, an 'ideal type' process in any realm of governance will be constrained by these realities. This is particularly the case with regard to any aspect of defence and the international arms trade, which is subject to many national security-related abnormalities and specificities (for example, in relation to transparency).

Nevertheless, it is possible to make some general observations which arise from this study and which may assist in moving towards an ideal type of arms procurement decision making in the South African context.

1. Parliamentary oversight in relation to acquisition could be further strengthened. The days when Armscor and the Cabinet (or more usually the State Security Council) were free to make and implement acquisition decisions (even including those relating to nuclear weapons) secretly and without public

[82] *Citizen*, 25 July 1997; and *Business Day*, 25 July 1997.

[83] Liebenberg, I., 'A socio-historical analysis of national decision-making behaviour', SIPRI Arms Procurement Decision Making Project, Working Paper no. 107 (1997), pp. 9–13.

accountability are over. The establishment of the various defence oversight committees has been a significant step forward.[84] However, the committees have had to find their way with few resources, both in terms of developing expertise and in working out an appropriate relationship with the civil service, the public, the military and the executive. The establishment of the NCACC also indicated a government commitment to take political control over the arms trade and defence industrial issues. (Although it is mostly focused on arms exports the NCACC has taken on broader responsibilities, including some aspects of acquisition policy, as is evidenced by its commissioning of the White Paper on the Defence Industry.) However, there is no formalized link between the NCACC and the parliamentary committees, although the NCACC is obliged to submit an annual report to Parliament. Nor has the principle of an independent inspectorate for arms trade issues, called for by the Cabinet in August 1995 when the NCACC was set up, led to any institutional arrangements. The Cabinet authorized an inspectorate to 'ensure that all levels of the [arms control] process are subjected to independent scrutiny and oversight and are conducted strictly in accordance with the policies and guidelines of the NCACC'. It also called for the inspectorate to make reports to the parliamentary committees.[85]

2. Transparency and public accountability in regard to acquisition could be substantially improved, building on the basis of the constitution, which in Section 32(1)(a) states that 'everyone has the right of access to any information held by the State'.[86] One way to strengthen this would be to formalize the processes of parliamentary oversight and approval and clarify the reporting relationship between the various committees and the executive, in particular the NCACC. The principle of obtaining parliamentary approval for cardinal acquisitions should in practice lead to the establishment of a mechanism for ensuring that these decisions are put before the National Assembly and the National Council of Provinces, as they involve substantial public moneys and may have foreign and other policy implications. The functioning of the JSCD with regard to acquisitions would be greatly enhanced by the provision of expert technical advice, possibly administered by a secretariat. It is also not clear at what stage in the acquisition cycle the JSCD and Parliament as a whole are expected to be informed and provide oversight. To do its job properly, Parliament would probably need to consider cardinal acquisitions at the specifications stage, the tendering stage and when the tenders are evaluated.[87]

3. The Defence Review process and the subsequent realization that there was insufficient money to pay for the force design arrived at indicated the unsound relationship between the defence planning process and the budget cycle, and more generally between the internal processes of the DOD and the political process. It is clearly unsatisfactory (not to mention a waste of time and money)

[84] The JSCD has oversight over the DOD: this anomaly needs to be addressed so that it also has oversight over Armscor.

[85] South African Cabinet Office, 'Introduction to press conference', 30 Aug. 1995.

[86] Constitution of the Republic of South Africa Act, Act no. 108 (1996).

[87] Calland (note 78), p. 33.

when a two-year force planning exercise which finally gives rise to a force design proves to be inappropriate in that the money is not available. The introduction of medium-term budget planning over three years may improve such planning, but the DOD evidently needs to pay more attention to budget realities when it undertakes its planning exercises. This means that there should be much closer coordination between the political process and defence strategic management, which could be instituted through the JSCD.

The disjunction between the Defence Review's envisaged force levels and the realities of funding may also eventually give rise to a reassessment of the appropriateness of the SANDF's roles and functions, and hence a reassessment of procurement needs. While the core force concept is one of a scaled-down SANDF, it nevertheless calls for a balanced all-round capability for conventional defence. This may not be affordable in the long run, and the DOD may start to configure the defence forces for their actual tasks, now regarded as secondary—border protection, assistance to the police and peacekeeping.[88] As the SANDF becomes more involved in peacekeeping, as it is certain to do, procurement for peacekeeping operations is likely to become a more important issue. The SANDF has indicated that it sees its role in African peacekeeping operations in terms more of providing equipment and logistical and communications support than of providing troops. Even without SANDF involvement, with its considerable acquisition experience, Armscor could provide a useful service for UN peacekeeping operations in Africa or more widely.

4. One issue which has not been considered in any detail, but which is essential for the evolution of common security in Southern Africa, is the question of relations with the other 13 members of the Southern African Development Community (SADC) with regard to arms procurement.[89] The SADC member states are committed to a wide-ranging set of regional confidence- and security-building measures as well as a more ambitious programme to coordinate peacekeeping, carry out conflict resolution and build mutual defence structures. Transparency is an essential aspect of such confidence building. The MOD does not appear to have given much thought to the effect of South Africa's arms procurement or the process it follows on its neighbours' perceptions: while corvettes are unlikely to be regarded with any alarm, jet strike aircraft or tanks may be a different matter. The reaction of some SADC states to orders placed by Botswana for Leopard main battle tanks and F-5 aircraft in 1996 illustrates how unexpected or unexplained acquisitions can be regarded with alarm by neighbours, even in the context of a common security regime. Armscor has, however, mooted the idea of putting its resources at the disposal of other SADC states and of the South African defence industry becoming the primary supplier to the SADC, in part by donating some of its outdated and redundant equipment.[90] This may be seen as hegemonic behaviour, but it is evident that South

[88] Williams (note 14), p. 26.

[89] The members of the SADC are Angola, Botswana, Democratic Republic of Congo, Lesotho, Malawi, Mauritius, Mozambique, Namibia, Seychelles, Swaziland, Tanzania, Zambia and Zimbabwe.

[90] Buys (note 8), p. 20; and Omar (note 39), p. 20.

Africa would be well served by beginning a dialogue with its neighbours around arms procurement and by consulting within the SADC security structures about its procurement intentions as a confidence-building measure.

5. Improvements in the formal decision-making process around South African arms procurement should be accompanied by enhancements in the public's understanding of and information about acquisition decisions. The capacity of the South African media to deal with these issues is limited as a result of the distortions of the past. In particular, expertise and capacity on defence matters are lacking in the black-oriented press, especially since the demise of some of the 'alternative' or community newspapers of the anti-apartheid struggle.[91] While a few NGOs have established a niche in the defence arena, the interest of civil society in arms procurement issues remains limited, although, as with the case of the corvettes, this can change when large amounts of public money are involved and stark choices appear to be on the agenda. Armscor has made efforts to seek partnerships with NGOs, at one stage hoping to formalize an Armscor–NGO forum,[92] but some NGOs feared being co-opted while others had few resources and were able to focus on arms procurement issues only for limited periods. It is probably unrealistic to expect them to be subjected to ongoing analysis and attention from NGOs and civil society as they are in the USA, for example; nevertheless, capacity-building initiatives, for example by international aid donors, could assist.[93]

6. In the public service, a process of institutional rearrangement of responsibilities has been initiated with regard to procurement. The powers of Armscor, which was once both player and referee in the arms procurement process, have been substantially reduced and the Defence Secretariat, the MOD and the NCACC have taken over most of its powers of authorization and approval. The White Paper on the Defence Industry may make further recommendations in this regard in order to enhance public accountability and transparency and ensure cost-effectiveness.

7. Integration of procurement decisions with national economic, industrial, science and technology priorities could also be improved. Major capital expenditures, particularly if placed outside the country, could provide an opportunity for leveraging strategic and trade advantages. The implementation of the National Industrial Participation Policy is contributing substantially to this process, but, as the White Paper on Science and Technology has identified, there is greater scope to seek synergies between the military and civil technology sectors. This is particularly true with regard to IP policy and international partnerships, which the government, through agreements with other governments, can influence in order to ensure an appropriate exchange of technologies.[94] Under

[91] Liebenberg (note 83), p. 16.

[92] Omar (note 39), p. 17.

[93] E.g., the Group for Environmental Monitoring has managed to sustain public interest in defence industrial issues by holding public workshops to seek inputs into the White Paper on the Defence Industry and, with the Defence Management Programme at the University of the Witwatersrand, organizing a course in Defence and Development which focused on defence industrial conversion issues.

[94] Hatty (note 33), p. 29.

apartheid there was no effective link between arms procurement, which was driven by strategic military considerations, and the technology development process, and there is still considerable room for improvement.[95]

In a broader context, affirmative procurement and the restructuring of the defence industry can contribute to the realization of socio-economic objectives. It is therefore essential that in the pursuit of value for money procurement decisions continue to take into account wider political, economic and social implications. The exact way in which the government supports the domestic industry, and which sections of it it supports, needs to be determined, but it is clear that there cannot be an absolutely 'free market' in defence procurement, if only because other governments are subsidizing their industries. The thriving process of diversification and commercialization in the South African defence industry plays an important role in this as it makes the domestic industry less dependent on public money and leads to greater integration with civil technology and industrial development.

[95] Truscott *et al.* (note 12), p. 43.

7. Taiwan

*Chih-cheng Lo**

I. Introduction

The Republic of China (ROC) on Taiwan is one of the world's main arms-importing countries.[1] Since the 1950s it has maintained sizeable armed forces to protect itself against the military threat from the People's Republic of China (PRC). From 1949 to the 1960s, 'armed liberation' was the PRC's main strategy towards Taiwan and it has not renounced the possibility of using force against Taiwan. Given the security threat the PRC posed, a strong and capable national defence is of the greatest importance for Taiwan. The purchase of arms to guard against any military attack from across the Taiwan Strait has been and continues to be the main concern of its national security policy. Defence in 1998 accounted for 22.4 per cent of government expenditure.[2] In financial years (FYs) 1992–96, approximately 39 per cent of the annual defence budget was spent on arms procurement.[3]

Despite its importance, the subject of arms procurement decision making has hardly been touched upon by scholars. Any discussion of the process of security decision making in Taiwan, particularly relating to arms acquisition, is exceptionally difficult because of the threat from the PRC. The protection of critical national security secrets from public access (and theoretically from enemies) is clearly justified.[4] The need to withhold information relating to military operations, military personnel, weapon technology and arms procurement has rarely been questioned, particularly since the PRC continues to seek to cut off foreign

[1] Over the 5 years 1995–99 Taiwan ranked 1st in the world as a recipient of major conventional weapons in terms of SIPRI trend-indicator values. Hagelin, B., Wezeman, P. D. and Wezeman, S. T., 'Transfers of major conventional weapons', *SIPRI Yearbook 2000: Armaments, Disarmament and International Security* (Oxford University Press: Oxford, 2000), p. 368.

[2] Taiwanese Ministry of National Defense, *1998 National Defense Report, Republic of China* (Li Ming Cultural Enterprise Co.: Taipei, 1998), p. 132.

[3] Cheng-yi Lin, 'Taiwan's threat perceptions and security strategies', SIPRI Arms Procurement Decision Making Project, Working Paper no. 115 [1998], pp. 13–14.

[4] Chih-cheng Lo, 'Secrecy versus accountability: arms procurement decision making in Taiwan', SIPRI Arms Procurement Decision Making Project, Working Paper no. 116 [1998], p. 1.

* The author gratefully acknowledges the help of Dr Hung Mao Tien for advising on and guiding the research for this chapter. It also draws extensively on 11 working papers commissioned for the SIPRI Arms Procurement Decision Making Project which were presented at a workshop in Taipei on 28 April 1997, in collaboration with the Taiwanese Institute for National Policy Research. The contribution made by these papers to the research is gratefully acknowledged. They are not published but are deposited in the SIPRI Library. Abstracts appear in annexe B in this volume.

arms supplies to Taiwan.[5] However, unnecessary secrecy handicaps the public in knowing whether officials have engaged in any corrupt, illegal or improper conduct, and can impair good governance and decision making and even damage national security. Defence officials' use of the excuse of secrecy in the interests of national security has hampered the rational formulation and effective implementation of arms procurement policies. It is believed that open debate and public scrutiny could produce better policy decisions. The process of democratization in Taiwan initiated in the late 1980s and numerous scandals in arms acquisition[6] have generated greater public interest in opening the 'black box' of defence policy making in general and arms procurement decisions in particular.[7]

After these procurement scandals, the general public expects greater openness and accountability on the part of government. Striking a satisfactory balance between the competing interests of military confidentiality and accountability is a part of consolidating democracy in Taiwan. A rationally designed and institutionalized arms procurement process which reconciles the values of democratic accountability and secrecy in the interests of national security should be a priority for Taiwan on its way towards a consolidated and secure democracy.

Sound policy recommendations cannot be made without a clear understanding of the existing process of arms procurement decision making. This chapter examines that process, with a focus on the following aspects: (*a*) the characteristics of the processes, the organizational structures, and the major actors and influences in making national security and arms acquisition decisions; (*b*) the defence budget processes and constraints; (*c*) the domestic research and development (R&D) and defence production capability; and (*d*) the limitations and deficiencies in the process that impair legislative oversight and accountability. Section II describes the organizational structures of national security, the actors involved and the influence of Taiwan's predominant supplier of arms, the USA. Section III examines the procedures for arms procurement, section IV the process of defence budgeting, financial planning and audit, and section V the system of domestic R&D and the defence industrial base. Section VI looks at issues of democratic accountability and legislative oversight in Taiwan's arms procurement decision making, and section VII presents conclusions.

[5] See, e.g., Bristow, D., 'Taiwan looks beyond USA', *Jane's Intelligence Review, Pointer* (monthly supplement), Dec. 1998, p. 7; 'Chirac: conciliation in China', *International Herald Tribune*, 17–18 May 1997; and 'China warns US on Taiwan arms sales', *Interavia Air Letter*, 10 Jan. 2000, p. 4.

[6] See section III in this chapter.

[7] On arms procurement scandals, see sections III and VI in this chapter; and Chen, E. I-hsin, 'Security, transparency and accountability: an analysis of ROC's arms acquisition process', SIPRI Arms Procurement Decision Making Project, Working Paper no. 114 (1998), pp. 12–15.

II. Formulating the national security and defence strategy

Institutional structure, actors and processes

Taiwan's national security policy and defence decision-making system operates within the National Security Council (NSC), functioning under the presidency, and the ministries under the Executive Yuan (the highest administrative organ of the state) such as the Ministry of National Defense (MND), the Ministry of Foreign Affairs (MOFA), and the Mainland Affairs Council (MAC). The NSC is the advisory body to the President and relies on its subordinate, the National Security Bureau (NSB), for the collection and analysis of intelligence. The President's policy and strategy statements, prepared by the NSC, establish basic conceptual guidelines that assist the MND, the MOFA and the MAC in developing threat assessments and strategies. In accordance with the principle of civilian control, the MND is in charge of defence affairs and the Minister of National Defense must be a civilian. However, all defence ministers since 1949, with two exceptions, were serving officers immediately before they took office. The majority of officials in the MND are in fact also former military.[8]

The MND is responsible for formulating military strategy, deciding on and carrying out military procurement, setting military personnel policies, devising draft and mobilization plans, defining logistics and supply policies, arranging for R&D of military technology, compiling data for the national defence budget and so on. The defence minister is the head of the military administration system and in charge of all defence policy decision making. For the military administrative system the Chief of the General Staff (CGS) reports to him and is therefore responsible to the Prime Minister; however, in the military command system and for operational matters he reports to the President.

The General Staff Headquarters (GSH), headed by the CGS, is in charge of: planning and supervision of joint war activities; political warfare; personnel; military intelligence; operations; education and training; logistics, organization and equipment; communications; military archive management; and medical services. In practice, it is the CGS who makes the final decision in deciding which arms are to be purchased and from which sources. (During a hearing in the Defense Committee of the Legislative Yuan in December 1993, Sun Jen, then Minister of National Defense, admitted that he had no control over arms procurement decisions.[9] Very few arms procurement projects actually came to his office.[10]) Under the GSH come the offices of the Deputy Chiefs of the General Staff for Intelligence (J-2), Operations (J-3), Logistics (J-4) and Planning (J-5), and the Military Intelligence Bureau (MIB), which have played a

[8] Discussion at the SIPRI–Taiwanese Institute of National Policy Research workshop, Taipei, Apr. 1997.

[9] *Legislative Gazette*, vol. 83, no. 4 (1993), p. 99.

[10] E.g., it is reported that the final decision in 1992 to buy the French Mirage 2000-5 was made not by the Minister of National Defense but by the CGS. Yann-huei Song, 'Domestic considerations and conflicting pressures in Taiwan's arms procurement decision-making process', SIPRI Arms Procurement Decision Making Project, Working Paper no. 124 (1998), pp. 20–21.

particularly important role in threat assessment and joint capabilities planning in Taiwan.[11]

Dual chains of command

Tensions sometimes arose in the past in arms procurement decision making as a result of an inherent conflict between the military administrative system and the military command system. Ambiguous and sometimes contradictory legal arrangements of government control over the military created problems in arms procurement decisions. It is therefore important to understand the differences between the military command and military administration systems.

The MND has jurisdiction over defence policy and budget formulation but under Article 36 of the constitution the GSH is responsible to the President in the military command system and makes the final decisions on arms procurement.[12] This dual and parallel system of control over the military was less problematic during the period of authoritarian rule since the supreme leaders controlled both lines of command. As Taiwan turns into a democratic polity, however, it has enabled some aspects of procurement to be kept secret and not encouraged transparency and accountability in arms procurement. Among the major difficulties that can arise as a consequence of divided government control over defence policy making are: (*a*) problems in the coordination of, or even confrontation between, the two lines of command; (*b*) inadequate interaction between the President and the Prime Minister, which can affect the role played by the military; and (*c*) the relative independence of the military in its command function from the Executive Yuan. This also limits checks and balances by the Legislative Yuan.

As a result, the Taiwanese Cabinet approved a National Defense Law and the revised Organic Law of the Ministry of National Defense on 26 August 1999. They identify four elements of the national defence system—the presidency, the NSC, the Executive Yuan and the MND. The President as Supreme Commander is now empowered to call the NSC, and in that capacity gives direct orders to the Minister for National Defense, who then entrusts the CGS with specific tasks.[13]

Taiwan's defence strategy

Preparing for any form of military attack from the PRC is the dominating principle guiding Taiwan's defence planning. The assessment of the PRC's military offensive capacities thus defines Taiwan's national security goals and its military strategy for achieving those goals. The various elements and dimensions of the PRC threat define the national defence posture, which in turn decides arms acquisition priorities, the type of weapons to be acquired and the sources of

[11] Cheng-yi Lin (note 3), p. 4.
[12] Yann-huei Song (note 10), p. 20.
[13] Chang, F., 'Cabinet takes steps to unify national defense systems', *Free China Journal*, 3 Sep. 1999, p. 1.

arms supply. The consolidation of Taiwan's fighting capability and the maintenance of sufficient and credible deterrence have become the bedrock of its national security strategy.[14]

Taiwan's national defence strategy calls for the balanced development of the three armed forces, but naval and air supremacy have the priority. This is set out in the 1992 *National Defense Report,* Taiwan's first defence White Paper, which states that 'the defence operations in the Taiwan area should firstly lay stress on air domination and sea control'.[15] In short, Taiwan's military build-up is based on three guiding principles: (*a*) to maintain air and naval superiority over the Taiwan Strait; (*b*) to maintain counter-blockade capabilities; and (*c*) to be able to win the fighting at the beachhead.[16] In particular, Taiwan is concerned about improving its anti-submarine warfare (ASW) capability. It is therefore not surprising that the modernization of naval and air forces has been given priority in recent years and that the lion's share of the defence budget has gone to procurement for them.[17]

Taiwan has made great efforts to maintain its military deterrence by acquiring more advanced weapons and improving the quality of its human resources. Its defence strategy also involves the Ten-Year Plan for Restructuring of Defense Organizations and Armed Forces 1993–2003 (called the Chinshih Plan), prepared by the GSH, to restructure the armed forces, streamline levels of command, renovate logistical systems, merge or reassign military academies and senior staff units, and reduce the total number of men and women in uniform.[18]

Threat perceptions influencing force posture

On 1 May 1991, the Taiwanese Government announced the end of the Period of National Mobilization for the Suppression of the Communist Rebellion. It recognized the PRC regime as an unfriendly political entity effectively governing the Chinese mainland and renounced the use of force as a means for settling cross-strait disputes. Taiwan now asserts that it and the PRC are two equal political entities. The possibility of armed conflict arising from China's military

[14] Yang, A. Nien-Dzu, 'Arms procurement decision-making: the case of Taiwan', SIPRI Arms Procurement Decision Making Project, Working Paper no. 123 [1998], p. 1.

[15] Taiwanese Ministry of National Defense, *1992 National Defense Report, Republic of China* (Li Ming Cultural Enterprise Co.: Taipei, 1992), p. 83.

[16] Taiwanese Ministry of National Defense, *1996 National Defense Report, Republic of China* (Li Ming Cultural Enterprise Co.: Taipei, 1996), p. 62.

[17] E.g., in FY 1996, out of a budget of NT$58.75 billion for procurement of major weapon systems, 36.85% was allocated to the purchase of aircraft, 49.86% for the purchase of naval vessels and 12.44% for missiles and air defence systems. Wen-cheng Lin, 'Taiwan's arms acquisition dependence and its effects', SIPRI Arms Procurement Decision Making Project, Working Paper no. 120 [1998], pp. 4–5.

[18] The total number of troops in the ROC armed forces was reduced from 600 000 between 1950 and 1979 to 470 000 in the early 1990s. Taiwan's armed forces will stand at 400 000 by the year 2003. The army accounts for 50% of the armed forces and the navy and air force 25% each. For details of the Ten-Year Plan, see Taiwanese Ministry of National Defense, *1993–94 National Defense Report, Republic of China* (Li Ming Cultural Enterprise Co.: Taipei, 1994), pp. 74, 153; 'Taiwan army changes focus', *Jane's Defence Weekly,* vol. 26, no. 15 (9 Oct. 1996), p. 21; 'Taiwan wants lean, combat-ready army', *Straits Times*, 24 Feb. 1997, p. 13; and 'Military set for major restructuring', *China News,* 8 Apr. 1997, p. 2.

adventurism still overshadows the Taiwan Strait.[19] Cross-strait tensions were heightened after Taiwanese President Lee Teng-hui's semi-official visit to the United States in June 1995. They reached a peak in March 1996 when the People's Liberation Army (PLA) fired four M-9 missiles into waters about 20–30 km off the coast of Taiwan.[20] These exercises showed Taiwan's vulnerability to missile attack from the PRC.

Taiwan's leaders believe that the PRC may consider using force against it in the following circumstances: (*a*) if Taiwan declares independence; (*b*) if foreign powers intervene in Taiwanese security affairs; (*c*) if Taiwan continues for an extended period to refuse to negotiate for reunification; (*d*) if domestic chaos erupts on the island; (*e*) if Taiwan's armed forces are found to be so far weaker than those of the PRC that they would be unable to withstand a PRC offensive; and (*f*) if Taiwan develops nuclear weapons.[21] On various occasions the PRC leaders have stated their intentions to use force in the first three cases.

Security perceptions of the major political parties[22]

The divergent attitudes of the major political parties towards the issue of reunification or independence have created somewhat different stances on national security, military objectives and arms acquisition policy. The official position of the long-ruling Kuomintang (KMT) is 'one China with two political entities'. It seems that the KMT has taken a middle-of-the-road approach towards the issue of reunification or independence. Its leaders tend to believe that, so long as Taiwan does not declare independence, the USA will extend its assistance to the island if mainland China attacks. In contrast, the New Party (NP) stands very firmly for reunification. Although it does not agree with the PRC's claim that Taiwan is part of the PRC, it does insist that Taiwan is part of China. It opposes independence for Taiwan in the strong belief that it would only bring disaster. The Democratic Progressive Party (DPP) has determined to seek independence and views mainland China as a hostile foreign country. It has a more provocative policy towards the PRC, believing that the USA will help defend Taiwan even if it declares its *de jure* independence.

Notwithstanding their divergent views in this respect, the three major political parties show no great difference in their positions on the actual military threat to Taiwan.

Some DPP legislators advocate introducing the Theater Missile Defense (TMD) system in Taiwan, believing that it will contribute to strengthening the country's defence. The NP, on the other hand, believes that an arms race and provocative actions, such as introducing the TMD, would only jeopardize cross-strait relations. The KMT, which currently still dominates security and defence policy, takes a position somewhere between the two. The inputs of the

[19] Wen-cheng Lin (note 17), pp. 2–4.
[20] Wen-cheng Lin (note 17), p. 4.
[21] Taiwanese Ministry of National Defense (note 18), p. 62.
[22] This section is based on Chen (note 7), pp. 5–7.

opposition parties will increase as Taiwan democratizes further and as the KMT gradually loses its grip on power. Nevertheless, stronger ties with the USA and an effective deterrent force will remain the cornerstones of the island's security.

Guiding principles and approaches to arms procurement[23]

The general principles guiding Taiwan's arms procurement policy as stated in its defence White Papers are the following. First, in relation to the operational requirements for weapons and equipment, and consistent with the Chinshih Plan, the decision-making process is required to include systematic analysis, compare force levels with those of the PRC, decide the types of weapons and equipment required, and then comply with the annual Administrative Plan of the MND. Second, arms procurement policy stresses the principle of multiple-purpose applications of equipment and 'one system being utilized by three services of the armed forces'.[24] Third, it takes into account the need to develop the country's military R&D and arms industry. Whenever possible, the Taiwanese Government seeks to acquire manufacturing know-how along with arms purchased in order to build up the domestic defence industry and upgrade the country's military R&D capabilities and achieve a certain level of self-sufficiency in weapons production and maintenance. Fourth, if weapons and equipment have to be purchased abroad, this should be done in accordance with the state's external economic and trade policies. Fifth, procurement of arms from foreign sources should be made directly from the manufacturers. Sixth, sources of supply should be diversified as far as possible.

The influence of the United States

Outside the formal institutions of national security planning, the most important external actor capable of influencing Taiwan's national security, military objectives and arms procurement policy is the United States. It has played the most important role in shaping the island country's national security and arms procurement policy since the nationalist government fled to Taiwan in 1949. The USA's need for strong partners in the Asia–Pacific region to contain the communist expansion allowed Taiwan to purchase high-quality tactical weapons at very reasonable prices.

Despite its 'hands-off' policy towards the civil war in China, the administration of President Harry S. Truman supplied Taiwan with vast quantities of arms for its defence. On the outbreak of the Korean War in June 1950, the USA decided to intervene actively in cross-strait affairs by dispatching the Seventh Fleet to the Taiwan Strait. In August 1950 arms valued at US$14 million under formal military assistance were delivered to Taiwan.

[23] This section is based on Wong Ming-Hsien, 'Influence of the ROC's foreign and security policy on its arms procurement decision making', SIPRI Arms Procurement Decision Making Project, Working Paper no. 121 [1998], p. 9.

[24] Taiwanese Ministry of National Defense (note 15), pp. 157–58.

Before 1954, the United States feared being drawn into a conflict not only by a possible PLA invasion of Taiwan but also by a Taiwanese offensive against mainland China. In exchange for the US security guarantee, Taiwan agreed not to take military initiatives against the mainland without US consent and changed its offensive policy, of attacking to regain mainland China, with the Mutual Defense Treaty with the USA of 1954. It was under US pressure that Taiwan renounced its 'counter-attack the mainland' strategy in 1962 and modified its policy to recovering the mainland through a strategy of '70 per cent politics and 30 per cent military'.[25] After 1960 the USA gradually changed its method of arms supply to Taiwan from direct aid to Foreign Military Sales (FMS), and both aid and FMS were terminated when it broke off diplomatic relations with Taiwan in December 1978.[26]

The question of arms sales to Taiwan was not a major barrier in the normalization of relations between the USA and the PRC during the 1970s. The enacting of the Taiwan Relations Act (TRA) in 1979 under the Carter Administration marked the most important milestone of the US–Taiwanese military relationship.[27]

Events in 1982 diluted the effects of the TRA. The PRC threatened to downgrade its diplomatic relations with the United States if it continued to sell arms to Taiwan. The result was the Joint Communiqué of 17 August 1982, whereby the USA promised that 'its arms sales to Taiwan will not exceed, either in qualitative or in quantitative terms, the level of arms supplied in recent years since the establishment of diplomatic relations' between the USA and the PRC on 1 January 1979.[28] The USA also expressed its intention 'to reduce gradually its sales of arms to Taiwan, leading over a period of time to a final resolution'. As a result of this communiqué, the USA reduced its arms sales to Taiwan every year and controlled the quality of weapons supplied so as not to exceed the level of 1979. However, in March 1983 the Reagan Administration announced that future arms sales to Taiwan would be indexed for inflation. This permitted the USA to claim that it was complying with the 1982 Joint Communiqué while still increasing arms sales to Taiwan. In September 1992 the PRC protested at the US decision to sell Taiwan 150 F-16 combat aircraft, stating that it violated the terms of the communiqué. In May 1994, the US Congress voted to increase US arms sales to Taiwan, and this was believed to have removed the restriction of arms sales to Taiwan provided in the 1982 Joint Communiqué.[29]

There are several reasons for Taiwan's concentration on arms purchases from the United States. First, the TRA guaranteed the provision of sufficient defensive weapons to Taiwan. Section 2(b) of the act states that 'it is the policy of the United States to provide Taiwan with arms of a defensive character and to maintain the capability of the United States to resist any resort to force or other

[25] Wen-cheng Lin (note 17), p. 2.
[26] Yang (note 14), p. 2.
[27] Wen-cheng Lin (note 17), p. 9.
[28] *New York Times*, 18 Aug. 1982, p. A12.
[29] See, e.g., Yann-Huei Song (note 10), pp. 13–15.

forms of coercion that would jeopardize the security, or the social or economic system, of the people of Taiwan'.[30] Second, the United States is the only country capable of defying PRC pressure. It also has major national interests on Taiwan because of the island's strategic importance in the Western Pacific and strong economic ties with the USA. Third, the long-standing military ties with the United States make it very difficult to cut off the US logistical and supply systems. Taiwan and the USA were allies from 1954 to 1978. US advisers helped to restructure the Taiwanese military and formulate its strategy after 1949. Many senior officers in Taiwan were trained by or educated in the USA. They feel more comfortable using US weapons.[31]

Despite dedicating significant resources to modernizing and increasing its military might, Taiwan still believes that it is not capable of defending itself alone from a PRC attack. If the PRC initiates military confrontation, and provided Taiwan has not provoked it by declaring independence, Taiwan hopes that the USA will come to its aid in accordance with the Taiwan Relations Act.

The implications of dependence on the USA

Taiwan's dependence on the USA has significantly constrained rational calculation in arms procurement decision making. It can only buy from the USA those weapon systems that the USA is willing to sell. The USA is reluctant to sell the sophisticated high-technology systems that Taiwan badly needs, and there is almost no alternative: other countries are less able than the USA to withstand pressure from the PRC. The examples in table 7.1 illustrate the extent of Taiwan's dependence on the USA in the period 1990–98.

Consequently, the military equipment acquired by Taiwan mainly reflects the USA's global and regional strategic considerations. Their threat assessments and strategies for dealing with the threat are not always congruent or compatible. As far back as 1982, Taiwan's Tien-ma (Sky Horse) project, which aimed to develop medium-range surface-to-surface missiles (SSMs) with a range of up to 1000 km, and thus capable of attacking cities on the Chinese mainland, was suspended under pressure from the USA. With the collapse of the Soviet Union and the end of the cold war, the international security environment has undergone tremendous change. With the disappearance of the common Soviet threat, the loose anti-Soviet alliance has become obsolete. US relations with the PRC were at a low ebb after the 1989 Tiananmen Square incident, but have improved, particularly with the summit meeting between US President Bill Clinton and Chinese President Jiang Zemin in October 1997. The two countries claim to have a strategic partnership. US policy on arms transfers to Taiwan has changed accordingly. Without doubt, US decisions to sell arms to Taiwan are based on its evaluation of the security environment in the region and its own economic and political considerations. Taiwan is apprehensive of any possible shift by the USA from its existing policy.

[30] Wen-cheng Lin (note 17), p. 9.
[31] Wen-cheng Lin (note 17), p. 11

Table 7.1. Select major arms orders by Taiwan, 1990–98

Date of contract*	No.	Items purchased	Source	Value (US $m.)
1990	10	S-70B/SH-60B Seahawk ASW helicopter	USA	. .
1991	6	La Fayette Class frigates	France	2 400
1992	3	Knox Class frigates	USA	230
1992	150	F-16AM combat aircraft	USA	5 800
1992	60	Mirage-2000-5 combat aircraft	France	2 600
(1992)	(960)	MICA-EM AAMs	France	(part deal)
1992	(480)	R-550 Magic-2 AAMs	France	(part deal)
1992	26	Bell-206/OH-58D(I) combat helicopter	USA	367
1992	42	Bell-209/AH-1W combat helicopter	USA	(FMS deal)
1992	600	AIM-7M Sparrow AAMs	USA	(part deal)
1992	900	AIM-9S Sidewinder AAMs	USA	(part deal)
1993	(4)	C-130H Hercules transport aircraft	USA	. .
1993	40	T-38 Talon jet trainer aircraft	USA	49
1993	4	E-2T Hawkeye early-warning aircraft	USA	760
1993	200	Patriot missiles	USA	1 300
1994	3	Knox Class Frigates	USA	230
1994	160	M-60A3 Patton-2 main battle tanks	USA	91
1996	300	M-60A3 Patton-2 main battle tanks	USA	223
1996	4	C-130H Hercules transport aircraft	USA	200
1996	1299	RMP Stinger SAMs	USA	125
1997	54	Harpoon anti-ship missiles	USA	95
1997	700	DMS Stinger SAMS	USA	200
1997	11	S-70B/SH-60B Seahawk ASW helicopter	USA	. .
1998	2	Knox Class frigates	USA	. .
1998	28	155-mm M109A5 self-propelled howitzers	USA	. .
1998	1000	Apilas anti-tank weapons	France	. .
1998	728	DMS Stinger SAMs	USA	180
1998	58	Harpoon air-launched anti-ship missiles	USA	101
1998	9	CH-47SD helicopters	USA	486

Notes: *Or date of notice of contract. . . = Not available or not applicable. ( ) = uncertain data or SIPRI estimate. ASW = anti-submarine warfare; AAM = air-to-air missile; SAM = surface-to-air missile.

Sources: Chung Yang Jih Pao, 12 May 1997, p. 4; *Jane's Defence Weekly*, 12 Feb. 1997, p. 17; 15 Apr. 1998, p. 20; 22 July 1998, p. 14; 23 Sep. 1998, p. 14; and 21 Oct. 1998, p. 21; *Flight International*, 10–16 June 1998, p. 6; and SIPRI arms transfers database.

Although the TRA stipulates the provision of defensive arms and equipment to Taiwan, the quality and quantity of such weapons have been affected by the changes in the USA's perceptions of its security interests and in its relations with the PRC over time.

Given these circumstances, Taiwan also has problems in negotiating arms sales prices or good offset deals. The prices it pays for the same or similar types of US arms have been much higher than those paid by other foreign buyers.[32]

[32] Yann-huei Song (note 11), p. 15.

The compulsion to acquire its arms from the United States places Taiwan in a less favourable negotiating position.

The USA–Taiwan arms transfer process

A fairly formal and institutionalized arms procurement procedure is established between Taiwan and the USA, partly because of the long-standing arms transfer relationship. Under the TRA, technical sections form part of the representative offices in the respective capitals, Taipei and Washington. The technical section of the American Institute in Taiwan is responsible for assessing requests for arms acquisition from the Taiwan MND and forwarding the case to the US Department of Defense (DOD) and State Department for consideration. The technical section of the Taiwanese representative office in Washington (TECRO) then liaises with the DOD and the State Department. The proposed arms procurement list is presented at the annual unofficial defence meeting with the USA. The decisions on arms procurement made at this meeting are then referred back to the respective governments.

A similar procedure is followed by the Taiwanese military and French manufacturers with respect to supply of spare parts and logistical support for the French weapon platforms and systems which Taiwan has bought.

The diversification strategy

Diversification of foreign suppliers is an important principle of the arms procurement policy. The USA is still Taiwan's most important arms supplier and security provider and continues to provide it with defensive weapon systems, but, particularly given the PRC's relentless criticism of the US supplies of arms to Taiwan and its efforts to deny Taiwan access to the international arms market, diversification is only prudent. Ideally this would also give Taiwan greater bargaining power vis-à-vis the suppliers. The biggest problems arising from buying arms from different sources are (*a*) integrating them with the operational inventory of weapon systems, and (*b*) training personnel to operate and maintain weapons acquired from different countries with different operating manuals translated from different foreign languages into Chinese.[33]

Taiwan has bought advanced weapon systems from European countries. Contracts were signed with France in 1991 for six La Fayette Class frigates and in November 1992 for 60 Mirage-2000-5 aircraft, and with a Dutch shipyard in 1981 for two Zwaardvis diesel-powered submarines after the USA repeatedly turned down Taiwan's requests for submarines. Taiwan has also approached Australia, Singapore, South Africa, Sweden and East European countries. However, the PRC's relentless efforts to sabotage its arms procurement plans have made diversification difficult. The Mirage-2000 deal was the last major French arms sale to Taiwan: under pressure from the PRC, France declared in

[33] See, e.g., Yann-huei Song (note 10), p. 20.

January 1994 that it would not authorize any further transfers to Taiwan.[34] Only after great effort was an agreement on general logistics and support signed with France to ensure that it fulfils its obligation to deliver the arms purchased and continues to supply parts needed for maintenance and operational purposes.[35]

So far, however, Taiwan's efforts to diversify its arms sources have been in vain because countries have not been prepared to jeopardize their economic and political ties with mainland China. Taiwan's dependence on the United States for arms is likely to continue in the future.

III. The arms procurement decision-making process

The officials in the armed services and in the MND play the most important roles in the arms acquisition process. They coordinate the armed forces' needs, assess weapon acquisition programmes required by particular security considerations, evaluate possible alternatives, and identify budget needs and available resources. They serve as channels of information to the legislature and implement the projects approved.

The principal organizational actors[36]

New weapons and equipment requirements are selected according to Taiwan's strategic concepts and operational guidelines. There is a hierarchical structure consisting of several tiers of actors in the arms procurement decision-making process. The main organizations involved are shown in figure 7.1.

At the top in the first tier are the Executive Yuan and the Legislative Yuan. The former is responsible to the latter. The Prime Minister chairs the Cabinet meeting on policy and budget integration and the Speaker of the Legislative Yuan chairs the meeting to approve the defence budget.

The second tier is the GSH and the MND. The principal organization responsible for arms procurement in the GSH is the General Staff Department (GSD). Its J-5 department is the key department responsible for arms procurement functions and coordination. Also in this tier is the Military Procurement Bureau, created in 1995 (initially under the GSD) to integrate the purchasing units of the armed services and be responsible for the overall planning and purchasing of major weapon systems and equipment for the armed forces.[37] In March 1998 it was placed directly under the MND as part of the reform of arms procurement procedure.[38] Various military purchasing units were integrated into the Procurement Bureau, under which there are five departments, two sections and one foreign procurement unit stationed abroad.

[34] Wen-cheng Lin (note 17), p. 10; 'France: no more new arms to Taiwan', *China Post*, 13 Jan. 1994, p. 1; and 'Tension with France over arms sale has ended: Qian', *Straits Times*, 12 Jan. 1997, p. 16.

[35] Leung, A., 'The reinforced fortress', *Military Technology*, Mar. 1996, p. 74.

[36] This section is based mainly on Yang (note 14), pp. 4–5.

[37] *Republic of China Yearbook 1997* (Government Information Office: Taipei, 1997), p. 124.

[38] See also below in this section. On the relationship of the Procurement Bureau to the departments of the GSD, see Yang (note 14), p. 5.

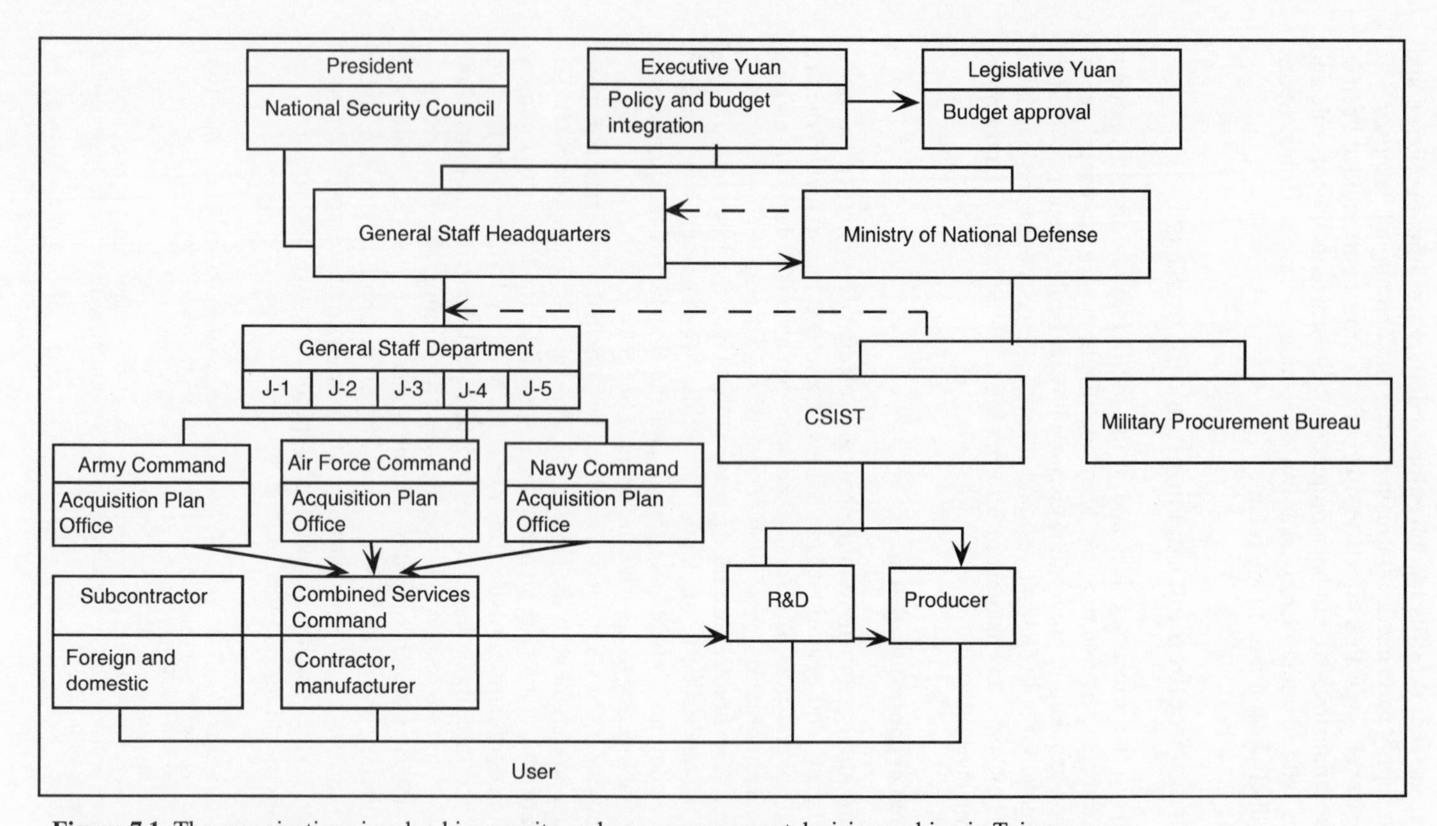

Figure 7.1. The organizations involved in security and arms procurement decision making in Taiwan

Source: Taiwanese Ministry of National Defense, *1996 National Defense Report, Republic of China* (Li Ming Cultural Enterprise Co.: Taipei, 1996).

Table 7.2. The stages of arms procurement in Taiwan

Stage	Body responsible
1. Establishment of long-term procurement plan (from defence policy and 10-year force development plan)	Requirement Committees of the the respective armed services
2. Review of long-term procurement plan	MND System Analysis Committee
3. Studies of operational, technical and financial aspects of weapon systems proposed	GSH, with assistance of CSIST
4. Compilation of request for funding	MND
5. Review of request for funding	MND Planning Committee
6. Drafting of annual defence budget	MND Accounting Bureau
7. National budget drafted	General Accounting Office of the Executive Yuan
8. Approval of budget	Legislative Yuan
9. Permission for procurement to go ahead	Cabinet

Note: CSIST = Chung Shan Institute of Science and Technology.

The third tier includes the acquisition planning offices of the armed services, which make the initial assessment of equipment acquisition plans and establish priorities. The procurement planning and acquisition offices of the respective armed services set up inspection teams to carry out foreign procurement.[39] The fourth tier consists of the major defence manufacturing units and prime defence contractors, which are responsible for implementation of defence contracts. The fifth tier is made up of the defence manufacturers and R&D institutes that carry out R&D and production projects and programmes according to defence contracts. They are not involved in procurement decision making. This tier also includes the user services, which conduct trials and field tests and report the shortcomings of weapons under development. The first three tiers mainly deal with decision making, analysis and planning, the last two with manufacturing and R&D.

The formal procedures of arms procurement

In order to define responsibilities clearly and develop effective decision making, several committees were established in the MND for each stage in the process of arms procurement. They include the Requirement Committee, the System Analysis Committee, the Decision-Making Committee and the Acquisition Reviewing Committee. These committees team up to supervise the acquisition projects of critical weapons and equipment. The stages in the arms procurement process according to the new procedure are illustrated in table 7.2 and figure 7.2.

[39] The on-site inspection team consists of representative of the user organization, a PPAO officer, a technological adviser and a logistic engineer to monitor production schedule and pre-production tests and trials. Yang (note 14), pp. 11–12.

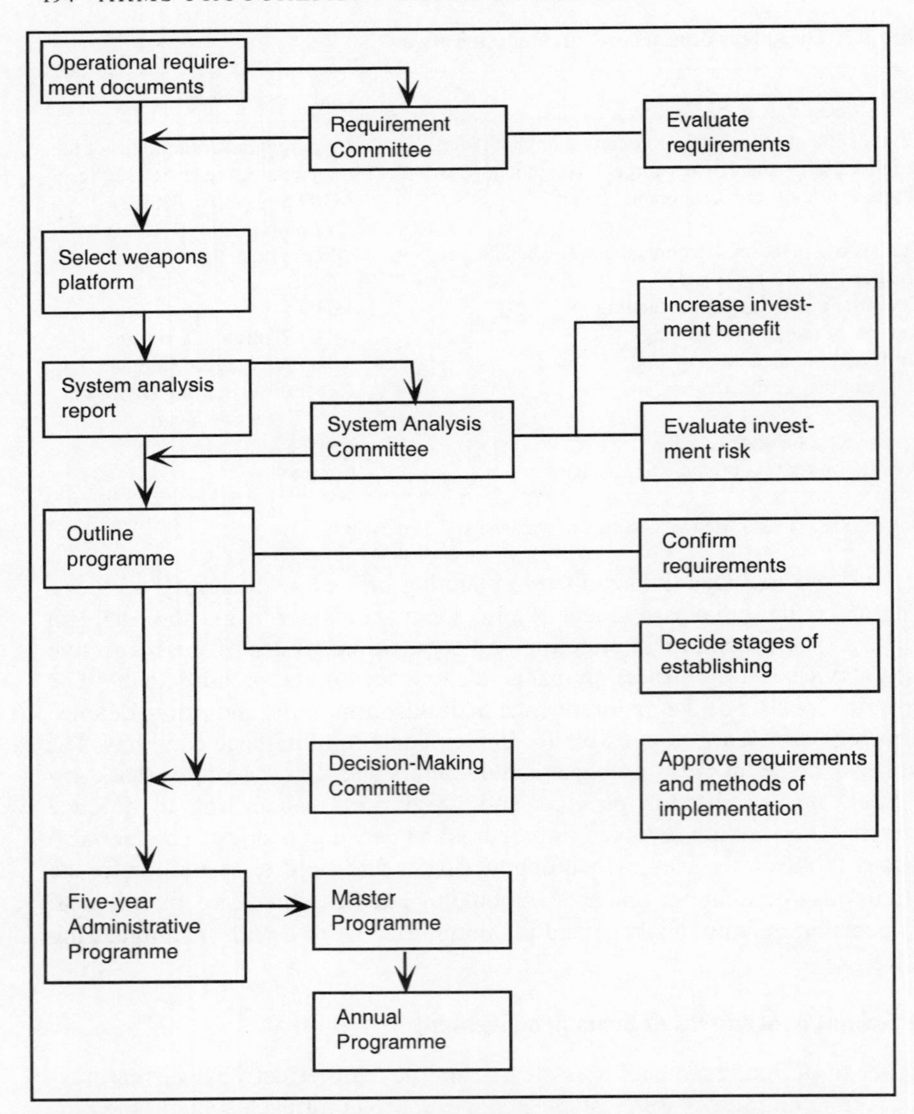

Figure 7.2. Investment outline programme and system analysis process of weapon acquisition in Taiwan

Source: Chin-chen Yeh, 'Arms acquisition decision making in Taiwan', SIPRI Arms Procurement Decision Making Project, Working Paper no. 117 (1998), citing Taiwanese Ministry of National Defense, [Procurement regulation of military materials] (Ministry of National Defense: Taipei, 1995) (in Chinese).

The Operating Procedure and Regulations on Reviewing Arms Acquisition and Major Engineering Constructions introduced by the MND in July 1995 involves two main stages.[40] The process by which the procurement plans are incorporated in the defence budget and the national budget is described in section IV of this chapter.

Planning application

The first step is system analysis, which is done in the J-5 division of the GSH and in the respective headquarters of the three branches of service.[41] When considering arms procurement projects, the Requirement Committee of each branch of service must first submit a procurement plan to be reviewed by the MND's System Analysis Committee and ratified by the Policy-Making Committee. The latter is made up of various deputy chiefs of the general staff.

Next, on the basis of medium- and long-term weapon development programmes, the armed services formulate plans for arms procurement according to the type, specifications and quality of weapons and equipment required. The procurement planning office of each service then works out proposals for new weapons which define the purpose and main combat performance and technical specifications, and provide planning schedules and budget estimates. This is done after thorough studies have been carried out of the operational, technological and financial aspects by the GSH. While doing this, each service has to compromise between its operational requirement and technical and financial feasibility. The Chung Shan Institute of Science and Technology (CSIST) also has the responsibility to evaluate the operational, technological and financial feasibility of the plans submitted by the Acquisition Planning Offices.

An item of equipment or weapon is listed in the annual financial programme and funds allocated only if the system analysis report is favourable.

The next stage is requisition of the items selected.

Requisition, acquisition and execution[42]

This is the most complex part of the procedure. It includes the following steps:

1. Evaluation of application. A team of experts from the relevant agencies examines the planning application for a particular weapon system to identify unsuitable or infeasible aspects or anything that might cause failure of the programme. If it identifies such possibilities, the units applying may be asked to explain or modify their plans. The application is also evaluated using nine main considerations: (*a*) the threat from the enemy; (*b*) consistency with the current defence policy; (*c*) consistency with the strategic plan and war-fighting principles; (*d*) the importance of the weapon to each force's mission; (*e*) the readiness of the weapon system; (*f*) the possibility of technology transfer and

[40] The procedure and regulations are not published.

[41] Lin Chi-Lang, 'The policy analysis of land force arms procurement: the case of the Republic of China's army', SIPRI Arms Procurement Decision Making Project, Working Paper no. 118 (1998), p. 3.

[42] This section is based mainly on Wu, S. Shiouh Guang, 'Problems in Taiwan's arms procurement procedure', SIPRI Arms Procurement Decision Making Project, Working Paper no. 122 (1998), pp. 5–6.

advantages to domestic production; (*g*) the available financial resources; (*h*) the time-frame and plans for phasing out the system which is to be replaced; and (*i*) cost-efficiencies. The services applying for procurement of a weapon system have to provide further information or modify their plans if requested. Only after the evaluation team has approved the plan can the project be included in the annual budget proposal for next year.

2. *Contracting.* Contracting can be divided into two types: (*a*) 'common-place' or general purchases; and (*b*) foreign purchases. In the case of the former a standard form of contract is usually used. It includes clauses on 'guarantee of durability', penalties for damage or breach of contract, payment and so on, but can be modified to suit special situations. Legal advisers are usually consulted at this stage. When purchasing arms from the USA, the quotation documents issued by the US Government are generally used. Sometimes a special clause on warranty, penalties for delay in delivery, constant supply of spare parts and ammunition, and so on is added. The procedure for acquisition from European countries is less standardized. This may be part of the reason why corruption is much easier in Taiwan's purchase of arms from European countries.[43]

3. *Auditing supervision.* If the price of a purchase is NT$50 million (US $1.5 million at 1998 rates of exchange) or over, according to Article 5 of the Supervision Rules on Governmental Constructions, Purchases, Ordering and Sales of Properties, officers from the Ministry of Audit will be asked to supervise the purchase. The procurement offices stationed in foreign countries are also bound by Article 28 of these rules and are required to send copies of their comparison of quotations, the results of the bidding and the contracts for verification by the auditing agencies concerned.

4. *Delivery and acceptance.* Once the item has been delivered, it is examined for compliance with the contract. Quality, quantity, delivery time and place are carefully checked before payment is made. If any part of the contract has not been properly carried out by the supplier, then the process of asking for a penalty is initiated.

Under the new system of arms acquisition, military hardware is procured by a centralized management system but authority to make a purchase can be delegated to lower echelons. For domestically produced items costing less than NT$50 million, the individual branches of service have the authority to make the purchase. If the price of a domestically produced item is over NT$50 million, the Military Procurement Bureau takes it over. In the case of foreign procurement, items costing under US$1 million can be purchased by each military service itself. When the price is more than US$1 million, the case is submitted to the Military Procurement Bureau, which then hands it over to its procurement office abroad.

[43] Wu (note 42), p. 6.

Changes in the decision-making process in the 1990s

Before 1995, Taiwan's major overseas arms procurement projects were handled by the various services of the armed forces, the Taiwanese Military Procurement Mission to the USA and the Division of Materials of the Combined Services Command. The system was replete with examples of improper decisions and misconduct on the part of officials, scandals, waste and cover-ups. For example, in April 1994 the naval chief, Admiral Chuang Ming-Yao, was forced to step down because of irregularities discovered in the navy's arms procurement process. Eight retired senior military officers were censured by the Control Yuan in 1994 when a legislator from the opposition DPP accused them of tailoring a bid to favour Grumman in the procurement of aircraft.[44] As democratization took root in Taiwan, the opposition parties began to criticize the arms procurement scandals and demand more openness in the management of defence programmes. In 1993, opposition members of the Legislative Yuan questioned the rationale of buying medium-sized rather than light tanks for Taiwan's ground defence and stalled the original plan. Delivery was not completed until 1996.[45]

The situation changed after the murder in 1993 of the former Director of the Navy General Headquarters Weapons Acquisition Office, Captain Yin Ching-feng, over a scandal related to the navy's purchase of foreign-made minesweepers.[46] In 1994, the MND began to study ways to correct the defects in the arms acquisition process and presented a report to the Legislative Yuan, entitled *Review and Improvement on the Purchase of Military Hardware*.[47] In January 1995, eight study groups were called together to study the arms procurement process in depth with regard to personnel, education, purchasing, planning, political warfare, the audit function, the role of the judge advocate and administrative support. The Military Procurement Bureau was established in July 1995 under the GSH to institutionalize and professionalize the defence acquisition process and to make it as transparent and accountable as possible.[48] In March 1998 the Bureau was placed directly under the MND as a measure of damage control in response to pressure from the Legislative Yuan and the general public after yet another procurement scandal was exposed in February 1998.[49]

[44] '8 censured over planes purchase', *China News,* 3 June 1994, p. 1. On the Control Yuan, see sections IV and VI in this chapter.

[45] 'ROC quest for tanks persists', *China Post,* 1 Feb. 1994, p. 15; Opall, B., 'US Government finds tough customer in Taiwan', *Defense News,* 17–23 Jan. 1994, p. 1; and Chen Kao, 'Taiwan's military is learning to play by new rules of the games', *Straits Times,* 15 June 1995, p. 38.

[46] According to former Prime Minister Hau Pei-tsun, who had served as defence minister and CGS, conflict of business interests over the purchase of parts for the maintenance of these minesweepers might have resulted in the murder. *China News*, 19 Mar. 1998, p. 3. For details, see Lee Mei-qei, [Who killed my husband Yin Ching-feng?] (Ta-tsun: Taipei, 1994) (in Chinese).

[47] Chen (note 7), p. 9.

[48] Chen (note 7), p. 9.

[49] 'Taiwanese ministry moves to control purchasing', *Jane's Defence Weekly*, 1 Apr. 1998, p. 11. A former French Foreign Minister, Roland Dumas, admitted in Mar. 1998 that bribes of $500 million had been paid to facilitate the French Government's approval of the sale of 6 La Fayette Class frigates to Taiwan in 1991. Chen (note 7), pp. 12–13.

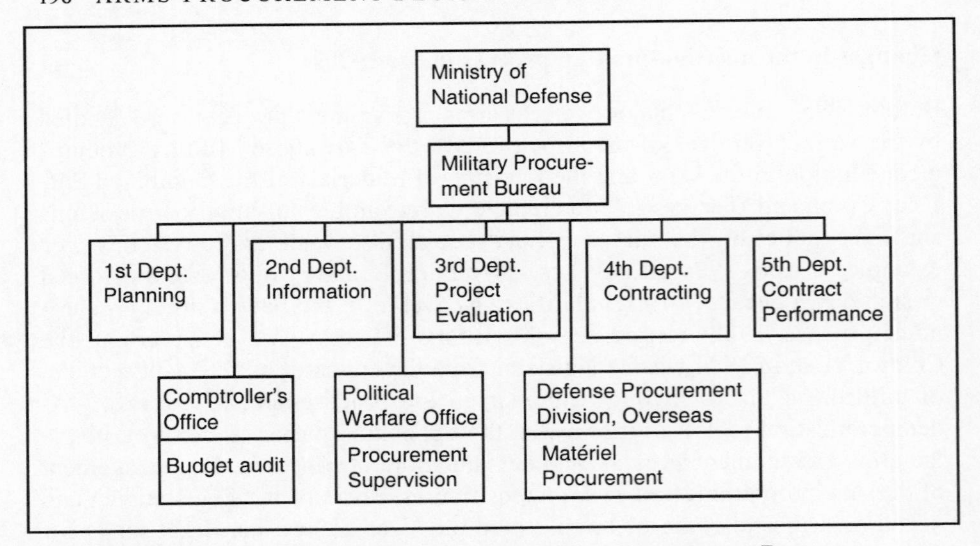

Figure 7.3. The organization of the Taiwanese Military Procurement Bureau

Source: Wu, S. Shiouh Guang, 'Problems in Taiwan's arms procurement procedure', SIPRI Arms Procurement Decision Making Project, Working Paper no. 122 (1998), p. 4.

The Military Procurement Bureau is also responsible for recruitment and training of procurement officers and overall streamlining of the whole arms acquisition process. The purpose of the reorganization was to fine-tune the purchasing procedure and to professionalize the personnel responsible for arms acquisition. It is hoped that misconduct will be reduced as a result of the reforms. In order to prevent corruption, ideally, personnel who are responsible for arms purchase are to be rotated every three or four years, but there are practical difficulties in implementing this policy since most of the important weapon programmes are long-term ones and it is not easy for the armed forces to replace key personnel frequently without causing serious disruption to programmes under way.[50] While J-3 is still responsible for drafting operational plans and operational requirements for arms procurement, the Military Procurement Bureau now has the responsibility to monitor implementation, for instance, by reviewing and assessing the qualifications of contractors. It presides over the bidding procedures and controls the payment process. Figure 7.3 shows the present organizational structure of the Military Procurement Bureau.

A Procurement Commission was also established in April 1997 for the consultation and evaluation of arms procurement operations. Its recommendations are limited to internal organizational reforms of the army.

The introduction by the MND of the 1995 Operating Procedure and Regulations[51] was one of the most important achievements of the reform of arms pro-

[50] Wu (note 42), pp. 3–4.
[51] See note 40.

curement. The ministry even published a handbook entitled *Questions and Answers on How to Participate in the Purchasing of Military Hardware* to help manufacturers and businessmen interested in doing business with the military, and an information centre has been set up to answer enquiries from manufacturers and businessmen.[52] All these policy and procedural changes were introduced to meet the public demand for more openness and transparency in arms acquisition decision making.

IV. The budget planning and programming process

In the 1990s Taiwan has also undergone rapid social, political and economic change. Together with external changes resulting from the end of the cold war, these have had a significant impact on the political system in general and the MND in particular. However, budget constraints have been a major factor affecting all forms of government procurement. Taiwanese defence planners have to assess the impact of budget constraints when deciding on the choice of weapons and choice of sources. The military budget, no longer a sacred cow, has been trimmed to make room for welfare spending.

The defence budget has risen steadily in real terms over the past decade but has fallen consistently as a share of GDP and as a share of the government budget,[53] and it has been increasingly opened up to public scrutiny. Before the 1970s, defence expenditure accounted for approximately 75 per cent of government spending. It dropped to 50 per cent in the 1970s, to below 40 per cent in 1981, and to less than 30 per cent in 1992.[54] In the 1990s the government reduced defence expenditure to 21 per cent of the total national budget in 1999.[55] Of the defence budget, 70 per cent is spent by the General Staff, 25 per cent goes to pensions, and only 5 per cent is under the control of the MND.[56]

Budgeting covers not only immediate operational requirements but also the long-term development of national defence in the future. The budget items show size of force objectives, weapon systems, the actual situation of training, strength of logistics, and direction of integrated national defence force.

Financial planning and budgeting

The national defence plans consist of a strategic programme for long-term force building, usually covering 10–20 years, intermediate five-year programmes of arms procurement, and annual budgets.[57] The defence budget is based on a comprehensive strategic analysis, which includes assessment of threats and

[52] Chen (note 7), p. 9.
[53] Taiwanese Ministry of National Defense (note 2), p. 132.
[54] Cheng-yi Lin (note 3), pp. 12–13.
[55] SIPRI military expenditure database.
[56] Chen (note 7), p. 14.
[57] Chin-chen Yeh, 'Arms acquisition decision making in Taiwan', SIPRI Arms Procurement Decision Making Project, Working Paper no. 117 (1998), p. 2.

resources, leading to an integrated political and military strategy, operational concepts, defence technology and industry, defence financial assessment and arms procurement options.

The units that apply for procurement have to present their procurement plans to the GSH based on the financial plan approved by the Legislative Yuan, hitherto usually in June, so that the budget needed for implementation does not exceed the allocations approved by the Legislative Yuan. A copy of the application plans should also be submitted to the Ministry of Audit, which comes under the Control Yuan, after being validated by the relevant agencies.

The defence budget is drafted concurrently with and derived from the national budget, which is prepared by the General Accounting Office of the Executive Yuan. The coordination work for drafting the annual defence budget is done by the Accounting Bureau of the MND. It submits the draft plan for defence expenditure to the General Accounting Office of the Legislative Yuan (the Executive and the Legislative Yuan have separate General Accounting Offices) which holds intensive consultations with other accounting agencies in various government departments. After general consensus has been reached the defence budget plan is sent to the Executive Yuan meeting for deliberation and finally to the Legislative Yuan for approval.

The system also allows the MND to formulate special budget plans for important procurement. This is given priority and special funding when overseas arms procurement deals are being confirmed. These special budget plans are formulated by the GSD and submitted to the Executive Yuan for decision. The Legislative Yuan then holds secret meetings to decide the special budget allocation.[58]

If funds allocated are not spent during the fiscal year concerned, they have to be returned to the Treasury. In 1996 one of the reasons for Taiwan's decision to switch from purchasing French-made Mistral portable surface-to-air missiles (SAMs) to buying US-made Stingers was this time constraint.[59] If money has to be returned, the MND has to fight for approval of the budget again for the same purchase the next year, and if it is approved again it affects other procurement planned for that year.

Inter-service competition for budget share is reflected in the balance of the defence budget. Taiwan's defence strategy has changed from an offensive to a defensive doctrine, giving priority to air and sea defence over land defence in long-term force building. As a result, a larger share of the defence budget has been allocated for the air force and navy since the late 1980s, while the army's force modernization programme has been modest in comparison. On completion of the major arms procurement programmes for the air force and the navy, it is expected in some quarters that the army's share of the defence budget will be increased, as the MND is dominated by the army. However, it is likely still to be squeezed out by the purchase of combat aircraft and frigates.

[58] Yang (note 14), p. 11; and *Legislative Gazette*, vol. 83, no. 32 (1994), pp. 102–48. The Dutch submarines which Taiwan has purchased, Mirage and F-16 combat aircraft, La Fayette Class frigates and Cheng Kung Class frigates (4100 tons) were all acquired under special budget plans.

[59] Yann-huei Song (note 10), p. 19.

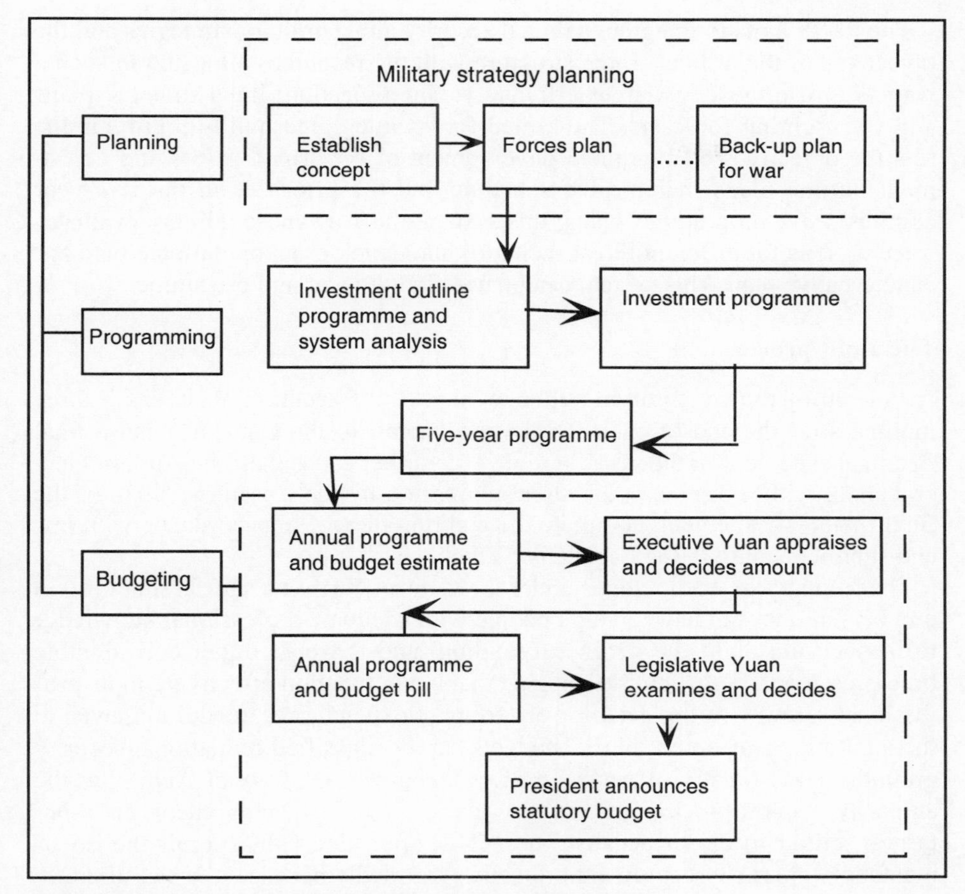

Figure 7.4. Flow-chart of the military budget process in Taiwan
Source: Chin-chen Yeh, 'Arms acquisition decision making in Taiwan', SIPRI Arms Procurement Decision Making Project, Working Paper no. 117 (1998), p. 6.

Because of Taiwan's strong economic growth and the increasing awareness of the military threat from the PRC, the MND's combat aircraft and frigate procurement projects were supported and approved by the Legislative Yuan. A special budget was accordingly allocated for the purchase in the early 1990s.

The Planning, Programming and Budgeting System[60]

Since 1975, the MND has used the US concept of the Planning, Programming and Budgeting System (PPBS), combined with basic concepts from the original budget system and standard budget laws. This system introduces objectivity to planning, programming and budget execution, and is linked with achieving Taiwan's national defence strategy (see figure 7.4).

[60] This section is based mainly on Chin-chen Yeh (note 57), pp. 3–4.

The PPBS aims to integrate national security aims, military strategies and the objectives of the military force structure with the resources allocated in such a way as to use those resources efficiently. The procedure links strategic plans and war-fighting forecasts. The armed forces integrated build-up programme sets the objectives of integrated development of the armed forces and investment outline plans indicate the schedule and the priorities in the five-year administrative programme. Using this system, the Taiwanese military evaluates force options for different threat scenarios and develops an operational plan and an alternative plan. This system can be used in both war and peacetime.

The audit process

Article 60 of the constitution stipulates that 'the Executive Yuan shall, three months after the end of each fiscal year, submit to the Control Yuan a final financial statement of the year'. Article 105 further states that 'the Auditor General shall, within three months after submission by the Executive Yuan of the final financial statement, complete the auditing thereof in accordance with law and submit an audit to the Legislative Yuan'.

The Ministry of Audit comes under the Control Yuan. Its staff is small (40 in number), it does not have enough people with adequate professional knowledge to inspect and audit the defence spending, and it works under considerable time-pressure. It is difficult to imagine that it can function effectively in its professional scrutiny of the detail of government expenditure. In addition, even in the auditing stage, some 'black' budgets can be classified on national security grounds. The Ministry of Audit, in accordance with the Law of Audit, has the authority to audit procurement of arms and other military equipment, but it has largely failed to do so because the MND often does not submit the documentation on arms procurement projects, purportedly because it is classified. In addition, under Article 29 of the Statute for Inspection Procedures Governing Construction Works, Procurement of Products, and Disposal of Properties by Government Agencies, whenever considerations of confidentiality, 'emergency' or 'ensuring the quality of military equipment' are involved in the procurement process, military units are allowed to bypass certain requirements provided for in the law. However, the MND is required to submit afterwards the reasons for the purchase, which will then be checked by the audit agencies. It has used its own interpretation of the word 'afterwards' to keep arms procurement decisions secret for a period of time.

As a result, it is very difficult to meet the requirement of accountability in arms procurement. A legislator has complained that it is difficult for the legislators to gain access to defence budget information but, ironically, arms sales dealers have been able to acquire classified documents relating to arms procurement.[61]

The processes of audit and programme review that are internal to the MND are shrouded in secrecy. The most obvious weakness of the control mechanisms

[61] Yann-huei Song (note 10), p. 22, quoting *Legislative Gazette*, vol. 83, no. 38 (1994), p. 18.

in general is that those who are responsible for supervising and those who are supervised are mostly military officers who may have connections or have worked together at one time or another.[62] The traditions of 'old boy' connections and 'mutual cover-up' create strong group cohesion within this closed professional community.

Offsets and industrial cooperation

In recent years, Taiwan has been demanding offsets when negotiating arms sales deals with foreign contractors in order to develop its defence-related industries and reduce dependence on foreign weapon supply.[63]

In 1993 the Executive Yuan established the Steering Committee Office to direct industrial cooperation and offsets for arms acquisition contracts with the following objectives: (*a*) to develop strategies for industrial cooperation for military equipment around the possible procurement alternatives; and (*b*) to prepare and review the industry cooperation plan based on the development priorities of industry and technology to be acquired.[64]

A procurement project costing over NT$5 billion is required to include an industrial cooperation clause in the contract, which should be worth at least 10 per cent of the contracting price or negotiated price. Industrial cooperation can take the forms of: (*a*) cooperative production; (*b*) common investment; (*c*) technology transfer; (*d*) cooperation in R&D; (*e*) personnel training and education; and (*f*) other types suitable for the Taiwanese investment environment.

When arms are to be bought abroad, the Legislative Yuan directs the foreign supplier to submit offset plans such as technology transfer or co-production plans for spare parts. Examples of offsets are: (*a*) the automation project for weapon production machinery—industry cooperation is 35 per cent of the price of the contract; and (*b*) the project for procurement of a navy missile system; here industry cooperation is 30 per cent of the value of the contract.[65]

Recognizing the importance of offsets to the development of local industries, the Legislative Yuan retroactively demanded Taiwan's first-ever offset from the US Lockheed Corporation in connection with the purchase of the 150 F-16s in 1992. Payments were to be stopped if Lockheed failed to provide Taiwan with technology and production contracts related to the aircraft. Under pressure, Lockheed signed a 10-year industrial cooperation agreement worth US$1.1 billion which ensures the production of some of the aircraft parts and the creation of maintenance depots in Taiwan.[66]

It seems that Taiwan's efforts to negotiate offsets have borne some positive results. In June 1997 it was able to negotiate offsets worth US$24 million when

[62] Wu (note 42), p. 8.

[63] Projects such as acquisition of the Mirage 2000-5 aircraft and the Perry Class and Lafayette Class frigates have included 15–25% offset package in the contracts. Yang (note 14), p. 11.

[64] Chin-chen Yeh (note 57), p. 20.

[65] Huang Hui-Chia, 'Promoting ways for industrial cooperative planning and implementation', Taiwanese Industrial Development Bureau, Ministry of Economic Affairs, 1996.

[66] See, e.g., Yann-huei Song (note 10), p. 28.

Table 7.3. Percentage of offsets in selected arms transfers, 1980–92

Recipient country	Supplier country	Type of aircraft	Date of deal	Offsets (% of price)
Canada	USA	F/A-18A	1980	*58*
Greece	France	Mirage-2000	1985	*150*
Saudi Arabia	USA	E-3A Sentry	1981	*35*
South Korea	USA	F-16C	1981	*30*
Taiwan	USA	F-16AM	Sep. 1992	*10*
	France	Mirage-2000-5	Nov. 1992	*10*
	USA	E-2C	1993	*10*
Turkey	USA	F-16C	1984	*24*
UK	USA	E-3D Sentry	1986	*130*

Source: Yann-huei Song, 'Domestic considerations and conflicting pressures in Taiwan's arms procurement decision-making process', SIPRI Arms Procurement Decision Making Project, Working Paper no. 124 (1998), p. 40; and SIPRI arms transfers database.

purchasing the M-3 amphibious bridging and ferry system at a cost of US$60 million from the German corporation EWK.[67] In January 1998, three Taiwanese companies were able to produce helicopter components for the US-based Sikorsky under cooperative production agreements linked as an offset requirement to Taiwan's earlier purchase of 10 Sikorsky S-70(M)-1 helicopters. However, the value of offsets it has achieved is low compared with those obtained by other major arms-importing countries. Table 7.3 compares the offsets negotiated by Taiwan and some other countries over the period 1980–92.

V. Defence technology and industrial considerations

Establishing and upgrading an indigenous weapon R&D and production capability has been a top priority in Taiwan's arms procurement agenda, given the uncertain nature of foreign arms supply. Before 1975, state-controlled ordnance factories had acquired the capability to produce infantry weapon systems, artillery and various types of ordnance through technology transfers from the USA and European countries.[68]

Indigenous R&D structures and process

Although Taiwan has achieved some significant results after three decades of effort, its reliance on domestic defence production has varied with the availability of foreign weapons. As access to foreign weapons and equipment became more difficult after 1982, the balance between indigenous development

[67] 'German firm offers offset credit for arms contract', Taiwan Central News Agency (in English), 12 June 1997, in Foreign Broadcast Information Service, *Daily Report–China (FBIS-CHI)*, FBIS-CHI-97-163, 12 June 1997.

[68] Yang (note 14), p. 2.

and import was inevitably tipped in favour of domestic production. Recognizing the need to enhance indigenous research and design capability for developing advanced weapon systems, the MND established the Aviation Industry Development Center (AIDC) and the CSIST (under the GSD) in 1969.

Before the CSIST was established, arms procurement decisions were made primarily on the USA's recommendation. The CSIST's role gradually became more significant as foreign suppliers, especially the USA, began to accommodate the objections raised by the PRC after the early 1970s. The CSIST focused on developing advanced technology weapons with multiple functions. It has four divisions: (*a*) the Institute of Mechanical Engineering; (*b*) the Institute of Chemical Engineering; (*c*) the Institute of Electrical Engineering; and (*d*) the Institute of Aeronautical Engineering. The Institute of Nuclear Energy Research was transferred from the CSIST to the civilian sector in 1980.[69]

The CSIST also has responsibility for technology and scientific assessment of R&D polices and decisions; for collaborating with other agencies for developing special advanced weapon systems such as missile technology, radar, communications and fire control systems; for material science and nuclear science research; and above all for weapon system integration. It is responsible for weapon upgrading, technology testing, design, type approval, trial production, test batch processing, and outlining policies, priorities and targets for the development of defence science, technology and manufacturing. The CSIST was put directly under the MND in April 1998, with the result that this very secretive institute is for the first time partly open to legislative scrutiny.

As mentioned, the CSIST carries out operational, technical and financial evaluations of arms procurement plans submitted by the Acquisition Planning Offices of the armed services.[70] It employs over 12 000 scientists, engineers and support staff, of whom 80 per cent are ranking military personnel. Of the 6800 scientists and technicians, 90 per cent have a PhD in a specialized disciplines and many have years of experience in overseas defence industries.

Not all military R&D is managed by the CSIST. The AIDC was responsible for developing the Indigenous Defense Fighter (IDF); and R&D and production of the Cheng Kung Class frigate were carried out by the United Ship Design Center (USC), a government-financed institute established in 1971, and the China Shipbuilding Corporation (CSBC). However, most military R&D is done by the CSIST's research institutes and the funding comes from the annual defence budget.

The Combined Services Command, which is part of the armed services command system, is responsible for design, development, procurement and manufacturing of weapon systems which use lower-level technologies. It is also responsible for ammunition and logistics support.

[69] Lung Kwang Pan, 'Weapon acquisition and development under foreigner influence: trajectory of Taiwan's highest military research institute', SIPRI Arms Procurement Decision Making Project, Working Paper no. 119 (1998), p. 3.

[70] Yang (note 14), p. 6.

Table 7.4. Taiwan's decentralized scientific R&D institutions and agencies

Parent body	R&D institution
National Science Council	1. Science-based Industrial Park Administration 2. Science and Technology Information Center 3. Precision Instrument Development Center 4. National Laboratories
Ministry of Economic Affairs	1. Industrial Technology Research Institute 2. Non-profit institutes of private organizations and state-run corporations (AIDC, CSBC)
Ministry of National Defense	Chung Shan Institute of Science and Technology
Ministry of Transport and Communications	1. Telecommunications Laboratories 2. Data Communications Institute 3. Research and Development Center of the Central Weather Bureau
Atomic Energy Council	Institute of Nuclear Energy Research
Department of Health	1. National Institute of Preventive Medicine 2. National Health Research Institute

Notes: AIDC = Aviation Industry Development Center; CSBC = China Shipbuilding Corporation.

Source: Republic of China Yearbook 1999 (Government Information Office: Taipei, 1999), p. 314.

Taiwan's R&D institutions are shown in table 7.4. Billions of dollars have been allocated to them since the mid-1970s. The MND has used the National Defense Industrial Development Fund to assist public and private enterprises in cultivating qualified technical personnel, purchasing facilities, transferring advanced technology and developing a more sophisticated production base.

Taiwan's collaborative R&D network

Apart from the agencies directly involved in weapon development, R&D on new weapons is occasionally based on cooperation between the military and civilian research organizations, notably the universities and special scientific institutes. The CSIST coordinates its R&D, System Manufacturing and Integration divisions with research academies and establishments and the various manufacturing entities. The technologies for the new type of missile-equipped corvette (the Kwang-hua III project), the IDF interceptor and guided missiles were developed with contributions from both civilian and military industries with different product specializations and R&D capabilities. Collaborative R&D has been carried out by subcontracting research projects to individual scientists or research teams.

The MND issued the Defense Science and Technology Development Plan in 1986 to strengthen cooperation between the academic and industrial sectors[71] and, along with several cabinet-level institutions such as the National Science Council, the Ministry of Education, and the Ministry of Economic Affairs, has set up the Executive Committee for the Development of Defense Science and Technology. The indigenous development and production of the IDF, which became operational in 1995, the Cheng Kung Class frigate and Tien-Chien air-to-air missiles (AAMs) are notable examples of success.

'Make or buy' decisions

Arriving at a balance between foreign procurement and domestic production is critical in Taiwan's arms procurement process. Operational urgency and the military's preference for foreign weapons have often had a negative effect on the development of the domestic defence industry. After contracts were signed in September 1992 with the USA for the purchase of F-16s, and in November 1992 with France for 60 Mirage-2000-5s, fewer IDFs were produced by the AIDC for the Taiwanese Air Force.

Domestic defence industrial capacities

Table 7.5 shows the major weapon systems produced by the CSIST. Taiwan has produced the Ching-Feng (Green Bee, Lance-type, with a range of *c*. 75 miles, 120 km) anti-ship guided missile. The domestic defence industry is now able to produce a great variety of modern weapons including artillery, tanks, helicopters, tactical missiles and jet combat aircraft. With the joint efforts of the CSIST, the CSBC and other public- and private-sector factories, conspicuous results have been achieved.[72]

The CSBC in Kaohsiung is capable of building frigates and fast attack craft. Under the Kwang-hwa I programme, the first of seven domestically produced Cheng Kung missile frigates, modelled on the US Perry Class, entered service in May 1993. The Kwang-hwa III programme is another indigenous project which aims to build 12 500-ton patrol boats.

Despite this progress, the Taiwanese defence industry still lags behind world standards. In fact, foreign technology, especially US technology, has been critical to those sophisticated weapon systems that are made in Taiwan. For instance, the Tien-Kung SAM, the Hsiung-Feng air-to-ship missile (AShM; originally a ship-to-ship missile) and the Cheng Kung Class frigate all rely on foreign technology or sub-systems. The IDF was developed with the help of US companies. Taiwan's dependence on foreign supplies will only increase over time.[73]

[71] Chung Shan Institute of Science and Technology, *The Thirty-Year Anniversary of the CSIST* (CSIST: Taoyuan, 1999).

[72] Lin Chi-Lang (note 41), pp. 7–8.

[73] Wen-cheng Lin (note 17), p. 7.

Table 7.5. Weapons produced by the Taiwanese CSIST, 1980–98

Weapon	Use and principal parameters	Similar to
AT-3 jet aircraft	Training/attack jet; max. speed 1.05 mach; service limit 15 000 m.	
IDF jet aircraft	Air superiority combat aircraft; max. speed 1.8 mach; digital fly-by-wire; advanced 9-g cockpit	F16/J79 (USA)
Ching-Feng	Surface-to-surface missile; range *c.* 160 miles (257 km)	
Tien-Chien I missile	Infra-red guided short-range air-to-air missile; all-aspect; 'fire and forget'	AIM-9L (USA)
Tien-Chien II missile	Advanced medium-range air-to-air missile with mid-course navigation and terminal guidance: multi-target engagement	AIM-120 (USA)
Hsiung-Feng I missile	Ship-to-ship missile; max. range 35–40 km; semi-active radar guidance system	Gabriel (Israel)
Hsiung-Feng II missile	Equipped with various-launched platforms; max. range 120–50 km; active radar homing system	Harpoon (USA) Exocet (France)
Tien-Kung I missile	Surface-to-air missile; single-stage, dual-thrust solid-propellant rocket motor guided by mid-course inertial reference and radar in the terminal phase; max. speed mach 3.5; max. range 60 km	Patriot I (USA)
Tien-Kung II missile	Surface-to-air missile equipped with advanced active seeker; 'fire-and-forget'; max. speed 4.5 mach; max. range 100 km	Patriot II (USA)
Kung-Feng 6 MLRS	MLRS, 117 mm calibre; range 1–15 km; 2.1 m long; 42 kg; solid-propellant rocket motor	
Chang-Bei radar	Electronic scanning, multi-function phase-array radar; capable of target searching and tracking	Aegis [AN/SPY-1A] (USA)
CS/MPQ-78 radar	Mobile fire-control radar; can be incorporated with both gun and missile	
CS/UPS-200C radar	Surface search radar; can be equipped with both TWS and IFF systems for surveillance missions	

Notes: MLRS = multiple-launch rocket system; TWS = tracking-while-scanning; IFF = identification friend or foe.

Source: Lung Kwang Pan, 'Weapon acquisition and development under foreigner influence: trajectory of Taiwan's highest military research institute', SIPRI Arms Procurement Decision Making Project, Working Paper no. 119 (1998), p. 5.

Of the 1996 arms procurement budget, 73.7 per cent went to imports and 16.7 per cent to domestic development.[74] This highlights Taiwan's weakness in defence R&D and production capabilities. Even though it has put great effort

[74] Lin Chi-Lang (note 41), p. 5. The remaining 10% is accounted for by maintenance and logistics.

in to developing its own defence industries through technology transfer or offset agreements, its arms development and production projects, dubbed indigenous, are actually licensed copies of foreign systems or assemblies of imported components.[75]

The successes of the defence industrial sector have been mostly confined to machining components and light manufacturing. Taiwan's indigenously produced weapons have high maintenance requirements and are of inadequate quality compared with those imported.[76] Military users are not confident about domestically produced weapons and are not sensitive to their economic and political importance. Moreover, indigenous development slows down acquisition time. Experience indicates that it usually takes 10–20 years to develop a weapon system at home. Taiwan's domestic R&D and manufacturing capacities therefore do not usually have the opportunity to demonstrate what they can do, especially when foreign supplies are easily available. Finally, the country's indigenization efforts have been constrained by limited technical expertise, funding and domestic industrial infrastructure.

VI. Legislative and public-interest monitoring

The democratization of Taiwan has increased the influence of the political parties on national security policies and defence policy making. In addition, the increasing demand for the introduction of checks and balances is reinforcing the oversight power of the Legislative Yuan in the policy-making process.[77] Still, the final decision is chiefly if not exclusively in the hands of the president and his institutional subordinates.

There is growing tension in Taiwan's arms procurement between the demand for openness and transparency and the requirements of secrecy in arms acquisition deals. Information on defence policy-making processes has long been restricted, known exclusively to insiders. Before the 1990s there was no tradition of public debate on arms procurement. The public awareness resulting from Taiwan's rapid democratization since the late 1980s created new pressure in the Legislative Yuan for greater disclosure of government information in regard to defence policy decisions.

During the period from 1949 to 1985, Taiwan did not function as a democratic country and the defence procurement decision-making process was a 'black box'. Anti-communist concerns resulted in high levels of political and military secrecy, backed up by various domestic intelligence services with wide-ranging powers of arrest and detention. The political leaders used the all-encompassing martial law and the intelligence services to censor and control criticism of their performance, including in arms procurement.[78] Defence procurement policy was largely unchecked by the legislative branch. Few legis-

[75] Yann-Huei Song (note 10), p. 27.
[76] *Jiefangjun Bao* (Beijing), 10 Sep. 1999, p. 5.
[77] Chen (note 7), pp. 14–15.
[78] Chen (note 7), p. 2.

lators took the issue of the supervision of defence procurement seriously enough before the murder of Captain Yin Ching-feng in 1993.

Until very recently, the MND consistently ignored demands by the Legislative Yuan for the CGS and Commanders-in-Chief of the three branches of the armed forces to appear before the Defense Committee on the floor of the Legislative Yuan to report on defence affairs.

When the MND began to conduct studies on ways to improve the arms acquisition policy in 1994 and the Military Procurement Bureau was created in 1995, it was generally agreed that the Taiwanese military had made progress in terms of accountability and transparency in the arms acquisition process as a result of the MND reforms. However, the level of corruption revealed by the series of procurement scandals that broke out in early 1998 surprised the people. The public has begun to think that if the defence budget is to be put to better use then transparency and accountability will be indispensable, at the cost of confidentiality.

The legislative branch believes that more accountability and transparency of a nation's arms procurement process leads to more rational choices. It also views these characteristics as necessary to prevent corruption. On the other hand, the executive branch believes it prudent to maintain a low profile in arms procurement in order to avoid interested parties exercising undesirable influence. It is encouraging that open debate on defence procurement processes has increasingly attracted public attention. In practice there is no absolute transparency in the country's arms acquisition process. Enhanced bureaucratic accountability and well-developed legislative procedure and regulations for scrutinizing arms procurement are badly needed.[79]

Secrecy and accountability

No national defence White Paper was published until 1992.[80] Although long overdue, it represented a significant move in the direction of transparency in defence policies. Of late, responding to requests by the legislators, the Legislative Yuan has held closed sessions to examine proposals for arms procurement. In general, however, the MND has demonstrated reluctance to admit, and even hostility towards, the public demand for greater openness on defence issues in general and arms procurement in particular.

The inclination of the government (or more precisely the MND) to withhold arms procurement decisions from the public on supposed national security grounds is theoretically and generally accepted by many people in Taiwan. The reasons for keeping weapon acquisitions secret are quite evident. The first is the PRC's strategies, including economic and political strategies, to deter potential suppliers from selling arms to Taiwan and to compel existing suppliers to cut or stop their supply. Accordingly, in negotiations for more sensitive and high-

[79] Chen (note 7), pp. 8, 13.
[80] See note 15.

technology military purchases, confidentiality is deemed necessary. The second reason is the possibility of crisis or even a conflict between the two sides of the Taiwan Strait. The 1995 military exercises by the PRC and missile tests showed how serious the security dilemma is. The government is thus bound to protect defence secrets of which the disclosure could threaten national security.

The third reason is that public participation in decisions on arms procurement will not self-evidently produce more rational decisions. It is difficult for the general public to understand and compare the technical merits of rival weapon systems. Not even the legislators can assess different procurement programmes. In addition, open debate on and public scrutiny of defence procurement could prolong the process of decision making and transactions, and would therefore increase the economic and political costs for Taiwan. It is therefore argued by some that there is a legitimate need for secrecy in the arms procurement process in order to ensure its efficiency.

Proper access to official records and government information is critical to the idea of public accountability. Without adequate information, government cannot be properly scrutinized or held accountable, whether legally, politically or financially. Nevertheless, arms procurement decisions in Taiwan are usually claimed to fall into the category of secrecy for national security.[81]

Evidently, defence officials sheltering behind the argument of secrecy for national security have hampered the rational formulation and effective implementation of arms procurement policies. Such attitudes in the long term may even harm national security interests—the very element secrecy is intended to protect. Secrecy is maintained at the expense of accountability and responsibility. Scandals and reports of waste in military budgets imply that the current mechanisms of internal audit and programme review have failed to achieve the goal of making arms procurement more responsible and accountable.

Institutional limitations

Some problems can be easily identified with regard to the existing institutional framework for arms procurement decision making in Taiwan. The existing institutional design is biased in that there is an information asymmetry as between the executive and legislative branches, with the latter in a very disadvantageous position. The executive branch (or, more specifically, the MND) has enjoyed exclusive discretionary authority to decide on the classification or disclosure of information. Without any statutory foundation for classification of documents, the legislature and the general public simply have no ways to oversee or scrutinize the decisions and conduct of defence officials and therefore cannot ensure that officials in government are answerable for their actions.

Without sufficient knowledge and information, the public and the legislators simply have to accept the decisions made by the military elite. The Legislative Yuan does not have the capability to monitor the arms procurement procedure

[81] Chih-cheng Lo (note 4), p. 1

either. It can approve the budget sent by the Minister of Defense, but the Ministry of Audit is responsible for the auditing of the budget. The Legislative Yuan can only obtain information from the reports provided by the Minister of National Defense and by the Ministry of Audit. The reports do not mention the details of the implementation of arms acquisition decision or any inadequacies identified. It is therefore extremely difficult for members of the Legislative Yuan to scrutinize acquisition programmes.

Additionally, the Ministry of Audit is not capable of supervising arms acquisition activities. Its Second Department is responsible for the auditing of the defence budget, but there are not enough staff in this department to evaluate the more than 400 000 projects per year. Professional expertise is needed in several fields, and it is next to impossible for the department to send officers to each of those 400 000 projects and conduct careful auditing. As a result, again, most of the auditing jobs end up as mere paperwork.

An organizational culture of 'follow the order' and 'obey the superior' may also contribute to corruption. In Taiwan it is very difficult for a military officer to resist pressure from a superior. Because of the prevailing culture, many months of professional assessment by an evaluation team can be easily reversed by an ad hoc judgement of a high-ranking officer.[82]

The involvement of organized crime is yet another factor that works against transparency. As arms sales involve huge profits, they become a natural field for organized crime. The secrecy requirement serves as a perfect cover for illegal activities. Organized crime has apparently penetrated into the arms procurement process in Taiwan. On 18 March 1998, the Combined Services Command reported that a lieutenant-colonel and a major-general (a former director-general of the CSF Public Construction Service) had been kidnapped and intimidated by gangsters. According to the news report, they were forced to sign contracts with companies that are controlled by the gangsters. The murder of Captain Yin Ching-feng in 1993 may also have been the result of the involvement of organized crime.[83]

Constitutional limitations

There is no doubt that the legislative branch should play the central role in ensuring the political and financial accountability of the executive. Among the constitutional limitations to accountability in Taiwan the first is the predominance of the executive and its power of discretion over legislative checks and balances.

As regards political accountability, Article 3 of the Additional Articles of the constitution stipulates that 'the Executive Yuan has the duty to present to the Legislative Yuan a statement on its administrative policies and a report on its

[82] During the ROC Navy's assessment for the purchase of a 2nd-generation battleship, the South Korean Waisan emerged as the front-runner. However, the decision was reversed in favour of the French La Fayette Class frigate after a Taiwanese admiral visited France. It is believed that changing this decision cost Taiwan billions of dollars without comparable increase in its security. Wu (note 42), p. 9.

[83] See note 46; and Wu (note 42), p. 9.

administration. While the Legislative Yuan is in session, its members shall have the right to interpellate the president of the Executive Yuan and the heads of ministries and other organizations under the Executive Yuan'.[84] Additional Article 3 states that 'should the Executive Yuan deem a statutory, budgetary, or treaty bill passed by the Legislative Yuan difficult to execute, the Executive Yuan may request the Legislative Yuan to reconsider the bill. Should the Legislative Yuan not reach a resolution within the said period of time, the original bill shall become invalid'; and 'the Legislative Yuan may propose a no-confidence vote against the president of the Executive Yuan' but if that fails it may not initiate another no-confidence motion against the same president of the Executive Yuan for at least a year.

For the executive branch to be held accountable for its policies, examination of public officials appears to be the only tool the legislators can use. Other mechanisms, such as votes of no-confidence or overriding the executive veto, are scarcely feasible in Taiwan's current political setting. The use of the constitutional powers of the Legislative Yuan to oversee the work of the executive through questioning ministers and cross-examining the relevant officials is not seen as feasible in Taiwan at present. Without the power of impeachment, censure, appropriation and auditing, the Legislative Yuan is very much like a dog barking at the train.

The Control Yuan is Taiwan's watchdog agency, which has the authority to investigate and indict officials. However, its efforts at examining arms procurement decision-making methods are in most cases not as successful as they ought to be. According to Additional Article 7 of the constitution, it 'shall be the highest control body of the State and shall exercise the powers of impeachment, censure, and audit'. Article 95 further stipulates that 'in exercising its power of control, the Control Yuan may request the Executive Yuan and its ministries and commissions to make available to it any orders they have issued and all other relevant documents'. It appears that the transparency of government decisions is ensured, since the Control Yuan may request 'all relevant documents' which it considers necessary. However, in practice, because of the absence of any statutory foundation to the security classification system, the executive branch has the exclusive authority to decide what information can be disclosed.

In the area of financial accountability, the legislative branch is also handicapped in holding the executive answerable for the defence budgets. Article 59 of the constitution stipulates that 'the Executive Yuan shall, three months before the beginning of each fiscal year, submit to the Legislative Yuan a budgetary bill for the following fiscal year'. Article 70 states that 'the Legislative Yuan shall not propose any increase in the budget estimates submitted by the Executive Yuan'. Given the shortness of the time available to them, it would be extremely difficult for the legislators to examine the proposed defence budget thoroughly. While the MND has a large staff to work on compiling the data and

[84] Yann-huei Song (note 10), p. 20.

proposing the budget, the Defense Committee of the Legislative Yuan has no research staff to analyse the defence budget. As a result, faced with an uncooperative or even a hostile attitude on the part of the MND, legislators sometimes have to rely on 'whistle-blowers' to uncover the hidden budgets or scandals in arms procurement decisions. Once the budget bill is passed, the Legislative Yuan, without appropriation power, simply has to wait for the Auditor General's report.

Opportunities for waste, fraud and abuse

Lack of accountability opens the door for corruption and abuse. The military has been charged with numerous irregularities in arms procurement.

The Chief of the General Staff and the Commanders-in-Chief of the three armed services are exempted from appearing in the Legislative Yuan to answer questions raised by the legislators. The armed forces are authorized by law to decide not to submit arms procurement projects to the Ministry of Audit for inspection and audit for a certain period of time if they consider it necessary to keep a purchase secret. In addition, retired senior officers and relatives of active-duty high-ranking officers have been able to exert influence on arms procurement in exchange for payments or personal interests.[85] Because of these practices, there have been irregularities in Taiwan's arms procurement process.

In exchange for favours, active-duty officers have in the past leaked confidential arms procurement documents to arms dealers. In other cases, retired senior military officers taking advantage of their connections and experience have started new careers in influential positions in defence-related industries and engaged in influence peddling. The 'old boy' network extends to the MND and agents representing arms manufacturers in Taiwan and other countries. In addition, relatives of the active-duty high-ranking officers working for foreign defence contractors have engaged in activities which helped their firms obtain contracts.[86] The costs of these irregularities are very high.

The corruption problem has been so serious that it not only results in the waste of valuable public resources but also seriously undermines Taiwan's security and damages public trust in the military and the government. These defects in the arms procurement decision-making processes can be attributed to structural as well as human factors. Without first identifying and clarifying these major and, more importantly, interlinked causes, any suggestions for improving the defence acquisition decision-making process will prove fruitless.

Taiwan relies very heavily on advanced weapon systems purchased from foreign countries. Diversification of arms supply has meant reliance on agents for information and connections, some of whom will try all methods, including bribes, to win contracts. Former Prime Minister Hau Pei-Tsun has argued that Taiwan's purchases of weapons from European countries are more likely to be problematic because such transactions are not adequately overseen by govern-

[85] Yann-huei Song (note 10), pp. 22–23.
[86] Yann-huei Song (note 10), p. 23.

ment agencies.[87] Most of the recent arms scandals in Taiwan had to do with purchase from European countries. The lack of transparency leads to inadequate scrutiny mechanism and poor monitoring and management of arms procurement decision making.

The processes of internal audit and programme review in the MND are still shrouded in secrecy. Because the reports of these reviews need not be made public, their application remains at the discretion of MND officials. The most obvious and serious weakness of the supervision mechanism is that those who are responsible for supervising and those who are supervised are mostly military officers who may have connections or have worked together at one time or another. The traditions of 'old boy' connections and 'mutual cover-up' create a strong group cohesion within this closed professional community. Without independent scrutiny and external check on the executive branch's arms procurement, the supervision mechanisms which it introduces are not likely to achieve any significant results.

Possible remedies

External checks and balances might be made more effective by restructuring the current institutional framework. This would involve improving the capacities of the Ministry of Audit to scrutinize the arms procurement reports and strengthening the Legislative Yuan's capability to supervise the arms procurement process. It would also mean facilitating access to information other than that available from the executive branch. All government organizations have a distinct tendency to control access to information they possess and this attitude is particularly strong in the defence field. The MND enjoys much greater discretionary authority than other state agencies in withholding information from the public. There is therefore a compelling need to review the present constitutional provisions that give the executive branch arbitrary power to decide on government secrecy. This is the key to producing more open and accountable arms procurement decision-making processes.

Any proposal for appropriate freedom of information legislation should therefore be welcome. There should be clear and strict rules ensuring that information is released except where disclosure would cause harm to a limited number of specific national security interests. Such legislation on government secrecy could also control the current problem of information leaks, which are inherently liable to political abuse and manipulation.

Given the unimpressive record of the legislature, it is difficult to be optimistic about the chances for legislative action in this area. There have been attempts by opposition legislators to initiate legislation on public access to government information through the regulated release of information on budget details and arms procurement decisions, but the executive branch and the military have been resistant to such reforms.

[87] Chen (note 7), p. 14.

VII. Conclusions

Arms procurement decisions in Taiwan have long been made in a very protective and secret process. Given the tremendous military and diplomatic threats from the Chinese mainland, the Taiwanese authorities have political and security justifications for keeping arms acquisitions confidential. The absence of open debate and public scrutiny of arms procurement methods and processes has resulted in a lack of accountability of the military and the misuse of defence resources. The current institutional framework of Taiwan's defence decision-making processes has created difficulties in ensuring the legal, political and financial accountability of the defence community. Attempts to initiate procurement reforms have encountered resistance from the military and some defence officials.

After the recent arms procurement scandals and the resultant public outrage, the MND did seek to improve its acquisition process within its internal administrative structure. The current mechanisms of internal audit and programme review have, however, failed to make arms procurement processes more rational and accountable. Without independent professional capacities for scrutiny and institutionalized external checks on the executive branch's arms procurement activities, the corrective mechanisms initiated by the MND will only produce very limited results. Significant changes in the arms procurement decision-making process are not likely to occur unless changes in the political framework take place—for instance, a shift of power from the ruling to the opposition parties or a strengthening of the legislative branch.

In reality there are many difficulties in the way of Taiwan's achieving an institutional framework which incorporates both the values of democratic accountability and secrecy in the interests of national security. The PRC's strangulation of foreign arms supply to Taiwan has created enormous constraints on the rationality and efficacy of Taiwan's arms procurement decisions. Selling countries' own political, economic and security considerations always condition its arms procurement from foreign sources. In most cases the weapons Taiwan needs most, even when proposed after comprehensive security assessment, are either late or under-supplied. The general prospects for diversifying sources are not bright. Understanding fully the risk of relying on a single source for arms purchase, Taiwan has made great efforts to improve domestic arms production and to expand the pool of supplier states. This has achieved some results. More importantly, the uncertainty of and fluctuations in foreign arms supply have had a significant negative impact on the development of Taiwan's arms industry.

8. Comparative analysis

Ravinder Pal Singh

I. Introduction

The comparison of arms procurement decision making in the countries covered by the first volume published by the SIPRI Arms Procurement Decision Making Project was helpful in focusing on the major aspects of public accountability in arms procurement processes. The research questions asked in the first part of the project and the propositions identified as the basis for comparison have not been changed in the second volume. This has made it possible to extend the sample of countries for which common problems can be examined. In addition, this concluding chapter addresses the motivations for and factors that contribute to secrecy in security bureaucracies, which is the primary barrier to public accountability. The challenge before the international community is to design ways of arriving at a balance between the need for public accountability and the need for confidentiality in military matters.

Of the countries participating in the second phase of the project, four—Chile, Poland, South Africa and Taiwan—have been making the transition to more open and representative forms of government during the 1990s. In developing political systems that are in keeping with the requirements of good governance, they have been able to introduce more genuine, practical public accountability to their arms procurement processes than is to be found in some countries with longer experience of democratic politics. However, Chile is an example of the military ensuring its continued political influence and autonomy. It negotiated itself out of power from a position of strength and succeeded in enshrining provisions in the Chilean constitution to preserve its own position.

Five of the countries covered by the project in both phases—Greece, Japan, South Korea, Taiwan and Thailand (until the mid-1970s)—were part of the US security alliance system. Four of them had political systems which were either led by or under the influence of the military during the cold war. Except in the case of Japan, democratic control of the military was not effectively encouraged by the USA, and this allowed opportunities for the military in the remaining four countries to eschew public accountability norms in their security sectors.

This chapter identifies mainly those relevant elements in the research which have not been included in the country chapters.

II. Military and politico-security issues

This section highlights the political and military security characteristics of national security planning and arms procurement decision-making structures in the countries covered by the study. The comparison between the decision-making processes is based on three elements, against which each country's arms procurement decision making is examined in the sections which follow: (*a*) the quality of definition of threat assessment methods and long-term defence planning, which is necessary if national strategy and arms procurement policies are to be coherent; (*b*) coordination between foreign and defence policy-making processes and between the armed services; and (*c*) the influence or autonomy of the military in national security and arms procurement decision making, the military's role in domestic politics, and in some cases the influence of a dominant arms-supplying country or military alliance.

Threat assessment and long-term planning

Among the first initiatives taken by the democratically elected government in *Chile* after 1990 was an attempt to bring coherence into defence planning at the Ministry of Defense level by formulating a five-year defence policy and taking a 'global political strategic approach' which examines the political, military, economic and diplomatic aspects of threats and alternative countermeasures, and indicates military capability needs. It is formulated through an interactive process involving Ministry of Defense officials and the military. The armed forces do not generally carry out such a broadly based assessment. This process places the arms procurement needs in a broader perspective.[1] Notwithstanding these initiatives, the tendency of the three armed services to operate independently of each other preserves a gap between long-term threat assessment, planning and implementation. It adds to the problems of coordinating and defining an explicit defence policy. The three armed services enjoy a fair amount of autonomy in defining their defence policies and decision making, not least in arms procurement.[2] The military, moreover, sees the functions of the ministry as having more to do with the administration of the armed forces than with threat assessment and defence planning. The latter are considered to be outside the purview of the ministry and the congressional defence committees.

Threat assessment in *Greece* has traditionally been concerned with its ally in NATO but long-standing rival, Turkey. It defines equipment needs, priorities and application in the context of the Turkish threat. The US influence on Greece's security decision making and threat assessment was paramount under the military junta and up until the late 1980s, except for a brief absence from the military 'pillar' of NATO. However, since the end of the cold war the US influence has been reduced. Long-term planning has laid greater emphasis on building up the ground forces at the cost of the air and naval forces, as air and naval defence were to be provided by the USA. Since 1974 the role and influ-

[1] Castro, C. S., 'Effects of threat perceptions, security concepts and operational doctrines in the planning of forces', SIPRI Arms Procurement Decision Making Project, Working Paper no. 61 (1997), pp. 2, 6, 7.

[2] Navarro, M., 'The influence of foreign and security policies on arms procurement decision making in Chile', SIPRI Arms Procurement Decision Making Project, Working Paper no. 66 (1997), pp. 6, 9.

ence of the political leadership in arms procurement decisions have increased. However, the political culture of Greece is such that those decisions may be based on personal and political preferences as much as on technical, strategic or economic considerations.[3]

In the absence of an immediate external threat, in *Malaysia* capability building rather than threat assessment is the main criterion in force design. The experiences of the communist insurgency, the race riots of 1969 and ethnic tensions have combined to heighten the priority given to regime building and social cohesion. This has impeded the development of transparency in any kind of security-related decision making and resulted in weak capacities for oversight. The details of the long-term plans of the Malaysian Armed Forces are known only to a few select officials and political leaders in the executive branch and are not verified by the elected representatives of the people. The need for the armed forces to develop their own long-term perspective plans which harmonize with the broader requirements of Malaysian society remains unaddressed.[4]

Integration with NATO has been the key element of *Poland's* national security policy in the post-cold war order. Poland having accepted all the obligations of NATO membership, its defence priorities are aimed at building up the capacities for performing its new missions. Its long-term planning involves contributing to collective defence and modernizing its forces for integration with NATO's military structures and operational missions. The 15-year plan for the modernization and restructuring of the armed forces for 1998–2012 aims to reduce manpower and achieve compatibility with other NATO forces. Poland, in the changes it has made in its defence priorities in preparation for membership of NATO, provides an interesting contrast with Greece, another NATO member. Poland has given priority to its NATO commitments and is raising and equipping special units and formations to be available to NATO and interoperable with the forces of other NATO countries. Greece still maintains that its prime security concern is Turkey (a NATO ally): its defence planning is primarily engaged with the Turkish threat and in this it does not rely on NATO for its homeland defence.

The end of the apartheid regime in *South Africa* gave the new government an opportunity to redefine the strategic assessment and defence planning processes on sound working principles of civil–military relations. Long-term planning was introduced with the White Paper on National Defence of 1996, which identified a set of guiding principles that based the building of a 'core force' on threat-independent considerations in order to maintain a modern military capability.[5] Theoretically this should give a defensive orientation to South Africa's

[3] Meletopoulos, M., 'The sociology of national decision-making behaviour', SIPRI Arms Procurement Decision Making Project, Working Paper no. 74 (1998), p. 14.

[4] Robless, R., 'Harmonizing arms procurement with national socioeconomic imperatives', SIPRI Arms Procurement Decision Making Project, Working Paper no. 83 (1997), p. 16.

[5] Williams, R., 'Effects of threat perceptions, security concepts and operational doctrine on force planning in South Africa', SIPRI Arms Procurement Decision Making Project, Working Paper no. 113 (1997), pp. 18–19.

force design. A major Defence Review followed, carried out through a broadly based consultative process which included elements from the Parliament and civil society and was concluded in December 1997. It laid down clear links between strategic planning and arms procurement priorities, which were later approved by the Cabinet and Parliament.[6] This made possible a coherent sequence of long-term force planning.

Although the three main political parties in *Taiwan* share a common understanding of the threat from the People's Republic of China (PRC), which dominates Taiwan's defence decision making, there are different interpretations of the ways in which it can be met. The existence of dual chains of command contributes to difficulties in coordinating the management of the national defence system and the elaboration of defence policy: threat assessment is made by military intelligence under the General Staff Headquarters (GSH) and considered an aspect of the military command, for which purposes the GSH falls under the presidential chain of command; the establishment of defence policy, however, is the responsibility of the Ministry of National Defense. Apart from the five-year budget programmes and the one-off Ten-Year Plan for Restructuring Defence Organizations and Armed Forces 1993–2003, the research for this project did not identify any long-term planning processes in Taiwan. The absence of detailed long-term force-building plans could be one of the reasons why Taiwan has difficulty in developing indigenous military capacities. The fact that its force building is reactive to the military capacities of the PRC and to relations between the USA and the PRC also complicates long-term defence planning.

Coordination between foreign policy and integrated defence policy

Decision-making powers are concentrated at the level of the three commanders-in-chief of the armed services in *Chile*. Coordination between the three armed services and between them and the foreign policy-making process is not well developed. Except for limited coordination, such as for communications, a lack of integration in operational plans is evident. Procurement plans are developed by the individual armed services and projects are decided not on the basis of comprehensive acquisition of military capabilities but according to accretions needed to enhance the operational capacities of each branch of service.

There has been no tradition of coordination between foreign and defence policy making in Chile. Under military rule before March 1990, when officers of the armed forces staffed the Foreign Ministry, there was some blending of foreign and security policy, but formal processes of coordination between the two were not developed. As the democratic system has not yet matured, the lack of trust between civilian officials and the military reinforces problems in coordination. However, the same two guiding principles define the approach to

[6] Chandler, N., 'Armscor says doors open to "overlooked" arms bidders', *The Star* (Johannesburg), 29 Dec. 1998, p. 6. See also chapter 6, section III, in this volume.

foreign and security polices—the maintenance of military parity in the Southern Cone and the avoidance of dependence on a single source of arms supply.[7]

In *Greece* coordination between foreign and defence policy making is largely carried out at Cabinet level. However, in practice, the overriding concern with the Turkish threat has meant that no coordination process has developed at the functional levels. This has led politicians to articulate nationalistic policies and to emphasize military-related solutions to national defence problems. Diplomatic or other non-military approaches to security are not properly represented. Coordination between foreign and defence policies has been developed more to meet NATO requirements than to develop a coherent Greek security policy.

In *Malaysia,* the research suggests that civilian staff in the Ministry of Defence maintain contact with the Ministry of Foreign Affairs[8] but that organizational coordination between defence and foreign policy making has not been developed, apart from the functional coordination required for the 1971 Five-Power Defence Arrangement.[9] Inter-service cooperation is found mainly at operational levels in the armed forces. Despite the setting up of the Armed Forces Staff Headquarters in 1993, joint defence planning is not integrated at a higher level, as is indicated by the lack of a joint operational doctrine or a policy document.

In *Poland*, the coordination processes between foreign and security policies pay more attention to the requirements of membership of NATO as the guardian of Poland's security than to purely 'national' security issues. Defence cooperation with NATO members forms an essential part of the strategy of integration with NATO. Among the major elements of coordination with NATO are systems integration; adaptation of military infrastructure; interoperability in areas such as command and control, operations, air management and logistics; adaptation to NATO standards; the modernization of military equipment; and education and training.[10]

The four services of the *South African* armed forces have been integrated in the South African National Defence Force (SANDF), which defines operational requirements and priorities. A well-defined coordination process integrates their input with that of the Department of Defence, which ensures that decisions are made within the framework of national objectives and constraints. Coordination with foreign policy is better developed in arms export control than in arms procurement decision making. Close organic linkages between defence policy and foreign policy-making processes in regard to arms procurement have not been identified in the course of this research. Where arms procurement strategy is concerned, diversification of the sources of weapons is as important as building alliances with new countries by buying armaments from them.

[7] Navarro (note 2), p. 4.

[8] See chapter 4, section III, in this volume.

[9] The FPDA is with Australia, New Zealand, Singapore and the UK—Malaysia's only multilateral defence arrangement with other countries.

[10] Polish Ministry of National Defence, 'Report on Poland's integration with NATO', Feb. 1998, pp. 3, 11, 19, 22–25 (in English).

Coordination between the Ministry of National Defense of *Taiwan* and the Ministry of Foreign Affairs is not adequately developed. This lack of coordination handicaps political initiatives that could help produce a better balance in the country's security policy and arms procurement decision making. However, the operational plans of the three services are integrated by the GSH, so that there is coherence and coordination in arms procurement planning and priorities. This is evident from the changes being made in Taiwan's force designs, emphasizing air superiority, naval capabilities and missile defence for defence of the island, even if the army's influence remains strong (as is indicated by a larger army presence at senior levels in the Ministry of National Defense).

The political influence of the military and of predominant arms suppliers

The armed forces of *Chile* are extremely influential in the current political system as they have negotiated a considerable degree of autonomy under the new (1981) Constitution. In order to ensure that democracy did not dilute the military's autonomy in the future, constitutional safeguards to support right-wing political interests were put in place before the military departed from power. The powers of the President in nominating or removing commanders-in-chief are restricted by the constitution.[11] Checks on the military authority are probably not effective: for instance, of the 13 members of the National Defense Superior Council (CONSUDENA), which comes under the Ministry of Defense and whose function is to approve all arms procurement projects, seven are from the military. Where arms procurement is concerned, decision-making power is concentrated in two individuals—the President and the Commander-in-Chief of the service that is buying the equipment.[12]

Chile's arms procurement was heavily dependent on the USA in the 1950s and 1960s. US support was conditional on Chile's accepting a US military mission and abandoning its non-aligned posture. The military coup in 1973 was followed by US sanctions in 1976,[13] which led the military to follow a policy of avoiding dependence on a single source of arms supply while maintaining logistic compatibility. The return of an elected government resulted in the lifting of the arms embargo in 1990, but the army was reluctant to restore an arms procurement relationship with the USA—unlike the navy and the air force, who were more technology-dependent and pragmatic in making their decisions.

The 1953 agreement between *Greece* and the USA provided the legal basis of defence cooperation between the two countries. Since 1974, when Turkey occupied northern Cyprus and the Greek military junta fell, US influence on the Greek military has declined significantly. In particular, the conditions attached

[11] Robledo, M., 'Domestic considerations and actors involved in the decision-making process of arms acquisition in Chile, 1990–97', SIPRI Arms Procurement Decision Making Project, Working Paper no. 69 (1998), p. 6.

[12] Robledo (note 11), pp. 13, 16.

[13] Meneses, E., 'Chilean defence procurement: achieving balance between suppliers', SIPRI Arms Procurement Decision Making Project, Working Paper no. 65 (1997), p. 3.

to US Foreign Military Sales (FMS) loans are considered burdensome because payments have to be made in hard currency, there are no guaranteed delivery times for materials and there are no penalty clauses.[14] Even with gifts of US military equipment difficulties have been experienced: there is no transfer of technology to the Greek defence industry; the forces are heavily dependent on the USA for spare parts and maintenance; there are problems of interoperability with equipment from other sources; and procurement of US weapons results in operational doctrine being also defined by the USA.[15]

During the period when Greece was under threat from communist forces to its north, the government perceived US military aid as useful. However, the confrontation with Turkey led to the realization that the Turkish threat could not be countered within the framework of NATO. In order to reduce the long-term effects of dependence on the USA, the ratio of arms procured from the USA to those from other sources fell by value from 4 : 1 before 1974 to 0.9 : 1 during the period 1979–83.[16] During the 1990s the shift has been primarily towards buying arms of European origin.

Malaysia being a democracy with a strong central, civilian government, its military remains under civilian control. Political influence in matters such as arms procurement is resented by the military because the politicians bring in non-technical considerations. The military has a comparatively narrow professional perspective on security, while the political leadership and the civilian bureaucracy take a comprehensive view of national security requirements. In certain cases this has led to decisions being taken by the Cabinet without reference to even the senior officials in the Ministry of National Defense or the military leaders. For instance, despite the problems of interoperability, maintenance and training resulting from procuring major weapons from different sources, political considerations of avoiding dependence have led to the acquisition of fighter aircraft with very similar roles from the USA and Russia.

In *Poland* the authorities of the President and the Prime Minister over the military overlapped after the interim constitution was brought in in 1992. This led to problems in the accountability of the military as well as in defence policy making, and the military's desire to retain the autonomy it had enjoyed during the cold war also contributed to tensions. The 1997 Constitution established that the President had authority over the armed forces as Supreme Commander in time of war and the Minister of National Defence had authority over the Chief of General Staff in peacetime. The major functions of the integrated General Staff related to planning and not to the command function.[17] Membership of NATO implies that Poland must adjust its legal systems to NATO standards regarding transparency in defence planning and budget processes. It is therefore

[14] Giannias, H. C., 'Arms procurement and foreign dependence', SIPRI Arms Procurement Decision Making Project, Working Paper no. 73 (1998), p. 10.

[15] Giannias (note 14), pp. 11–14.

[16] Giannias (note 14), p. 6.

[17] Radio Free Europe/Radio Liberty, *RFE/RL Newsline*, 15 July 1999.

expected that over time the influence of the military will come to resemble the influence of the militaries in other NATO member countries.

In contrast to its predominant role during the apartheid regime, the new role of the military in *South Africa* is grounded in the principles of civil–military relations in democracies. It does not have political influence. Civil society organizations, notably the Anglican Church, work to promote development priorities and anti-militarism. In the absence of a direct military threat that might compel short-term decisions on procurement, South Africa enjoys something of a buyer's market in arms procurement. The new procurement plans are well diversified, so that no dominant arms supplier can gain influence.

The political influence of the military in *Taiwan* is a consequence of both historical and political factors. Taiwan maintains its independence from the PRC on the basis of military deterrence. Ironically, however, it cannot allow its military power to become strong enough to encourage the advocates of independence for Taiwan because of the reaction this would provoke from the PRC. The need to maintain this sensitive balance gives the Taiwanese military a rationale for maintaining confidentiality in its arms procurement plans and the resulting low level of public debate enhances the military's decision-making autonomy. Taiwan being weak in the foreign policy arena, the role of military strategy and consequently the military's influence domestically are enhanced.

Coordination is well developed between the military representatives of the USA in Taipei and of Taiwan in Washington. The US influence on Taiwan's security will remain as long as the USA remains the only power that can provide Taiwan with the sophisticated weapon systems it needs and withstand pressures from the PRC. Tension in US–Chinese relations tends to facilitate Taiwan's arms import initiatives, as was exemplified by its purchase of combat aircraft from the USA after the 1989 Tiananmen Square incident. Taiwan also retains an important place in the US security strategy for East Asia. However, the USA would not like to arm Taiwan to levels that would provoke China, particularly in view of the difficulties that followed the 1997 Guidelines for US–Japanese Defense Cooperation, interpreted by some as allowing Japan a role in the seas around it, including those north of the Taiwan Strait.

III. Defence budgets, financial planning and audit

Integrated defence budgets which are designed to indicate the costs of specific military functions, such as air defence, surveillance, logistics and so on, facilitate the evaluation of arms procurement decisions in relation to long-term priorities. On the other hand, defence budgets which divide up allocations by conventional cost heads such as pay and allowances, equipment, and operations and maintenance are less informative and inhibit cost–benefit evaluation.

This section examines aspects of accountability in defence budget making, financial planning for arms procurement, the capacities of legislative oversight bodies to monitor and review budgeting, and the role of statutory audit authorities. The analysis is based on the following elements: (*a*) defence budget planning and some aspects of financial planning such as life-cycle costs and offset policies; and (*b*) capacities in departmental and statutory audit functions which facilitate executive and legislative oversight in order to prevent fraud and inefficiencies in the system.

Defence budget planning and accountability

The defence budget in *Chile* is designed in terms of conventional cost heads such as salaries, operations and maintenance, welfare and infrastructure expenditure. It is controlled by the Ministry of Defense and governed by legislation. Funds for arms procurement are, however, separately appropriated through the 'Copper Law' of 1958 (revised in 1985) and not included in the defence budget; they are therefore outside the purview of legislative approval. The system is unique to Chile. The system of distributing the funds obtained under the Copper Law is not transparent because it is not covered by the annual state budget and not part of the general public-sector accounts.[18] This removes the military's decisions on arms procurement from public scrutiny in the political arena and from its logical base of strategic planning. It also makes the three services virtually independent in defining their operational needs and projects.[19]

Each service has tight control over its own money and treats its spending as a jealously guarded preserve. The situation is indicative of a division of assets on the basis of political balance rather than a military professional and political assessment of priorities. Politicians in both government and opposition show their support for the military budget and do not oppose it.[20] The Parliament has neither skills, time nor information to scrutinize the defence budget. Moreover, arms procurement expenditure is outside parliamentary control.[21]

The arms procurement budget in *Greece* is derived from a five-year medium-term plan. However, the servicing costs of long-term debt and the life-cycle costs of major procurement items have not been adequately factored in to the budgeting process. This is indicated by the fact that, of a budget of 4 trillion drachmas (reported elsewhere as $17 billion) for the arms procurement programme for the five years 1996–2000, half was to be used for repayment of older long-term debts. The bulk of the repayments for this programme will fall due in the years 2003–2007 and the costs of servicing new debts will continue

[18] Pattillo, G., 'The decision-making process in the acquisition of arms systems: an approach', SIPRI Arms Procurement Decision Making Project, Working Paper no. 67 (1997), p. 11. The defence budget for 1997 approved $1.158 billion (1.38% of gross domestic product (GDP)) and the Copper Law appropriated $240 million for arms procurement (0.28% of GDP). Additional appropriations are for pensions of retired personnel, police and Carabineros, and some research institutes, adding up to $2.958 billion (3.48% of GDP). 'Heavy investments in military sector', *El Mercurio* (Santiago), 28 June 1997, p. A1. The total sum available for arms procurement is made public in the CODELCO annual report.

[19] Pattillo (note 18), pp. 4, 6.

[20] Gaspar, G., 'Military expenditures and parliamentary control: the Chilean case', SIPRI Arms Procurement Decision Making Project, Working Paper no. 64 (1997), pp. 4, 5.

[21] Gaspar (note 20), pp. 7, 8.

until 2019. Opposition leaders have criticized the government for lack of transparency and for proceeding as if state decisions were private decisions.[22]

Greece has a well-defined offset policy and the organizational structure and legislation to implement it. The policy aims to increase the international competitiveness of the Greek defence industry by providing access to sophisticated technologies through co-production programmes. However, there have been shortcomings in its implementation. Private corporations have not been able to absorb offsets in a timely and correct fashion. This is mainly due to high investment costs; lack of infrastructure, skilled personnel and quality-control systems; lack of capability to estimate the technology transfer values of offsets; lack of coordination between official bodies and industrial corporations for implementing offsets; and the absence or weakness of penalty clauses in cases of failure to fulfil offset obligations.[23] Greece's approach to offsets has the twin aims of supporting the Greek defence industry and contributing to improvements in the technological infrastructure of the country.

A two-tiered defence budgeting system in *Malaysia* consists of five-year estimates to plan capital expenditure and annual budgetary allocations approved by the Parliament. A mid-term programme analysis and review is also carried out to determine if any revisions to allocations are required. In view of the need to measure technological options in financial terms, a need for financial planning in a 15- to 20-year time frame has been expressed in some quarters.

Inadequate project management, lack of clarity in defining project specifications and poor financial estimating by the military have resulted in some substantial cost overruns and delays[24] which might have been avoided had there been disaggregated budget information allowing scrutiny by the Parliament or independent financial experts.

Malaysia does not have a stated offset policy in arms procurement, but the approach it has taken during the 1990s indicates that, like Greece, it gives priority to industrial and technological benefits. Offsets in arms procurement contracts mainly take the form of joint ventures in the private sector for maintenance, support, the production of accessories, and the acquisition of training and design facilities or production technologies. The practice of seeking offsets has been imaginatively applied in the strategically important defence industries, with government retaining the controlling shares in these joint ventures.[25] The setting up of facilities for rapid prototyping and high-speed machining at the Standards and Industrial Research Institute Malaysia (SIRIM) and at the aerospace engineering school at Mara Institute of Technology are notable examples of the offsets achieved.

To an extent the defence budget of *Poland* can be described as an integrated defence budget. Allocations for arms acquisition are currently 10–11 per cent of

[22] Loukas, D., [Our expensive defence], *Ta Nea* (Athens), 14 Nov. 1996, p. 9.

[23] Antonakis, N., 'Offset benefits in Greek defence procurement policy: developments and some empirical evidence', ed. S. Martin, *The Economics of Offsets: Defence Procurement and Countertrade* (Harwood: Amsterdam, 1996), pp. 169, 173.

[24] Robless (note 4), p. 9.

[25] Robless (note 4), p. 10.

the defence budget, 63 per cent being spent on personnel and salaries and the balance on operations, maintenance and training.[26] The allocation for arms procurement is still well below the average in the NATO member countries that have more equipment-intensive militaries. In the Parliament the defence budget is overseen by the Commission of Public Finances as part of the overall national budget. Although the Parliamentary Defence Commissions are supported in their work by the Sejm Bureau of Research, which has been assertive in demanding a detailed draft defence budget, they still lack expertise or independent staff to scrutinize it.[27]

Poland does not seem to have developed a comprehensive offset policy in terms of priorities or ways of implementing the policy in different sectors of the national defence industry. A parliamentary proposal to seek up to 100 per cent offsets has been adopted.[28] Legislative and practical experience in this field is inadequate. Critics have noted Poland's failure to secure offset deals on various contracts, for example, the purchase of jet airliners from Boeing.[29]

South Africa's defence budget is divided into a General Defence Account and a Special Defence Account. The latter is used for capital costs for procurement of air, naval and ground systems and communications and some running costs for vehicles and weapons, communications, intelligence and research and development (R&D).[30] Experts believe that the current budget process does not take threat assessment as the point of departure, but starts from whatever budget has been allocated by the Finance Ministry.[31] The military is advocating higher expenditure on arms procurement on the grounds that the current allocations for capital expenditures are distorted in comparison with what could be regarded as typical budget allocations, seen in an international perspective.[32] Despite the participatory nature of the Defence Review, there is an opinion among parliamentarians that they are not given sufficient time to review the defence budget or exercise any influence.

[26] [Interview with Roman Musial, Chairman, Polish Chamber of Manufacturers for National Defense], *Polska Zbrojna*, 4 Apr. 1997, p. 20, in 'Poland: Defense industry fair advertised', Foreign Broadcast Information Service, *Daily Report–East Europe (FBIS-EEU)*, FBIS-EEU-97-079, 4 Apr. 1997.

[27] Stachura, J., 'Arms procurement decision making in Poland', SIPRI Arms Procurement Decision Making Project, Working Paper no. 94 (1998), p. 11.

[28] 'Ustawa o niektórych umowach kompensacyjnych zawieranych w zwiazku z umowami dostaw na potrzeby obronnosci i bezpieczenstwa panstwa' [Regulation on certain compensation agreements concluded as part of agreements concerning supplies for the defence and security needs of the state], 10 Sep. 1999, *Dziennik Ustaw Rzeczpospolitej Polskiej* [Journal of legislation], no. 80 (1999), pos. 903.

[29] 'Polish defence, arms sector seek more money', *Interavia Air Letter*, no. 13943 (5 Mar. 1998), p. 4; and *Rzeczpospolita*, 29 Jan. 1999, p. 11, in 'Poland: US, French plane offers for Poland viewed', Foreign Broadcast Information Service, *Daily Report–Arms Control (FBIS-TAC)*, FBIS-TAC-99-029, 1 Feb. 1999. The Polish airline LOT has been buying Boeings since 1988.

[30] [South African Parliament], 'National defence: Vote 29 1996/97 estimates. Explanatory Memorandum', p. 24.

[31] Sparrius, A., 'Quality in arms procurement', SIPRI Arms Procurement Decision Making Project, Working Paper no. 110 (1997), p. 22.

[32] According to a Department of Defence briefing given to the Parliamentary Joint Standing Committee on Defence, the allocations to personnel (57%), operating expenses (35%) and capital expenditure (8%) should be rationalized in terms of what would be a more typical distribution for countries with equipment-intensive armed forces, e.g., personnel 40%, operating expenses 30% and capital expenditure 30%. South African Department of Defence, *Bulletin,* no. 32/98 (2 June 1998).

Methods of assessing life-cycle costs are more advanced in South Africa than in other countries in this study, particularly with new technologies and costs of components purchased from abroad.[33] The offset package unveiled following the 30 billion rand arms acquisition programme in November 1998 indicates a high level of detail in identifying the industrial participation benefits.[34] However, there is no provision for scrutiny and monitoring of offset packages by experts other than those designated by the Department of Defence.

Taiwan's defence budget is developed using a systematic method of linking strategic plans with arms procurement programmes that define medium-term and annual budget plans. The budget breakdown indicates the shares allocated to specific military functions, described as air defence, 'counter landing' (as it is called), readiness support and sea control operations.[35] The method facilitates legislative overview of the country's defence planning. However, there is an unjustifiable lack of transparency in two aspects of budgeting. First, the legislators allow secret debate on special budget plans in the Legislative Yuan for special procurement projects.[36] Second, the Ministry of National Defense does not have oversight of nearly 70 per cent of the defence budget as it does not control the budget of the General Staff[37] and the GSH is not accountable to the ministry.[38] Because of the pressure China exerts on potential supplying countries, Taiwan has political difficulties in importing major weapons and thus in securing offsets. The legislature has, however, been assertive enough to demand offsets against arms imports. The offset policy aims to coordinate with and develop strategic industrial plans in key technology areas.

Departmental and statutory audit

In *Chile* the Controlaria General de la Republica (Office of the Comptroller General) is an autonomous body responsible for statutory audit of public expenditure.[39] It carries out routine audit of arms procurement in terms of financial probity but does not assess the relevance of projects or ensure that they are carried out. It is not known whether the staff of the Comptroller General includes technical experts and military specialists.

[33] Griffiths, B., 'Arms procurement decision making', SIPRI Arms Procurement Decision Making Project, Working Paper no. 105 (1997), pp. 9–11.

[34] South African Department of Defence, *Bulletin,* no. 85/98 (19 Nov. 1998), and *Bulletin,* no. 90/98 (26 Nov. 1998); and Ross, J. G., 'Beyond South Africa's arms deal', *Armed Forces Journal International*, Feb. 1999, pp. 24, 25.

[35] Taiwanese Ministry of National Defense, *1998 National Defense Report, Republic of China* (Li Ming Cultural Enterprise Co.: Taipei, 1998), pp. 130–31.

[36] Yang, A. Nien-dzu, 'Arms procurement decision-making: the case of Taiwan', SIPRI Arms Procurement Decision Making Project, Working Paper no. 123 [1998], p. 11.

[37] Of the defence budget 25% goes towards pensions and only 5% is controlled by the Ministry. Chen, E. I-hsin, 'Security, transparency and accountability: an analysis of ROC's arms acquisition process', SIPRI Arms Procurement Decision Making Project, Working Paper no. 114 (1998), p. 14.

[38] Chen (note 37), pp. 14–15. See also Yann-huei Song, 'Domestic considerations and conflicting pressures in Taiwan's arms procurement decision-making process', SIPRI Arms Procurement Decision Making Project, Working Paper no. 124 (1998), p. 21.

[39] Pattillo (note 18), p. 9.

The role of audit of arms procurement is small in *Greece* and its visibility low. The difficulty experienced by the Greek researchers on this project in identifying the agencies involved in and the functions of statutory and departmental audit of arms procurement decisions reinforces this conclusion.

In *Malaysia* the Federal Audit Department, a statutory organization, is answerable to the Public Accounts Committee of the Parliament, but because of the subservience of members of the ruling party to their political leaders there is neither a practice nor a spirit of executive accountability to the legislature. Consequently the role and capacities of the Federal Audit Department are very weak, as is evidenced by the lack of information on arms procurement audit processes available to the Malaysian experts on this project. The statutory audit staff are accountants in the government service and there are no multi-disciplinary audit teams with military and technical experts. However, internal audit is built in to all government departments, including the Ministry of Defence.

In *Poland* the auditing office is the Najwyzsza Izba Kontroli (Highest Chamber of Control, or NIK). It is an autonomous statutory authority which reports to the Parliament on defence budget implementation from the perspectives of legality, financial probity and the appropriateness of arms selected. Audits can be undertaken by the NIK on its own initiative or on request of the Parliament, the President or the Prime Minister. As the scope of its inspection is wide enough to include other military matters, its work is considered by the parliamentarians to be very useful.[40] The audit reports are public to a restricted degree,[41] but is reasonable to assume that some reports on waste and abuse have been classified as secret.

In *South Africa* remarkable changes have been made in opening up the defence budget to multiparty parliamentary defence and budget committees. However, South African experts maintain that if a substantive parliamentary check on the defence budget is to be possible, then a number of reforms are necessary: *(a)* forward estimates of defence spending are required one year in advance; (*b*) there is a need to develop support services necessary for scrutiny of the defence budget; and (*c*) capacities and methods for post-procurement performance audit need to be developed. [42] South Africa has a statutory audit authority which submits its report to the Parliament, but it does not function under the Parliament as its subordinate body.

The responsibility for audit in *Taiwan* lies with the Ministry of Audit, which is under the Control Yuan and not part of the executive. It is understaffed and lacks personnel with adequate training to evaluate arms procurement decisions and carry out performance audit of weapon systems. Furthermore, the GSH does not provide adequate details of the arms procurement programme, and

[40] Stachura (note 27), pp. 13–14.

[41] Stachura (note 27), p. 13.

[42] Calland, R., 'An examination of the institutionalization of decision-making processes based on principles of good governance', SIPRI Arms Procurement Decision Making Project, Working Paper no. 102 (1997), p. 40; and Sparrius (note 31), pp. 19, 21.

there is no statutory provision to enforce disclosure of classified elements of the budget at the auditing stage.[43] If it is true, as reports suggest, that only the Ministry of National Defense's departmental expenditure is audited and the GSH can avoid submitting the details of its expenditure to the Ministry of Audit, then only about 5 per cent of the defence budget is being audited.[44]

IV. Techno-industrial issues

This section focuses on the organizational capacities for defence R&D, arms manufacturing in the public and private sectors, the defence industry in relation to technology assessment and technology absorption, and the obstacles to parliamentary scrutiny, assisted by independent experts, of military technology and policy on the defence industry.

The defence industry, self-reliance and defence R&D

Despite the arms embargo, in *Chile* the defence industry and defence R&D have been driven by the criteria of cost, quality and availability rather than by objectives of self-reliance. The policy has been to acquire sophisticated weapons rather than strive to manufacture equipment domestically.[45] The government has not invested heavily in military R&D and Chile has consequently not developed the critical scientific industrial base required for independent evaluation of the arms manufactured. The three armed services maintain segregated defence industries to serve their needs, even though the defence policy aims at fostering coordinated R&D between the military, the private sector and specialized technical institutes in the universities.[46] In order to coordinate projects involving joint systems, a Committee on Analysis of Joint Product Development of Ministry of Defence was created in 1996.[47]

Greece's participation in the NATO Research and Technology Organization and the Western European Armaments Group (WEAG) Panel II gives it access to West European military R&D, while indigenous defence R&D is conducted at three of the research centres controlled by the General Directorate of Armaments (GDA) Technological R&D Directorate. Outside the government sector, R&D in advanced technologies is not sufficiently developed to support independent technological evaluation of Greece's military R&D.

[43] Chih-cheng Lo, 'Secrecy versus accountability: arms procurement decision making in Taiwan', SIPRI Arms Procurement Decision Making Project, Working Paper no. 116 (1998), p. 11.

[44] See note 37.

[45] See, e.g., Thauby, F., 'The decision making process in arms supply: the Chilean case', SIPRI Arms Procurement Decision Making Project, Working Paper no. 70 (1997).

[46] Thauby (note 45), p. 8.

[47] Porras, L., 'The influence of equipment modernizations, building national arms industry, arms export intentions and capabilities on national arms procurement policies and procedures', SIPRI Arms Procurement Decision Making Project, Working Paper no. 68 (1997), p. 15 and annexe.

The GDA aims to transform Greece's technological dependence into technological interdependence between the foreign arms suppliers and Greece.[48] The ideal distribution between the three functions of the long-term R&D budget that is suggested is 40 per cent for upgrading weapon systems, 45 per cent for development of new weapons and 15 per cent for 'breakthrough' research. It is also recommended that allocations to military R&D should be around 80 per cent for research and 20 per cent for development.[49]

Defence R&D in the public sector in *Malaysia* is conducted by the Ministry of Defence's Defence Science and Technology Centre (DSTC). Priority is given to the private sector for military industrial development, and the DSTC is starved of both funds and qualified manpower for carrying out any purposive military R&D. With only 18 researchers in its R&D unit, it is unlikely to carry out any meaningful R&D. Its functions are confined to quality control. Ironically, the DSTC has also been deprived of government funds under the Intensification of Research Priority Areas Programme (IRAP) because it has not been able to monitor security-sensitive defence research. The level of participation by engineers from the armed forces in the DSTC research staff is also very low, which has hindered the development of capacities for systems analysis and equipment development and testing.

Other reasons for the low salience given to R&D in the defence sector are the low probabilities of spin-offs to the civilian sector, the country's small R&D base, and the preference for acquiring technological know-how as opposed to indigenous development. Given the general shortage of skilled R&D personnel, Malaysian institutions are concentrating on the low-value-added segment of the defence industry.

In *Poland*, according to one newspaper, the military coordinates only 70 per cent of defence-related R&D work, and nearly one-third of funds earmarked for military R&D is outside the control of the Ministry of National Defence.[50] This has facilitated conversion in response to the drop in the demand for military goods, and therefore military R&D.[51] Although military R&D engineers have been moving to the civilian job market because of falling demand, broadly based relations between the civilian and military R&D systems have not yet developed.[52] Efforts by the institutes which do military R&D to market their products and services have yielded mixed results. Those which had products or

[48] Narlis, E. O., 'Arms development and defence R&D growth in Hellenic Republic', SIPRI Arms Procurement Decision Making Project, Working Paper no. 75 (1998), p. 2.

[49] Narlis (note 48), p. 4.

[50] Choroszy, R. (Maj.), 'A strategy for survival', *Polska Zbrojna,* 22 May 1998, pp. 20–21, in 'Polish strategic defense programs', FBIS-EEU-98-153, 2 June 1998.

[51] The value of arms procurement by Poland fell by 10% in 1991 and by 80% in 1992. Tarkowski, M., 'Balancing arms procurement with national socio-economic imperatives', SIPRI Arms Procurement Decision Making Project, Working Paper no. 96 (1997), pp. 1.

[52] Mesjasz, C., 'Restructuring of defence industrial, technological and economic bases in Poland, 1990–97', SIPRI Arms Procurement Decision Making Project, Working Paper no. 91 (1998), p. 21.

testing facilities that could be used in the civilian sector had better results in marketing than those with highly developed specialized military R&D.[53]

Military R&D in *South Africa* was formerly carried out largely in the public-sector corporations, while review and evaluation were carried out by Armscor. Opening up to international competition and the simultaneous decline in funding for defence R&D have compelled diversification and privatization. However, certain strategic R&D facilities that could not be commercially profitable have been retained in the public sector, such as testing ranges and laboratories for product evaluation. There is a division of authority in technology management, for instance, between the Armament Acquisition Steering Board and the Defence Research and Development Board.[54] The fact that they are separated from the technology developers in the private sector has prevented the development of a monolithic military R&D interest group. Executive oversight of R&D (in the shape of steering committees for specialist technology areas and Armscor) and some science and technology institutions in the private sector are independent of the end-users (the SANDF). However, experts believe that military R&D expenditure is shrouded in excessive secrecy and that public access to information is restricted.[55]

The primary responsibility for military R&D in *Taiwan* rests with the Chung Shan Institute of Science and Technology (CSIST), which is also responsible for TA. Combining both these functions in one agency works against the principles of checks and balances. However, growth in advanced industrial R&D, especially the high-technology areas required by the military—communications, aerospace, precision machinery, special materials, electronics and automation—will counterbalance the autonomy enjoyed by military R&D. It will also provide capacities outside government to monitor and review military R&D projects. By the year 2010 it is expected that Taiwan will have 75 000 researchers, 60 per cent of them with a master's degree or PhD.[56]

The defence industry, technology assessment and technology absorption

In *Chile* the rules of the market began to be applied to the defence industry in the 1990s in keeping with the broader national industrial policy of reducing state participation in the economy. As defence production was seen as risky by the private-sector engineering industry, certain selected private-sector defence

[53] Transcript of proceedings of the workshop held at the Institute of International Affairs, Warsaw, 26 Nov. 1997, within the framework of the SIPRI Arms Procurement Decision Making Project, p. 35; and Wieczorek, P. and Zukrowska, K., 'The influence of equipment modernization, building a national arms industry, arms export intentions and capabilities on national arms procurement policies and procedures', SIPRI Arms Procurement Decision Making Project, Working Paper no. 98 (1998), p. 10.

[54] Buys, A., The 'influence of equipment modernization, building a national arms industry, arms export intentions and capabilities of South Africa's arms procurement policies and procedures', SIPRI Arms Procurement Decision Making Project, Working Paper no. 101 (1997), p. 13.

[55] Cilliers, J., 'Defence research and development in South Africa', SIPRI Arms Procurement Decision Making Project, Working Paper no. 103 (1997), p. 14.

[56] Taiwanese Executive Yuan, National Science Council, *White Paper on Science and Technology*, (National Science Council: Taipei, Dec. 1997), pp. ii, iii.

industries, such as Cardoen, were encouraged by means of assured contracts to meet the military's product development requirements. However, the defence industry has remained comparatively small in scale in terms of weapon development. Its main functions are the manufacture of spare parts and maintenance and repair of equipment. Three large companies, which come under the army, the navy and the air force and are controlled by the respective undersecretaries, are involved in production of selected weapon systems and participate in concept development and determination of technical requirements.

The *Greek* defence industry still relies heavily on imported technology and know-how. Consequently a few technology priority areas have been selected on the basis of technologies available on the international market and existing Greek technological capabilities.[57] As there does not appear to be a quality assurance (QA) organization in the Greek Ministry of National Defence, the defence industry carries out its QA according to the standards of NATO's Alliance Quality Assurance Publications (AQAP) and the International Standards Organization (ISO). There are a large number of small and medium-sized companies in the private sector which allocate 20–80 per cent of their capacities to defence production.[58] The Defence Industry Directorate of the GDA also carries out international market research to improve the Greek defence industries' domestic and international competitiveness and promote exports.

Malaysia's defence industry was developed through joint ventures with foreign suppliers selected on the basis of their capacities to develop advanced products. This approach was necessary because demand from the defence sector was low and the techno-industrial base small. The government gave priority to the civil manufacturing sector as the engine of growth of advanced technological capacities in Malaysia, which led to the establishment of the Malaysian Technology Development Corporation (MTDC) in 1992 and Malaysian Industry Government Group of High Technology (MIGHT) in 1993. The objective of the MTDC is to make the R&D potential of Malaysian industry and the Malaysian academic world more visible; MIGHT is tasked with monitoring global technological developments for exploitation in Malaysia.[59]

In *Poland* the optimistic assessment of the potential benefits to the country's defence industry of joining NATO has been criticized by the Polish Chamber of Defence Producers, a voluntary organization of 177 companies. Two aspects have been highlighted: (*a*) the restructuring programme in the arms industry and modernization of the Polish military to adapt to NATO structures do not necessarily mean new opportunities for the indigenous defence industry—they will mean work on systems integration but imply arms procurement from NATO member countries; and (*b*) NATO membership will irrecoverably

[57] Narlis (note 48), p. 5. The priority areas selected are electronics, opto-electronics, telecommunications, fluid mechanics, aerodynamics and ballistics, advanced pyrotechnics and materials technologies.

[58] *Greek Defence Directory*, 4th edn (Delos Communications: Athens, 1998), pp. 103, 105.

[59] Supian Ali, 'Harmonizing national security with economic and technology development in Malaysia', SIPRI Arms Procurement Decision Making Project, Working Paper no. 87 (1997), p. 16.

deprive Poland of its military R&D and arms industrial potential.[60] At the end of the cold war, Poland had excess capacity in military production, with 128 companies engaged. Of those nearly 70 per cent were producing either dual-use or civilian equipment in order to maintain industrial capacities.[61] With a decline in demand for military goods and privatization, a highly qualified labour force and R&D engineering skills are likely to filter into the civilian sector. Among the intractable issues facing the old military industries is adaptation to market forces. This is being resisted by well-entrenched interests supported by local politicians and trade unions.[62]

The international competitiveness of the *South African* defence industry is indicated by its export performance. As all companies in the defence industry are commercial businesses, they have developed dynamic diversification strategies in order to survive the decline in defence business. Some significant commercial applications have emerged from defence technologies.[63] The defence industry is being encouraged to promote spin-offs to civilian industry in the National System of Innovation and civilian industry to spin on commercial off-the-shelf technologies. Other defence industrial cooperative initiatives include the National Research and Technology Foresight Programme, which aims to identify technologies and technological trends which will be important for South Africa's economic development; and the National Research and Technology Audit, which aims to assess the strengths and weaknesses of South Africa's science and technology system in order to understand the forces shaping long-term futures.[64] The focus of defence-related work at the Council for Scientific and Industrial Research (CSIR) is on selected core technologies.[65]

The TA capacities available to Armscor in the shape of private consultancies help its decision makers to obtain inputs from diverse specializations. However, post-procurement comparative evaluation, which would identify shortcomings in the weapon systems, track and analyse problems, and then improve weapon systems or procurement processes, is not being done systematically.[66]

Technological skills in both the defence and the civil sector are well developed in *Taiwan*, and this allows cross-fertilization between the two. The R&D organizations have successfully combined military experience with high levels of technology skills in their staff structure. As many as 80 per cent of the CSIST staff are from the military.[67] Taiwan's high-technology exports are developing. During the period 1990–95, Taiwan's technology-intensive exports in its industrial manufacturing sector rose from 25.7 per cent to 35.07 per cent

[60] *Polska Zbrojna*, 26 June 1998, p. 6, in 'Poland: Polish arms makers criticize industry restructuring plan', FBIS-EEU-98-177, 29 June 1998.

[61] Mesjasz (note 52), pp. 3.

[62] Plater-Zyberk, H., *Poland's Defence and Security: The Same Priorities, Different Approaches* (Royal Military Academy, Conflict Studies Research Centre: Sandhurst, May 1998), p. 16.

[63] Hatty, P., 'The South African defence industry', SIPRI Arms Procurement Decision Making Project, Working Paper no. 106 (1997), pp. 25–26; and Buys (note 54), pp. 15, 16.

[64] Buys (note 54), pp. 16, 18.

[65] Cilliers (note 55), pp. 12–13.

[66] Sparrius (note 31), pp. 22–23.

[67] Yang (note 36), p. 13.

and in the high-technology sector rose from 35.8 per cent to 45.5 per cent.[68] The ratio of scientists and engineers to auxiliary staff is also high in Taiwan[69] and there is active collaboration between the military R&D and technology research institutes and university laboratories in the civil sector.

In order to broaden the base of military and advanced industrial technologies in Taiwan, the inter-agency National Defence Technology Development Steering Committee has been strengthened; a specialized weapon development unit along the lines of the French Direction Générale d'Armements (DGA) has been set up; offsets and international joint ventures have been used to access military and maintenance technologies; interaction between the military, the advanced industrial laboratories and university R&D centres has been increased; the CSIST has been partly converted into an advanced centre for developing dual-use military and industrial technologies; and strategic alliances have been promoted with international expertise in advanced dual-use technologies.[70]

V. Organizational behaviour and public-interest issues

Organizational behaviour and public-interest issues have been the most challenging of the four themes of this analysis. The focus is on the limitations on and opportunities for the improvement of public scrutiny and oversight of defence policies and arms procurement decision making.

Public scrutiny of arms procurement decision making requires constitutional provisions, assertiveness on the part of the legislature and the availability to the public of sufficient information. In some cases, the government's resistance to legislative oversight is indicated by its reluctance even to issue White Papers or policy documents to identify defence policies or arms procurement guidelines. In such circumstances the military's autonomy in arms procurement decision making develops at the cost of the broader priorities of society.

The extent to which the legislative bodies demand security-related information is conditioned by a society's attitudes towards military security, traditional elite behaviour and the nature of a country's political organization. Since attitudes which encourage military autonomy and excessive confidentiality create barriers to public accountability, they can also allow inefficiencies to creep into the arms procurement processes, permitting waste, fraud and abuse.

The analysis in this section is based on: (*a*) the capacities and quality of legislative oversight of the military's arms procurement policies and decisions; and (*b*) social and elite attitudes that tend to exclude defence policy making from the purview of public policy oversight.

[68] *White Paper on Science and Technology* (note 56), p. 25. In the USA, high-technology products have been defined as those which have significantly more R&D than other products. Technology-intensive products are defined as those in which R&D expenditures exceed 2.36% of sales. US National Science Foundation, 'Science and technology resources of Japan: a comparison with the United States', Special report, NSF 88-318, Washington, DC, 1988, p. 38.

[69] In the CSIST it is 1 : 0.56 and in the Hsinchu Science-based Industrial Park (SIP) it is 1 : 0.64. Yang (note 36), p. 13; and *White Paper on Science and Technology* (note 56), p. 26.

[70] *White Paper on Science and Technology* (note 56), pp. 44–45.

Content and quality of legislative oversight

In *Chile* the Congress is not an actor either in the making or in the monitoring of arms procurement decisions: the decision-making autonomy of the military in this area is well established and is considered sensitive. The military, while negotiating the new constitution, ensured that the Congress does not have constitutional authority to check and monitor arms procurement decisions or the procurement budget. The Organic Law on the armed forces specifies that spending on military equipment and spare parts must be carried out in a confidential manner.[71] Apart from occasional departmental leaks, there is no way in which society can find out about waste and abuse in arms procurement by the military. Indeed, the level of interest shown by the Congress in arms procurement issues is low. Among the reasons for this are that it is not a major electoral issue; that the legislature does not have independent experts to assist its oversight of the military; and that in general the Congress does not receive adequate information.

The *Greek* arms procurement processes are rarely monitored or scrutinized either by the Parliament or by other mechanisms of democratic oversight. The processes are slow and marked by the procrastination, indecisiveness and inertia that characterize the Greek bureaucracy and political system in general.[72] The Parliamentary Committee on Foreign Affairs and Defence has limited importance. The executive branch gives it a largely symbolic role; it does not have specialized research staff; it does not receive the required information; and its does not examine financial details in its deliberations.[73]

The Greek experience illustrates one point that goes against the original assumptions of this study regarding the restraining effect of legislative oversight on arms procurement. Where public perceptions of military threat are heightened by a traditionally hostile relationship with another country, partisan politics willingly subordinates broader social interests to the resource requirements expressed by the military. In Greece, moreover, important security decisions are made by a small political elite led by the Prime Minister and his close personal advisers. This group has considerable autonomy and tends to promote personal political agendas.[74]

Despite the general belief among the *Malaysian* opinion formers and security experts in the region that transparency in arms procurement decision-making processes would help reduce insecurity in the South-East Asian region, in Malaysia this kind of transparency is seen by the ruling elite as a threat to inter-

[71] 'Ley no. 18.948 Orgánica constitucional de las fuerzas armadas' [Organic law on the armed forces], *Diario Oficial* [Official gazette], 27 Feb. 1990, Article 99; and Robledo (note 11), p. 14.

[72] Meletopoulos (note 3), p. 14.

[73] Valtadoros, C., 'The influence of foreign and security policies on Greek arms procurement decision making', SIPRI Arms Procurement Decision Making Project, Working Paper no. 76 (1998), p. 8; and Dokos, T. and Tsakonas, P., 'Perspectives of different actors in the Greek procurement process', SIPRI Arms Procurement Decision Making Project, Working Paper no. 72 (1998), pp. 3, 8.

[74] Meletopoulos (note 3), p. 7.

nal security.[75] Members of Parliament are not prevented from asking ministers questions on arms procurement, but the parliamentary rules indicate that such questions shall not seek information about any matter that is of its nature secret. However, analysis of the content of parliamentary questions on defence management and the executive's responses to them indicates that both the capacity and systematic methods for scrutiny and monitoring are lacking. The type of questions asked on security indicates that arms procurement is not a major issue either for the legislature or among non-governmental organizations (NGOs).

In *Poland* the redefined oversight powers of the Parliament, the Sejm, over the defence policy-making processes are both wide-ranging in scope and intrusive. The Sejm defines the powers of the various executive authorities in this respect, controls defence expenditure and exercises important control functions such as the decision to declare war and delegate powers in national emergencies.[76] Even so, it does not have the expert staff it requires if it is to exercise its oversight functions.[77] Both formal and informal contacts between the Ministry of National Defence and the Parliament have been developed. A post of Undersecretary of State for Parliamentary Affairs in the ministry was established in 1994. Officials from the ministry and the General Staff are invited to the meetings of the Parliamentary Defence Commissions as experts and advisers. Periodically, depending on the issue, the Minister for National Defence, Vice-Ministers and Chief of the General Staff also participate in discussions on new aspects of security issues or report on the work of the MoND.

On defence budgeting, security and foreign policy issues which are overseen by the Parliament and the parts of the defence industry that still come under the state, MPs are allowed access to confidential and classified documents and materials of the Ministry of National Defence and its subordinate institutions and to buildings where classified information is kept without special authorization.[78] A special parliamentary commission was set up in 1997 to monitor defence tendering and arms procurement decisions. A parliamentary deputy with special interest in the defence industry is allowed to attend the commission's meetings but has no vote.[79]

In terms of the accountability of the armed forces to elected representatives, Poland again provides an interesting to Greece. The organizational and constitutional changes made by Poland are far-reaching and have been made in a much shorter timeframe than is the case in Greece. Legislative oversight and

[75] Sharifah Munirah Alatas, 'Government–military relations and the role of civil society in arms procurement decision-making processes in Malaysia', SIPRI Arms Procurement Decision Making Project, Working Paper no. 84 (1998), pp. 3, 26.

[76] Response of the Mission of Republic of Poland to the OSCE to the Questionnaire on the Code of Conduct on Politico-Military Aspects of Security, Letter to all Delegations and Missions to the OSCE Forum for Security Co-operation and Conflict Prevention Centre, Vienna, 22 Sep. 1997, p. 2.

[77] *Congressional Quarterly*, vol. 56, no. 6 (7 Feb. 1998), p. 276.

[78] Kowalewski, M., 'The role of the Parliament in shaping civilian and democratic control of the armed forces', eds P. Talas and R. Szemerkenyi, *Behind Declarations: Civil–Military Relations in Central Europe* (Institute for Strategic and Defence Studies: Budapest, 1996), pp. 14–15.

[79] Stachura (note 27), p. 16.

statutory audit structures for monitoring and scrutinizing arms procurement decisions are being actively developed in Poland but remain weak in Greece.

In terms of constitutional change and introducing the processes and public capacities to monitor the military and the executive, the pace of change in *South Africa* has been faster in the five years since 1994 than it has in many of the countries participating in this project in the past 50. The establishment of the Joint Standing Committee on Defence (JSCD), which holds public hearings, and of the Special Parliamentary Committee on Intelligence is indicative of the development of the potential for legislative oversight.[80] However, South African experts have criticized: (*a*) the lack of an ombudsman to adjudicate in cases where the executive withholds documents from scrutiny; (*b*) the absence of specialized committees to focus on questions of finance, foreign affairs and the defence industry; (*c*) the rapid turnover of membership of the parliamentary committees, which makes it difficult for the legislature to develop interest in these matters; (*d*) the absence of official records of parliamentary committee meetings; and (*e*) the shortage of expert research staff, particularly technical experts. They have even recommended that the agreement of the JSCD should be required before Cabinet approval is given for major arms procurement contracts.[81]

In *Taiwan* the GSH is outside the purview of Legislative Yuan hearings. Although democratization in the past decade has led to an increase in legislative assertiveness and to more details of defence expenditure being published, legislative oversight is still weak. Barriers to oversight have been created by the argument for secrecy in the interests of national security. It is quite probable that the opportunities for corruption in Taiwan's arms procurement processes are one of the motivations for secrecy in its arms procurement decision making, besides the military threat from the Chinese mainland.

Social and elite attitudes and exclusivity of defence policy making

One of the notable characteristics of the *Chilean* political culture is the emphasis on consensus and avoiding issues that could cause disagreements.[82] Secrecy relating to arms procurement decisions has become a norm as there is no need for the armed forces to lobby the Congress for funds and no requirement to bring issues to public debate or legislative scrutiny. Even under the democratically elected regime that has been in place since 1990, no legislative checks exist on the military's decision-making power in arms procurement. The auton-

[80] The Special Parliamentary Committee on Intelligence in South Africa has a mandate to scrutinize the intelligence services: (*a*) to oversee their expenditure; (*b*) to consider and make recommendations on security legislation; (*c*) to order the investigation of allegations by a member of the public of abuse by the intelligence services; (*d*) to refer any abuse of rights to the Human Rights Commission; and (*e*) to report every year to the Parliament. Calland (note 42), p. 17.

[81] Batchelor, P., 'Balancing arms procurement with national socio-economic imperatives', SIPRI Arms Procurement Decision Making Project, Working Paper no. 100 (1997), p. 17; and Sparrius (note 31), p. 18. On the role of the South African Parliament and the JSCD, see Calland (note 42), pp. 19–27.

[82] Davila, M., 'Some aspects of the decision-making process in Chile', SIPRI Arms Procurement Decision Making Project, Working Paper no. 62 (1998), p. 8.

omy of the military in managing the defence of the state has been traditionally accepted. Tensions over that autonomy have surfaced on occasions, as during the investigations in 1990 into payments made to the son of General Pinochet,[83] but the elected government has avoided introducing legislation which would destabilize the delicate equilibrium in Chilean civil–military relations.

While the media have played a passive role in bringing issues to public attention, the academic world has over the years contributed to building up a mass of critical knowledge on security issues.

Two kinds of politician have been noticeable in the Chilean political system—the 'traditionals', whose activities are concentrated on political negotiations and bargaining, and the 'technocrats', professional experts on various issues, mostly younger people and by and large trained in engineering. There is friction between the two types, which also represent a generation difference. It is possible that with the number of technocrats increasing in the Chilean political system calls for military accountability will become stronger.

In *Greece,* society still places great importance on personal rather than institutionalized relationships and modern political practices often clash with traditional forms of patron–client relationships. Many features of traditional society persist to a greater extent than they do in West European countries,[84] including the influence of extra-institutional actors such as friends, relatives, middlemen and political advisers. Meletopoulos argues that the Greek social and political system has been conditioned by the experience of the Byzantine world followed by long rule by the Ottoman Turks. Professional and organizational behaviour is characterized by a powerful communal tradition, the influence of the Orthodox Church, patron–client relations between the leaders and the public, and a 'Byzantine–Oriental conscience'.[85] It is highly unlikely that abuse of public funds will draw overt public criticism. It is not uncommon even for civil servants to have a second profession. A large number of middlemen representing foreign-owned defence firms operate,[86] and there have been cases of fraud relating to arms procurement in the past.

The control of the United Malay National Organization (UMNO), the ruling party in *Malaysia*, has remained unchallenged since Malaysia became independent. The dominant political behaviour is characterized by feudal or patriarchal relationships, with the political elite expecting loyalty and a culture of deference resulting in deification of the political leadership.[87] Even more strongly entrenched in the military is the habit of avoiding questioning the political authority for fear of repercussions, which has led to an attitude of unquestioning obedience. A belief has gained currency that public accountability in arms procurement or defence issues undermines national security, and there is no political or professional motivation to improve oversight of and

[83] Robledo (note 11), p. 10.
[84] Meletopoulos (note 3), pp. 4–6, 14.
[85] Meletopoulos (note 3), pp. 5–7.
[86] Dokos and Tsakonas (note 73), p. 4.
[87] Sharifah Munirah Alatas (note 75), pp. 43–44, 45.

legislative checks on the decision-making process. Accountability and transparency in decision-making processes are seen as being at variance with the national preference for quiet, almost secretive, behind-the-scenes dealings. Such working norms do not allow the development of public accountability.

In *Poland* the military has traditionally been a prestigious institution with no tradition of control by democratically elected representatives. During the first half of the 1990s the military resisted increasing (civilian) executive and legislative control on the argument that it had the best and final judgement on military matters, and the new political elite did not have the practical skills in public policy and defence management to exercise control over the military. The second half of the 1990s saw developing confidence in managing security policy and the enactment of legislation on the dissemination of information.[88] The Ministry of National Defence still has a large number of departmental heads recruited from the military.[89] Of late, the Polish press has been active in bringing defence issues to public notice,[90] but it is still rather under-informed on technical issues, apart from sensational reports by the Sejm Commission on National Defence or from personal contacts in the military. It is believed that in order to eliminate extra-constitutional influences a comprehensive analysis of the entire arms procurement process is required, and that a system should be designed by systems analysts and legal experts working for the Parliamentary Defence Commissions before appropriate legislation was drafted.[91]

The *South African* political and bureaucratic elite still bears the legacy of the centralized decision making and authoritarianism of the apartheid regime. An 'affirmative' culture of accountability has not yet developed: prevalent norms, such as the idea that accountability in security decision making undermines secrecy and that military leaders are always right, still influence defence thinking.[92] A tendency to 'overdo' confidentiality remains, despite the remarkable progress that has been made in opening up the defence budget and decision-making processes to legislative scrutiny. The Department of Defence engaged in an unprecedented broadly-based consultative process in the drafting of the Defence Review in 1996–97, acknowledging the principle that control of the military by civil society, especially budget control, is fundamental to democracy.[93] Public criticism of the 30 billion rand arms acquisition programme of 1998, initially directed against it as an unnecessary burden on society, has of late also developed into criticism of its potential to fuel a regional arms race.[94]

[88] Private communication with Dr Andrzej Karkoszka, former Deputy Minister of Defence, Poland.

[89] Trejnis, Z., 'The sociology of national decision-making behaviour', SIPRI Arms Procurement Decision Making Project, Working Paper no. 97 (1998), p. 12.

[90] Tarkowski (note 51), p. 6.

[91] Miszalski, W., 'Characteristics of acquisition procedures in terms of the organizational structures involved', SIPRI Arms Procurement Decision Making Project, Working Paper no. 93 (1997), p. 10.

[92] Liebenberg, I., 'A socio-historical analysis of national decision-making behaviour', SIPRI Arms Procurement Decision Making Project, Working Paper no. 107 (1997), p. 7.

[93] Crawford-Browne, T., 'Arms procurement decision making during the transition from authoritarian to democratic modes of government', SIPRI Arms Procurement Decision Making Project, Working Paper no. 104 (1997), p. 22.

[94] 'Cancel arms purchase, says De Lille, MP, Pan Africanist Congress', URL <gopher://.anc.org.za/OO/anc/newsbrief/1999/news 1122>.

Increasing democratization in *Taiwan* may also in the long term challenge the paternalistic style of political leadership there, but a strong relationship orientation and the sense of obedience to superiors ingrained in the society tends to work against transparency. The military has been able to maintain relative autonomy because of the security threat from China. Secrecy in arms procurement planning is also considered essential because disclosure invariably results in a reaction by the leaders of the PRC against the probable supplying countries. The need for secrecy has allowed official abuse and corruption, and even allowed organized crime to influence arms procurement decision making. Mere structural changes in the interests of greater accountability will not help to address a grave problem such as this.[95]

VI. Good governance, public accountability and secrecy

In the course of this project various approaches could have been taken to examining the national arms procurement decision-making processes: for instance, technical and organizational efficiencies or value for money could have been the criteria against which the decision-making processes were examined. The project chose instead to examine whether national arms procurement decision-making processes enable balanced decisions to be made from the perspective of broader societal interests. It investigated the decision-making process in the context of public accountability, the assumption being that building up capacities for democratic oversight of security decision making will in the long run contribute to building checks and balances in the security sector and could lead to developing restraints on arms procurement. Such oversight will be more durable if it is institutionalized in a transparent system of checks and balances.

Among the basic characteristics of good governance and public accountability are: (*a*) a clear separation of powers and capacities to exercise those powers between the executive, the legislature and the judiciary—the principle of checks and balance; (*b*) a clearly expressed written constitution defining the separation of powers, framing rules and regulations which define the methods of scrutiny by the legislature, and specifying how the constitution can be amended and its misuse prevented or punished; (*c*) a transparent system of public financial accounting and legislative capacities to influence, monitor and review the budget-making process; (*d*) a political culture which acknowledges the public accountability of the executive, based on qualified access to information; and (*e*) a system of government which acknowledges the public right to information through instruments such as freedom of information legislation.

At a minimum, these require independent sources of information and the availability of expertise publicly so that legislatures and statutory audit authorities can objectively evaluate the executive policy-making process and

[95] On the successive scandals in arms procurement in Taiwan, see Chen (note 37), pp. 12–13; and *Free China Journal*, vol. 15, no. 12 (12 Mar. 1998), p. 1. The alleged bribe of $500 million paid to senior French officials for dropping objections to Taiwan's buying 6 La Fayette frigates from France is a strange case of a reverse bribe in the scandal-ridden arms procurement practices in Taiwan.

decisions made. The principles of good governance must guide every aspect of public policy making, including those relating to arms sales or procurement.

There is, therefore, a need to consider in what circumstances defence decision making can be treated as an exception and the use of secrecy be justified.

Security bureaucracies and secrecy in their decision making

The argument for secrecy is not unique to arms procurement. A range of motivations for secrecy in the public sector in general can be identified. However, military roles and functions cannot be excepted from the requirements of good governance mentioned above. Secrecy in arms procurement must be justified on grounds of (*a*) national security or (*b*) commercial confidentiality.

In the case of the former, secrecy can broadly be justified for the following reasons: (*a*) a need for secrecy of military holdings and stocks; (*b*) a need to withhold technical information which reveals the strengths and weaknesses of a weapon system; (*c*) a need to withhold operational information related to the employment and deployment of weapons; and (*d*) urgency, if rapid procurement is needed. Among the indicators given by the Chief of Defence Intelligence in the British Ministry of Defence are: (*a*) imminent aggressive action against or threat to the state; (*b*) activities of near neighbours pursuing a course prejudicial to the state's independence or security; (*c*) disruptive forces within the society; (*d*) terrorism; and (*e*) 'exceptional circumstances'.[96]

Arguments based on commercial sensitivity need to be handled with care. Companies must be fairly treated, but the argument of commercial sensitivity can be abused. A catch-all determination that no commercial information can be disclosed without companies' consent could also open up opportunities for lobbying and corruption.

In many countries, particularly in the developing world, the roots of secrecy are to be found in the vulnerability of the regime and lack of consistency in state policies on fundamental political, economic, social or ethnic issues. States may also perceive themselves as vulnerable because their borders are ill defined or not recognized, or because their state institutions are weak or are not legally established, or because of competing social interests. Such problems create tensions not only within the country concerned but also in the surrounding region and can lead to the development of an assertive national security policy.

Among the reasons why countries maintain secrecy in routine arms procurement decision making are the following.

1. *Lack of a clear information policy and a weak information dissemination process.* Particularly in developing countries, information collection and dissemination are underdeveloped, even between government departments. Policies on and procedures for handling or releasing information for the purposes of legislative oversight are in many cases unclear. Information policy and infor-

[96] Elworthy, S., 'Balancing the need for secrecy with the need for accountability', *RUSI Journal*, Feb. 1998, p. 5.

mation management receive very little attention. There is a lack of clarity in methods for releasing information or deciding on the classification of information. Classification can be used routinely for administrative convenience as well as to avoid accountability. The working papers and the workshop discussions in Chile, Greece, Malaysia, Taiwan and South Africa highlighted this.

2. *Lack of a legal obligation to disclose information.* There are countries with laws that forbid disclosure of any information related to military security. These laws are often cited by the military and bureaucracies to deny information even of a trivial nature to the elected representatives of the public. Adequate legal provisions have not been framed that can be used by legislators to gain access to and handle classified information. There is a need for legislative initiative to enact freedom of information provisions and address the constraints imposed by legislation enacted to enforce public respect for secrecy. The working papers and the workshop discussions in Chile, Malaysia and Taiwan highlighted this problem.

3. *A high degree of autonomy of the military.* The military in many developing countries enjoys a high degree of political influence and autonomy in many respects. On the other hand the military is reluctant to admit that any serious contribution can be made in defence matters by outside expertise and it distrusts civilians. As a result it rarely participates in public debate or is questioned on security matters. Public indifference on defence issues is encouraged by a common assumption that military professionals have the best and final judgment on security questions. This is a question of lack of political development, which is a long-term process. The working papers and the workshop discussions in Chile and Taiwan highlighted this problem.

4. *Lack of a tradition of transparency.* In many countries the norms of public access to information are underdeveloped because of a traditional lack of transparency in the society, which serves the purposes of the governing elite. Such countries and societies tend to have strong paternalistic belief systems. The public is not seen as being competent to understand or interpret decisions, there is a likelihood of misinterpretation, or there is simply no need or reason to inform the public. Countries that do not have strong democratic foundations are unable to produce a civil society that is assertive enough in expressing its right to information. Transparency is avoided by political elites which are concerned with consolidating their hold on the instruments of power. The working papers and the workshop discussions in Chile, Greece, Malaysia, Poland and Taiwan highlighted this problem.

5. *Ambiguity in the law.* The civil and military bureaucracies consider it safer from their career perspectives to interpret confidentiality, if the law makes this possible, broadly rather than narrowly. Officials may hesitate to make public policies or decision-making processes which are inadequately documented or internally contested, for which the rationale may be publicly criticized or which could cause embarrassment to the government. The working papers and the workshop discussions in Greece, Malaysia and Taiwan highlighted this problem.

6. *Commercial interests and lobbying.* Processes which are opaque or at best ambiguous can be manipulated by industrial lobbies. Commercial confidentiality creates opportunities for the defence industry to subsidize loss-making civil production lines, which are open to commercial competition and thus operate on tight profit margins, from the profits gained from the defence production line.[97] Confidentiality is promoted as a part of industrial lobbying because of the opportunities for gain that it creates. Discussions in Greece, Poland, South Africa and Taiwan indicated the presence of such attitudes.

7. *Bureaucratic behaviour.* Bureaucracies are often characterized by a culture of caution, secrecy and privilege in access to information. This attitude is habit-forming and leads to work methods that accept inertia and discourage information exchange with the public. Under-resourced public offices are often overwhelmed by the workload of processing information, and this can become a barrier in itself. Among the mutually reinforcing characteristics of bureaucratic tribalism are: (*a*) the assumption that control is exercised through a perception of competence, and therefore public criticism must be avoided by protecting information; and (*b*) the fact that in most countries absolute discretion is given to the executive to handle the secret affairs of the state. Such discretionary powers often lead to the misuse of official and legal provisions by bureaucrats in order to avoid accountability by classifying documents and discouraging public access even to low-level information. The working papers and the workshop discussions in Greece, Malaysia, Poland and Taiwan highlighted this problem.

8. *Weak democratic norms.* Legislators have the duty to monitor defence decisions on behalf of their electorates. However, they may be more concerned with their own careers or commercially lucrative issues. Politicians do not wish to be seen as overly critical of the military, particularly in countries under international sanctions or where a heightened sense of national security is embedded in the country's culture and history. In particular, politicians have a strong resistance to improvements to the legal framework for public accountability. The working papers and the workshop discussions in Chile, Greece, Malaysia, Poland and Taiwan highlighted this problem.

The effects of secrecy on a decision-making process can be twofold: first, it can lead to apprehensions on the part of other countries in the region, leading to an action–reaction spiral of arms procurement; and, second, it can allow corruption, fraud and abuse to creep into the system, which can encourage corporate interests to promote secrecy even further, thus leading to a vicious circle. The argument that public accountability in arms procurement is detrimental to national security because it implies transparency neglects the need to prevent abuse of power in policy making.

The negative effects of lack of accountability are equally important. It can, for instance, lead to unverified threat assessments being generated and conse-

[97] Author's discussions with Shazia Rafi, Secretary General, Parliamentarians for Global Action, New York, Apr. 1999.

quently needs for military equipment being exaggerated, which in turn can generate apprehensions and insecurity in neighbouring countries. Public understanding of the decision-making processes will enhance public confidence, and professional scrutiny by agencies other than those which have an interest in the decisions will benefit the military's decision-making capabilities in the ultimate analysis.

Democratic oversight of the military sector would, however, address only a small element of the larger problem—building up awareness in the society of citizens' fundamental right to know how the state is planning and applying policies for their security.

VII. Recommendations for the future

This project has revealed that questions still remain to be investigated, such as whether public accountability in security policy making serves the interests of consolidating peace; whether the requirements of public accountability only involve broadening public and parliamentary debate on arms procurement decision making; and how the public interest can effectively influence security policy making. It has suggested a method for developing restraint in arms procurement which could be more acceptable to the national defence opinion makers in various arms-procuring countries than conventional arms control initiatives, which are seen as being driven by the West. Arms procurement restraints combined with diplomatic initiatives for regional peace-building frameworks would also have greater durability against the criticism that military capability is the only guarantee of national security.

Among the elements that ensure that the military plays its proper role in a democratic society are: (*a*) the existence of proper constitutional and legislative structures with clearly defined responsibilities for the executive and legislative branches and a system of checks and balances; (*b*) coordination between foreign and security policy-making structures and processes, the primary role being played by the former in formulating a country's external policies; (*c*) a clear primacy of civilians in the ministry of defence, the military being ultimately accountable to the democratically elected representatives of the public; (*d*) substantive parliamentary oversight involving members of parliament trained in the techniques for and the responsibilities of holding the military authority accountable; (*e*) the presence of expert professional staff in national parliaments to keep the members fully informed on key security issues and related data; (*f*) the development of a cadre of security policy experts in the public domain, specializing in a range of security issues in order to generate public debate; (*g*) statutory audit structures to prevent corruption, fraud, abuse and neglect of public resources by the military, which remain unknown to the public because of military confidentiality; (*h*) transparency in the defence budget-making process in order to prevent the military's threat perceptions being driven by interest groups; (*i*) training and education in the armed forces about the role of the military in democratic society, including respect for human and civil rights; (*j*) a fair and effective military justice system that enforces established standards of conduct and discipline; and (*k*) an open and informed national debate preceding major decisions on national security and military matters. The commitment of armed force outside national borders should require broader endorsement by elected representatives.

The critical task, therefore, is to harness the opportunities presented by the present wave of democratization in order to address shortcomings in the public accountability of security policy decision making. Countries in all regions of the world have a role in managing domestic security and encouraging regional security in a democratic manner.

Quite often the criticism made by the military that civilian elected representatives do not sufficiently understand security rationales and technical requirements overlooks one essential element. Democratic control of the military does not imply that the elected representatives are necessarily better decision makers in security matters than the military, but they represent the popular will expressed through due constitutional process. The responsibility of the military has to be exercised through the elected representatives of the public.

A future research agenda

What can and should the international research community do to address such shortcomings in national security policy-making processes? Security must be seen in regional, international and human terms and in terms broader than conventional military security. The institutionalization of democratic oversight of security policy making would give an enduring quality to diplomatic and political alternatives to reliance on the military for security.

Structures of governance that should be examined in terms of the relationship with and accountability of armed forces are: the executive branch; the legislative branch; statutory audit bodies; the judicial system; and special constitutional authorities or commissions set up to carry out other public oversight functions.

Issues of democratic control and oversight that should be studied include: defence and security policy-making processes; the formulation of threat perceptions; public information; the intelligence and security services; financial planning and budget questions; defence industrial questions (where applicable); arms procurement processes; and human rights and juridical questions.

Each of these issues should be examined in the context of: (*a*) constitutional provisions; (*b*) organizational aspects; and (*c*) functional methods.

Constitutional provisions

Four essential aspects are: (*a*) the existence of a proper constitutional and legislative framework with clearly defined responsibilities for the executive and legislative branches and a transparent system of checks and balances applied by the legislature and statutory bodies; (*b*) the primacy of civilians in staffing the ministries of defence and constitutional provisions to ensure that senior military leaders are accountable to the elected representatives of the public; (*c*) a legislative review process to examine the legitimacy of military secrecy provisions in order to prevent misuse of confidentiality; and (*d*) ways of promoting informed national debate about the requirement for the elected representatives to monitor and scrutinize the major decisions on national security issues and the country's armed forces.

Organizational aspects

Barriers need to be identified in the following areas: (*a*) mechanisms for coordination between the foreign and security policy-making structures, the primary role being played by the former in formulating a country's external security policy; (*b*) methods of parliamentary oversight and the information available to members of parliament who are able and prepared to exercise the responsibility of holding the military authority accountable; (*c*) the availability of independent expert professional staff in parliaments or access to expertise; (*d*) the training and development of a cadre of defence policy experts in-country specializing in a range of defence-related issues, generating public interest in oversight functions and providing the multidisciplinary expertise needed to facilitate statutory audit functions; and (*e*) the availability of statutory audit structures to prevent corruption, fraud or abuse of public resources by the military.

Functional methods

Barriers and opportunities need to be identified in the following areas: (*a*) ways of encouraging transparency in defence budgets and accountability in budget-making processes to help the public judge the military's threat perceptions and financial demands; (*b*) ways of encouraging confidence-building measures such as regional codes of conduct on major conventional arms procurement decisions and arms procurement expenditures; (*c*) ways of encouraging accountability in arms procurement decision making and the responsiveness of the military to the information requirements of democratic oversight; and (*d*) the functions of agents and brokers in the arms procurement process, the methods used by them to exert influence, and the legal framework for checking extra-legal methods of marketing.

Research is going on or about to begin in some of the areas defined above. Independent research needs to be started in other areas as soon as possible. It is important for developing stable security structures as well as for good governance that governments, parliamentarians, the military, industrialists and the public be informed of important shortcomings and ways of overcoming them.

VIII. Conclusions

The public has a right as well as an obligation to participate in the security debate on major decisions made on its behalf if democracy is to work. Public accountability is facilitated if the legislature has access to independent experts. This will help to avoid conflicts of interest and organizational bias from being reflected in the executive's recommendations. Such expertise cannot develop if the public is denied essential information relating to defence management and policy making. A dynamic process of public accountability is not only cost-effective; it will produce better decisions.

This project has thrown light on some aspects of the question of civil–military relations. In the course of the research for the project and in the discussions at the workshops on which the results are based it has been assumed that popularly elected civilian leaders and civilian bureaucracies in national ministries of defence would promote democratic governance and that civilian officials in defence ministries are better able to harmonize the broader interests of society with those of national security than officials recruited from the military. These assumptions need to be validated.

The research in most of the countries in the course of this project indicates that only a small number of persons and institutions are prepared politically and intellectually to take on the responsibilities of national security planning that balances arms procurement requirements with broader public priorities. Decision making in the security area is by and large in the hands of the few and the decisions of the military are usually insulated from public scrutiny and accountability. Threat perceptions are manipulated to emphasize the military's decision-making autonomy in its areas of responsibility.

Even in functioning democracies some basic lacunae remain in the administration of security policy, and this is reflected in a country's external security relations. The concept of good governance when applied to the security sector at a minimum requires that the elected representatives of the public who are not in the executive branch have the possibility to scrutinize national security policies, defence budgets and arms procurement decisions in the context of comprehensive security and broader societal priorities.

Annexe A. Research questions

The workshop contributors were asked to highlight issues unique to their country. The instructions and research questions listed below were intended to assist them in preparing material which would facilitate a comparative analysis of national arms procurement decision-making processes.

The questions are arranged according to four themes around which the research was conducted: (*a*) military and politico-security issues; (*b*) defence budgets, financial planning and audit; (*c*) techno-industrial issues; and (*d*) organizational behaviour and public-interest issues. Some questions are deliberately repeated in the different themes so as to ensure that the varying perspectives of the contributors, who represent many different academic and professional disciplines and backgrounds, are reflected.

Contributors were asked to base their papers on strong empirical evidence and published data, but they were also encouraged to draw on their own experience and first-hand knowledge in refining their analyses.

Military and politico-security issues

Effects of security threats and operational doctrines on force planning

Discuss the effects of threat perceptions, security concepts and operational doctrines on force planning. How are military technologies tailored to the requirements of developing balanced force structures in terms of intermediate and long-term planning profiles? This topic should be addressed by someone with military experience.

Threat assessment

Discuss the methods and processes used for carrying out threat assessments, identification of strategic objectives, prioritization, and implementation and review of national security policies and alternatives. Describe and examine the efficacy of the arms procurement process as it develops from security policies into arms procurement plans and military capabilities.

Long-term forecasting

Examine the types of methodological research carried out on long-term forecasting for the development of balanced force structures. Discuss the methods of carrying out force structure analyses and examine such aspects as: (*a*) sequential analyses of operational scenarios; (*b*) the evolution of operational concepts; (*c*) the integration of service-specific threat analyses into defence force analyses; (*d*) operational and technical assessments of alternative systems; (*e*) estimates of resource availability; (*f*) budget simulation; (*g*) balancing defence plans with the available resources; and (*h*) balancing resource levels with required military capabilities.

Defence White Papers

Analyse the content of long-term planning guidelines or White Papers on national defence policy and the sequence of their evolution and development. Do such guidelines contribute to creating a comprehensive framework for policy planning and implementation, or to incrementalism and ad hoc accommodation? In the absence of a defence White Paper or long-term guidelines on national security, discuss any drawbacks experienced in equipment procurement prioritization. Does the absence of a defence White Paper allow non-defence factors to influence or inhibit monitoring of long-term defence planning and limit comprehensive analyses?

Procurement budgets and external threats

Do changes in procurement budgets reflect an increase in perceived external threats or vice versa? Or are changes in procurement budgets related to other factors? Is there a process for examining alternatives to procurement decisions that are made? Give examples if possible.

Responses to emergent military threats

Examine the criteria and planning considerations for the development of balanced force structures for meeting conventional threats. Discuss the following: (*a*) arms procurement responses to the mobilization requirements of emergent conventional military threats, low-intensity threats, small-scale conflicts in peacetime or other commitments such as UN operations; and (*b*) the effects of recent conflicts, other political/ military factors or technological changes that could affect procurement planning.

Constraints on arms procurement planning

Discuss the types and level of constraints on designing desired force structures or on arms procurement planning. Examples of such constraints include: (*a*) budgetary; (*b*) political (international/domestic); (*c*) arms or export control-related; (*d*) human resource-related; (*e*) technological or domestic industry-related; and (*f*) constitutional.

Political leadership and arms procurement planning

Examine the relative influence and control of civil and political leadership over arms procurement plans. To what extent do political guidelines, force design parameters and defence commissions contribute to developing a balanced force planning process? How do the security planning processes lend themselves to public accountability and to addressing dissenting opinion?

Influence of foreign and security policies on arms procurement

Examine the influence of foreign and security policies on arms procurement decision making. Discuss the following: (*a*) domestic arms procurement processes in relation to the country's position on international arms control initiatives; and (*b*) the impact of international technology export controls on the selection of arms supply sources. An expert in foreign and security policies or export control should address this topic.

Relationship to technology control regimes

Assess the country's relative position in technology control regimes and the level of acceptance of the export administration policies of major arms suppliers. Examine the experience of: (*a*) transferring generic technologies and manufacturing 'know-how'; and (*b*) developing 'know-why' capacities to enhance technological self-reliance.

Commitment to international arms control initiatives

Discuss the perceptions of various actors in the arms procurement decision-making process regarding national obligations towards international conventional arms control and transparency initiatives such as the UN Register of Conventional Arms. How are the relevant actors informed of continuing developments in international arms control discussions and related national commitments?

Risks and effects of export controls and embargoes

Examine the methods for political evaluation of the effects of export controls and UN or suppliers' embargoes. Analyse factors considered in decisions about procurement from foreign suppliers and the criteria governing the choice between suppliers. Discuss a formal or optimal model for the procurement of equipment and major conventional weapons from foreign sources with reference to joint ventures as well as direct 'off-the-shelf' imports.

Foreign supply vulnerability and risk assessment

Discuss the criteria for determining foreign supply vulnerability and acceptable levels of military/political risk in procurement policy. Analyse: (*a*) methods of risk assessment, including responses to disruptions in foreign supply; and (*b*) substitutability and alternative supply sources. What factors and actors are most important to this analysis?

Technology: isolation vs participation

Discuss the implications of technological isolation as opposed to participation in international technology transfer. Examine possible approaches to: (*a*) technology-related confidence building; (*b*) reducing problems in integrating with international science and technology initiatives; and (*c*) facilitating access to technology and learning.

National security, military security and military capability objectives

Discuss the perspectives of different actors in the arms procurement process concerning the relationship between national security, military security and military capability objectives. Examine the relevance of accountability and transparency in rationalizing arms procurement, inducing regional confidence and security, and restraining the use of extra-constitutional influences. Discuss ways of harmonizing the expectations of transparency with the military's legitimate need for secrecy. An expert on security issues or from the military should address this topic.

Conflicting security objectives

From the perspective of military, political and socio-economic development priorities, discuss the different interpretations of the participants in the procurement process of the broader objectives of national security, military security and military capability. Analyse possible approaches that could harmonize such conflicting interpretations.

Effects of public accountability on the arms procurement process

Examine the assumption that higher levels of public accountability in the arms procurement process could help to improve the quality of analysis and impede the use of extra-constitutional procurement methods that lead to delays, poorer performance or cost overruns in arms procurement programmes. Also present an opposing viewpoint.

Transparency in defence budgets and accountability in arms procurement

International arms control initiatives assume that transparency in military expenditure is a suitable means of promoting restraint of military build-ups and preventing the diversion of scarce national resources to the military. Compare the effectiveness of transparency in defence budgets with accountability in arms procurement plans as elements of arms control initiatives in terms of their measurability, verifiability and confidence-building value.

Security implications of transparency

What kinds of action, plan or policy relating to arms procurement could be discussed transparently in keeping with the legitimate requirements of military confidentiality? Analyse the implications of transparency for military security in relation to its application in regional confidence-building measures.

UN General Assembly Resolution 46/36 L

Discuss the implications of the transparency levels outlined in UN General Assembly Resolution 46/36 L for requirements of military confidentiality in relation to: (*a*) military holdings; (*b*) domestic arms production; and (*c*) arms procurement through foreign sources.

Organizational behaviour resisting public accountability

Analyse the organizational behaviour of military bureaucracies and factors contributing to their resistance to public accountability or legislative oversight.

The determinants of recipient dependence and their effects on autonomy

Arms procurement policies and practices have to a large extent been determined by predominant supplier–recipient relationships. During the cold war regional political and strategic necessity led to relationships of dependence. Examine the determinants of recipient dependence on a single or predominant arms supplier.

Determinants of recipient dependence

The determinants of recipient dependence on a single or predominant arms supplier could include the following aspects: (*a*) the relationship of threat perception and strategic support; (*b*) the degree of self-sufficiency; (*c*) the ability to increase domestic arms production; (*d*) the effects of diversification and availability of alternative suppliers; and (*e*) the domestic capacities for training, maintenance and availability of spare parts. Examine the consequences and effects of such dependence on political autonomy and foreign policy; domestic policy; strategic advantages or limitations; military–technological self-reliance; operational autonomy during armed conflict; and the opportunity costs of discontinuity in arms supply relationships.

Implications of financial concessions from a single or predominant arms supplier

Different modes of payment for arms could include: (*a*) grants for arms transfers; (*b*) military aid; (*c*) credit or cash sales; and (*d*) offsets or barter. While predominant suppliers might subsidize the procurement budgets of recipients, relationships of dependence can create distortions in long-term defence planning and capacity building. Concessionary financial terms restrict options to the supplier's major weapon platforms which, more often than not, are optimized for the recipient's requirements. It may be cheaper and more convenient to buy off-the-shelf equipment when domestic production is limited by national technical infrastructure or other considerations.

Strategies and countermeasures against recipient dependence

Examine the strategies and countermeasures against the development of recipient dependence in arms transfers, licensed production, and joint R&D and co-production projects.

Effects of arms dependence relationships

Through a specific case study, discuss: (*a*) the political and strategic necessity leading to the development of a relationship of dependence; (*b*) the influence of supplier capacities on the needs of the recipient; (*c*) the effect on public debate and legislative oversight; and (*d*) the effects of a large inventory of equipment from a predominant supplier on the military's operational autonomy in the recipient country.

Defence budgets, financial planning and audit

Budget planning

Examine the defence budget planning process and the influence of cost and the source of supply on the selection of weapon systems. Review the methodologies for procurement pricing negotiations, offset mechanisms and the establishment of priorities, and tendering and contracting methods. This topic should be addressed by an economist or an expert in international financial negotiations.

Long-term financial planning in defence budgeting

Discuss the process of long-term financial planning in defence budgeting. Examine the linkages between strategic and operational decisions and arms procurement budgeting necessary to achieve a given set of objectives. Is the arms procurement budget process integrated and mission-specific or is it programme-specific? Does the defence budget allocate funds separately for different services and agencies?

Methods of defence budgeting

Discuss different methodologies for defence budget planning. If budgeting practices are based on foreign models, examine the internal and external review processes and modifications that are introduced. Are the guiding principles for arms procurement based on monetary ceilings derived from national budgeting or are the equipment ceilings based on threat perceptions? As procurement budgeting requires long-term and multi-year allocations, how does parliamentary review of the annual national budget harmonize with long-term arms procurement commitments?

Cost assessment and price negotiating methods

Describe the elements of cost assessment and the composition of the price negotiating body in arms procurement from the state and private sectors, and foreign sources, including both direct purchases and cooperative projects. Examine the level and range of expertise available to the price negotiating bodies in carrying out sub-optimal planning.[1] Discuss the interaction between military and commercial costing and accounting practices.

Contracting procedures

Are contracting procedures standardized? If so, when and how do they operate? Describe contracting practices and alternatives to fixed-price contracts, cost-plus contracts or any other methods used. Are the contracting procedures and guidelines available to the public?

Offset policies

Analyse government policies in seeking offsets against arms procurement, if any. Discuss the offset policy's aims, strategies, priorities and characteristics, the coordinating agency involved and the methods used for implementing the offset policy. Examine the various methods of evaluation of products/services, the degree of compensation, the minimum size of agreements, and so on. Are offsets flexible, formalized, mandatory or written into government arms procurement regulations? Do offsets prioritize a technological approach (seeking access to specified technical capabilities), a market approach (primarily evaluating commercial prospects) or security considerations? Dis-

[1] Sub-optimization in the language of systems analysis implies breaking up decision making into component parts or sub-problems. Analysis and decision making are carried out in relation to different aspects of the problem in order to find optimum solutions. By analysing smaller sub-problems, greater attention can be paid to detail.

ncial or trading organizations in facilitating offsets and the
policies.

onomic resource allocations and arms procurement bud-
nodels that incorporate data on technology costing and
g manufacturing design costs. Assess recurring hardware
procurement costs, programme costs, life-cycle costs, and

gies and guiding principles used for financial evaluation of
ects by the ministries of finance and foreign affairs in relation
orthiness; (*b*) assessment of international financial support in
or direct funding; (*c*) evaluation of the impact of exchange rate
ial risk, and the value of offsets; and (*d*) the influence of inter-
titutions—for example, the International Monetary Fund (IMF)
—on defence budgeting.

countability in defence budgeting

ems with and impediments to introducing accountability and trans-
e budgeting. What are the acceptable thresholds for transparency in
ires that would be in keeping with the legitimate interests of military

iew

financial review process. Discuss the methodologies for assessing finan-
requirements and alternate resource levels through cost and operational
simulation of a given set of defence force alternatives.

g arms procurement with national socio-economic imperatives

c concerns the difficulty in balancing arms procurement with national socio-
c imperatives. Identify strategies for harmonizing the broader objectives of
security with technology acquisition from domestic R&D or foreign sources,
view to developing the national technology base. This topic should be addressed
conomist or a sociologist in the academic sphere or the national planning sector.

action of the military and economic development sectors

nology-intensive investments in the military sector could have useful applications
ational economic development and vice versa. For example, benefits in the areas of
mmunications and surveillance, advanced materials, marine technology, and signal
ocessing and sensors could be derived. Examine the institutionalization of structures
acilitating this process and the level of influence and interaction among policy makers
nd officials in the military and economic development sectors.

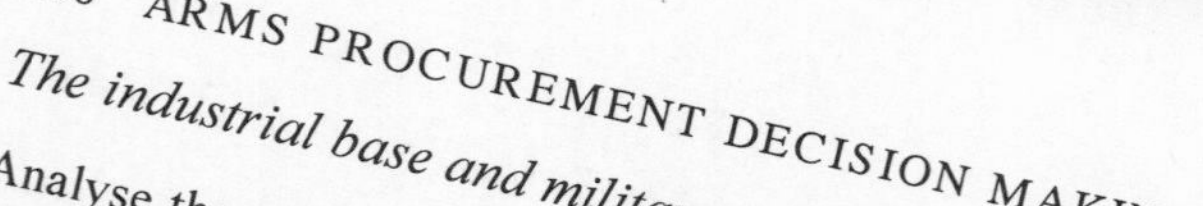

The industrial base and military technology

Analyse the assumption that, owing to escalating weapon system development costs and the accelerating pace of changes in military technologies, advanced weapon technology projects cannot be initiated by defence R&D alone. A number of studies have also indicated that military technologies increasingly appear to depend on advances in civil applications. Discuss the contention that a strong national industrial base is more conducive to developing military technology than vice versa.

Spin-off vs spin-on effects

Examine the assumption that technology transfer through licensed manufacture contributes to the national technology base and can have spin-off benefits for socio-economic development in the context of a case study and in comparison with other forms of technology transfer. Another case study could examine the relative spin-on effects of technology transferred from the civil to the military sector.

Effects of the military sector on the civil industrial sector

Examine the strategies for strengthening the civil industrial sector through the absorption of R&D, manufacturing and technical skills from the military sector. Discuss the components of civil–military integration strategies in terms of: (*a*) integration of R&D to foster dual-use technologies/processes critical to defence and techno-economic competitiveness; (*b*) integration of engineering, manufacturing and logistical support for cross-fertilization and efficient allocation of resources; (*c*) a shift towards flexible manufacturing and methods to increase inputs of commercially competitive technologies and components for military and civil products as well as production processes; and (*d*) balancing performance requirements with cost considerations.

Military auditing

Discuss alternative methodologies for military auditing in terms of the performance, operability and serviceability of the selected system.

Comparative review of arms procurement policies and practices

Examine the processes for reviewing arms procurement policies and for comparing decisions with practices. Discuss the relative merits of auditing the process as a whole and auditing specific procurement actions.

Performance auditing

Examine the methodologies for auditing costs, performance and serviceability against the initial objectives of arms procurement plans. Discuss the criteria for measuring efficient procurement and methods of testing and evaluating the criteria.

Arms procurement budget design

Discuss the arms procurement budget design in terms of its objectives. Is it integrated so as to indicate costs of specific military functions, such as air defence, surveillance,

erely divide up allocations by conventional cost heads
ipment, and operations and maintenance?

e evaluation

d if any evaluation of weapon systems and their pro-
systems have been introduced into service.

ance auditing is the quality and availability of data on
ted decisions. Analyse the availability and quality of
ry sources that are available to the public.

s

and a national arms industry

pment modernization, building a national arms industry,
l capabilities on national arms procurement policies and
ne from the perspective of a defence production organ-

apacities

chnological trends that seem to be emerging include:
of weapon systems; (*b*) an increase in development time
rising costs; and (*d*) improvements in overall quality and
omic trends indicate reductions in demand and increased
. Such indicators would suggest that military–civil con-
operation are likely, while the need to maintain com-
nced against the competing priorities of military self-
ons of national policies for: (*a*) technological self-reliance
chieving crisis independence in arms procurement.[2]

eeds

cal priorities in view of the heavy demands for subsidies
tries. Examine policy planning guidelines and methods for
uirements of domestic technological enhancement through
mport.

ary production processes

onding to the twin problems of escalating costs of defence
duce procurement budgets. For example, discuss: (*a*) the
s and practicalities of integrating civil and military produc-

meet either all or specified elements of its weapon and military hardware
riod of crisis or conflict.

tion processes; (*b*) concurrent engineering; (*c*) flexible manufacturing; (*d*) introduction of commercial practices in the defence sector; (*e*) budgets for the modification or development of equipment as an element of procurement budgets; and (*f*) domestic marketing (manufacturing advanced technology products for civil applications) and international marketing. What weight is given to such cost reduction criteria as: (*a*) modular design for reducing operations and maintenance expenditure; and (*b*) the development of interdisciplinary teams for R&D, manufacturing, marketing, and so on?

Implications of public and private ownership of defence companies

Examine the various criteria used for determining the extent of public/private sector control of defence production. Describe the structure of public/private ownership in defence companies engaged in the production of aerospace systems, ship systems, armament systems, electronics, and miscellaneous products and services. Discuss structural readjustment strategies for privatizing non-critical defence companies or expanding national or international cooperative initiatives such as joint ventures including: (*a*) R&D; (*b*) design and production; and (*c*) marketing collaboration.

Collaboration in technology acquisition

Discuss the implications of co-development, co-production, licensed production, subcontract production and other forms of cooperation in relation to strategies for technology acquisition, building defence industrial capacities and reducing the economic burden of the defence industry.

Joint ventures

Joint ventures are considered to be among the more efficient methods for facilitating technology transfer, as well as skills and resource sharing in the development of complex systems. Discuss the relevant criteria, priorities and types of joint venture or cooperative project. Describe the decision-making process within joint ventures and analyse the motivations governing such collaboration, the problems and benefits envisaged and the methods for managing competing priorities among the collaborators. Discuss military and national technological objectives with respect to joint-venture strategies.

Measuring relative levels of self-sufficiency and import dependence

Analyse the relative shares of imported and indigenously produced components in major weapon systems as a proportion of total arms procurement. This may include consideration of the relative indicators of: (*a*) imported complete systems; (*b*) complete systems produced under licence and the ratio of imported to domestically produced components; and (*c*) systems produced on the basis of indigenous R&D.

Influence of the defence industry on arms procurement

Discuss the level, scope and nature of the influence of the defence industry on the arms procurement decision-making process. What input does the defence industry have in

sely, to what extent do the priorities of the military sec-

tional defence industrial base

t experienced in building a national defence industrial
logy projects are becoming increasingly complex and
nd, examine the changes in industry–government rela-
oversight, forms of international cooperation and other
of constraints on the development of a national defence

hnology assessment (TA) and the selection of equipment.
corporating TA, systems analysis and costing method-
weapon systems. This topic should be examined by indi-
alysis and TA.

nology assessment

used for: (*a*) carrying out systems analysis and TA of
ecasting long-term technological development.

ne methodologies employed for conducting various types of
a typical procurement decision. These could include:
al maintenance assessments; (*b*) field trials; (*c*) assessment
; and (*d*) analysis of financial outlays and contractual offers.
setting credit rates and prices of services, training, the
bsequent technology upgrades, and methods for developing
delivery schedules, including any penalties for delays.

ures

es involving higher-level technology transfer, specify the TA
, provide examples for a comprehensive assessment of joint

sion-making process

alysis is a continuous cycle of formulating the problem, select-
g alternatives, collecting data, building better models, deter-
tiveness or satisfaction, questioning assumptions and data,
erformance, re-examining objectives, opening new alternatives,
limitations in the decision-making process, such as bias, sub-
of preconceived judgements, and so on.

Alternative methods of decision making

Examine various decision-making methods for developing a comprehensive analysis of the views of different experts or specializations. The committee method is one example. The objectives and criteria for measuring the efficacy of different methods should be clearly defined and a comparison of alternative courses of action should be made.

Building public capacities for policy analysis

Discuss the problems of developing competence in arms control and security issues in the society in general and professionalizing legislative oversight of arms procurement.

Trends in weapon systems development from an R&D perspective

New problems in arms development and procurement are being created by trends such as increases in performance and cost, an increasingly competitive market, decreasing development times, shrinking demand for weapons and decreasing military budgets. Weapon producers are resorting to transnational collaboration and other innovative approaches to meet these challenges. Discuss the implications of these changes for national defence R&D policies and the problems being faced by R&D organizations.

Implications of self-reliance for defence policy making

Examine the implications for defence policy making of a policy of technological self-reliance in relation to collaboration or procurement from foreign sources. Discuss the problems in developing cost- and risk-sharing methods, as well as technology linkages for the upgrading and replacement of equipment. Discuss the effects of levels of national competence and capacities for building components or complete major weapon systems ranging from semi-knocked down assembly to co-production.

Problems in developing defence R&D competitiveness

Discuss the problems in developing defence R&D competitiveness, for example: (*a*) the management of priorities between short-term project-specific research and long-term generic R&D; (*b*) protecting core competences; (*c*) greater reliance on continuous prototyping and design for produceability; and (*d*) developing human resources for specialized research.

Management of inter-organizational differences

Discuss how conflicting service and organizational philosophies are dealt with, particularly in the weapon conceptualization and project implementation phases. Examine possible options, for example: (*a*) the integration of scientists and engineers from the armed services into R&D projects and teams; (*b*) methods for keeping abreast of progress/changes relevant to service-specific technologies; and (*c*) the interface between weapon system developments and operational missions.

ɔpment vs the pull of operational requirements

ɔf the push of technology development and the pull of
ıe arms procurement process. Assess the influence of
ɛ media on determining the specifications of equipment
on.

h interaction

ng-term strategic research for future military applications
l research establishments. Examine the weapon develop-
to identifying spin-off and spin-on linkages, providing
ities of and the level of concern shown by the government
ıodies in facilitating long-term strategic research.

ry integration

hat a strategy for civil–military integration could include:
D to promote technologies for both national defence and
; and (*b*) the integration of engineering, manufacturing and
ote cross-fertilization and increased reliance on commercial
iability to reduce lead times and costs. Discuss the methods
tween defence laboratories, universities, industries and inde-
ıtions. Analyse the level of educational skills of personnel
in the state sector in comparison to the private sector. Assess
ıg and retaining qualified personnel in defence R&D labora-

system development process

development approach requires the integration of operational
ɔment, and logistical integration into the overall force structure
ıerations. Discuss the various stages and phases of the weapon
process and the roles of the interacting agencies and
ıould include an assessment of: (*a*) project identification and
eds; (*b*) concept exploration; (*c*) research and mathematical
ıratory development, preparation and evaluation of projects;
valuation of draft design and technical projects; (*f*) development,
ion and validation of prototypes; (*g*) full-scale engineering
ɛcisions to transfer to series production or a prototyping-plus
ɔduction, testing and deployment.

fence R&D establishments

sed for measuring and monitoring the industrial and human produc-
&D establishments. Examine the effects of innovations in enhancing

ı weapon system to the prototype stage, then successfully demonstrating and proof-
ınuing to the production stage. However, improvements to the system's components

productivity, including the delegation of decision-making authority to laboratory and project managers, and the development of a payment channel to the developer or to laboratories through a user service. Discuss methods of maintaining efficiencies in the user–developer relationship. Does the process include a periodical review of specific projects or of the entire defence R&D organization?

The private sector in the national defence industrial base

Examine the different concepts of and views concerning the composition of the defence industrial base. Analyse the level of national and international private-sector companies' participation in the arms procurement process, and their interaction and influence in decision-making processes in countries with private-sector involvement in military production. This topic should be addressed from the perspective of the private sector.

Defence industrial, technological and economic bases

Discuss the different interpretations of the defence industrial base, defence technological base and defence economic base. Discuss the methods for the measurement of techno-industrial compatibility and productivity in public and private enterprises in the defence industrial base.

Types of cooperative venture

Other than arms procurement-related cooperative projects, the private-sector defence industry in a number of countries is characterized by acquisitions, mergers and cooperative ventures with both domestic and foreign companies. Different business rationales offer an explanation of the need for new industrial ventures, ranging from lowering costs, risks and threats, and increasing capacities, markets, competitiveness, and efficiencies, to technology transfer. Discuss the types and objectives of cooperative ventures that private-sector military enterprises could have. Include an account of external influences on autonomy, problems in procurement plans and the legislative provisions available to the procurement agencies.

Problems of civil–military integration in defence production

Discuss the problems of civil–military integration in defence production. Are military and civil R&D mutually supportive of the development of tangible and intangible spin-offs and spin-ons? Examine the structural compatibility of the two sectors in terms of such factors as: (*a*) the application of dual-use manufacturing processes; (*b*) organizing combined research teams for specific projects; (*c*) organizing cooperative research associations in civil companies with specific military laboratories, projects or industries; and (*d*) developing cross-fertilization through a specialization-oriented network of military/government, industry and technical expertise.

Influence of national science and technology initiatives on military technology

Discuss the influence of national science and technology initiatives on military technology development. Analyse the effects of developments in military technology pol-

acities and productivity. Analyse structures facilitating
iencies, modes of control and public accessibility.

mpatible industrial cultures

ence industry seek to encourage private partnerships to
well as facilitate spin-offs and spin-ons. This public–
industry is hampered by technological differences in the
fence and civil industry as well as by their approaches to
ies in the state or state-supported sector are less exposed
access to funds and a lack of domestic competition. In
constantly needs to increase productivity and to adapt to
ercial competition. Discuss the problems of conversion in
tible industrial–technological paradigms and cultures.

ility in the national manufacturing sector

ntifying core and critical competencies in the defence indus-
pediments to the convertibility of the national manufacturing
es in terms of: (*a*) the demand for surge capacities; (*b*) the
uction at mobilization levels; and (*c*) the problems in
cific designs and specifications.

conversion strategies

privatization or conversion from military to civil production
structures, policy studies, the application of human resources,
d various techniques employed. Discuss the criteria, priorities
r impeding conversion initiatives.

ctor in arms procurement decision making

private sector in arms procurement decision making. How can
ribute to the development of more advanced and efficient mili-
ities? What are the impediments and potential problems that
contributions?

ptions of accountability and transparency

ncreased involvement of the private sector hinders or contributes
lity and transparency in the arms procurement process.

behaviour and public-interest issues

rations and elite motivations concerning equipment and

ce of domestic considerations and elite motivations on the choice of
rces of supply in arms procurement decision making. Examine the
res within the military sector, the bureaucracy, inter-service relations,

R&D organizations and the defence industry in relation to the arms procurement process. Ideally, this topic should be addressed by a social scientist or a media expert.

Information flow and decision making

Examine the institutionalization of feedback and the horizontal flow of information in the arms procurement process. Discuss the formal and informal modes of providing information for optimal decision making through the stages of concept definition, applied research, exploratory development and production.

Information assessment

Examine the process of selection, evaluation and acceptance of new information or assessments. Analyse the information-processing behaviour and degree of group conformity displayed by actors in the arms procurement process. Examine the relative levels of institutionalization and the influence of personal relationships in information flow. Discuss attitudes to new policies, dissonant information and re-evaluation methods.

Defining and coordinating military needs

Is the procedure for defining military needs and requirements for arms procurement separate and specific to the different armed services or is it related to general military security roles? How does the process coordinate the requirements of multi-service applications such as logistics, command, control, communications and intelligence, and space-based communications? Does it concentrate on allocation of resources and procurement and management of weapon systems or is there an overall rationale for building techno-industrial or operational capacities? Discuss the contribution of the process of integrating broader techno-industrial capabilities and military requirements.

Political culture and arms procurement decisions

Analyse the structural characteristics of the prevailing political culture and its influence on civil–military relations in general and arms procurement decisions in particular. How is the divergence between perceptions of national interest and the traditional role of the military sector reconciled? Discuss responses to divergent pulls, pressures and the influence of competing interests, as well as the constraints of bureaucratic and factional politics on the arms procurement process. What kind of influence could a defence White Paper have on the politics of decision making? What other kinds of structures or mechanisms are employed to achieve policy coordination?

Availability of technical and multi-disciplinary skills

Analyse the levels of scientific and technical skills and multi-disciplinary expertise available within different agencies and departments of the arms procurement decision-making structure. Does such expertise facilitate the making of sub-optimal studies, decisions and the cross-fertilization of ideas?

al priorities

ıethods for harmonizing intra- and inter-organizational nfluenced by or given more weight by a specific organ- view of the process reduce communication barriers, prove forecasting capacities?

ies and broader public interests

onization between organizational interests, such as those ublic interests and national policies. Discuss the concerns further the objectives of public accountability in public nd arms procurement in particular.

regulations

ormation, organizational politics, bureaucratic inertia and ajor factors contributing to dissonance in decision-making iality is needed concerning technical specifications or plans, esirable from the perspective of accountability. Discuss the egulations and legislation on public accountability.

esses, good governance and accountability

ization of decision-making processes based on the principles scuss the problems, apprehensions and barriers in building interest, transparency and accountability. This topic should be litician or a constitutional expert.

ountability and transparency

of various national interest groups which support or oppose nd transparency.

xpertise available to parliamentary committees

of expert advice and information available to parliamentary n monitoring arms procurement. To what extent do the members ticipate in debates in the legislature on defence policy making or anning? Discuss the role and influence of legislative oversight in rement planning.

overseas development aid and military expenditure

es and responses within legislative and administrative bodies to the eloped by international aid agencies between overseas development pient's level of military expenditure. How is public concern about itary expenditure or arms procurement on different levels of the ed and how does it influence the decision-making process?

Objectives of national security: perceptions of the legislature

Discuss the perceptions of different segments of the legislature concerning the broader objectives of national security as distinct from military security and military capability objectives. The manner in which public-interest priorities and public policy making can be harmonized should be examined in the context of sensitive issues such as arms procurement.

Effects of confidentiality on arms procurement policies

Analyse the effects of confidentiality on arms procurement policies, on procedures and guidelines that could enhance the influence of arms dealers and on the extra-legal dimensions of arms procurement. Are the decisions constitutionally valid?

Building public competence in the national security arena

One of the handicaps in promoting public accountability and debate in arms procurement processes is a lack of adequate capacities to engage the decision makers in an objective professional debate combined with insufficient levels of public awareness. Discuss alternative methods of building capacities and competence in society at large concerning national security and arms control issues.

Public concerns relating to the arms procurement process

Assume that the public interest regarding arms procurement relates to such concerns as: (*a*) that arms procured are essential from a national security perspective; (*b*) that governments pay a fair price that is appropriate to national capacities and needs; (*c*) that arms procured meet the expectations of the users; (*d*) that there is accountability in the process and that it is free from waste, fraud or abuse; and (*e*) that there is a legal basis for the decisions and actions. Discuss whether these assumptions are correct and analyse the effects of legislative oversight of the arms procurement process.

Arms procurement and organizational behaviour at the apex level

Discuss the characteristics of arms procurement processes in terms of the organizational structures involved. Are they competitive and do they incorporate a diversity of perspectives, or are they exclusive and insular, indicating a cultural or political bias? This topic should ideally be addressed by a military or civilian expert in public administration or organizational behaviour.

Dominant organizational attitudes and norms

In order to harmonize security policies with public-interest priorities, the arms procurement process needs to be examined in terms of: (*a*) the constitutionality of decision-making practices; (*b*) the levels of technical and analytical skills available for advising decision makers; (*c*) the levels of information flow; and (*d*) public accountability and interaction among various organizations and specialists. Do the dominant organizational attitudes and norms lend themselves to an internal audit of the arms procurement process?

onal performance

while rules and procedures are designed to regulate icy guidelines and to prevent waste, fraud and abuse, and initiative by producing rigidity and delay. How is ıal performance carried out?

pport transparency

ıuences that oppose or support transparency in arms pro- om an organizational behaviour perspective. Analyse the legislative oversight, constitutional and legal provisions, the bureaucracy and the military–scientific community,

nd non-delegation of authority

ety's dominant political cultures on the behaviour of differ- he arms procurement process. Discuss the situation in terms elegation of authority.

in large-scale national processes

d within major national processes, such as the arms procure- ot addressed. This is not because of a lack of innovative ideas because of impediments such as intransigence, resistance to rtia, extra-organizational factors, personal influences, systems eemingly incompatible positions of different interest groups. ents and barriers with those evident in the private sector or with arent and accountable systems.

onal marketing and the media

of marketing organizations and the international media on the erational needs and of threat assessments on procurement xtent are arms procurement requirements driven by long-term nfluenced by new information or organizational priorities?

ement in arms procurement decision making

ınizations have a greater role and influence in arms procurement ıan others? How have they achieved this influence? Provide zations expanding their influence in arms procurement decision

ional decision-making behaviour

es the attitudes, strengths and limitations in developing sub-optimal on-making structures. It should be addressed by a senior sociologist consultant.

Effects of sociological traits on decision-making behaviour

Discuss the effects of sociological traits, characteristics and culture-based codes on bureaucratic, military and political decision-making behaviour.

Effects of factional identities on decision-making politics

Examine the effects of small-group dynamics and factional identities on decision-making politics. Discuss the characteristics of the prevailing bargaining paradigm.

Cultural factors influencing elite behaviour

Discuss the cultural factors influencing the behaviour of the decision-making elite and the dominant psychological predisposition sustaining the inner circles of power. Consider the effects of transparency, public accountability and democratization or bureaucratization on such factors.

Management of dissent

Examine the effects of dissent and its management in the decision-making process.

Influence of different groups and interests

How is influence and power gained or lost in the decision-making process? Examine the levels of influence of different groups and interests.

cts of the working papers*

perceptions,
doctrines in
rms Procure-
orking Paper

at the highest
Defence High
fense develops
Appreciation,
ternal threats,
e those threats
ements for the
trine of dissua-
ch that seeks to
o—requires that
nd combat effi-

s of the decision-
SIPRI Arms Pro-
Project, Working

y in Chile has not
ocratic and partici-
decision-making
egime and the non-
party alliances have
the Congress, espe-
where disagreements
ition process and
e armed forces.

pplying and transpar-
SIPRI Arms Procure-
roject, Working Paper

bate on arms procure-
le owing to the auton-
s, the guaranteed mili-
ited influence of Con-
ntial system gives the
to control the military
ces have no coordinated
ecision-making process
rganizational and techni-

GASPAR, G., 'Military expenditures and parliamentary control: the Chilean case', SIPRI Arms Procurement Decision Making Project, Working Paper no. 64 (1997), 8 pp.

The preventive mechanisms designed by the armed forces during the transition to democracy in order to avoid civilian reprisals have limited the Congress' influence over the defence budget. It does not receive details from the regular budget for personnel and operational costs, and has no say in the budget for arms. Its political composition has also limited opposition and debate since it is dominated by two main coalitions—the government coalition and the centre–right coalition, which traditionally supports the military.

MENESES, E., 'Chilean defence procurement: achieving balance between suppliers', SIPRI Arms Procurement Decision Making Project, Working Paper no. 65 (1997), 16 pp.

The USA's dominance as an arms supplier in the 1950s and 1960s and the 1976 arms embargo resulted in a policy of arms procurement diversification in Chile which is further enhanced by the independent role of the armed forces in arms procurement.

NAVARRO, M., 'The influence of foreign and security policies on arms procurement decision making in Chile', SIPRI Arms Procurement Decision Making Project, Working Paper no. 66 (1997), 13 pp.

Formal coordination between the armed forces and the Foreign Ministry in defence policy and arms procurement is very limited owing to the considerable autonomy enjoyed by the armed forces and independent sources of funding for arms procurement. However, the armed forces do consider general trends in Chile's foreign policy and international engagements when making procurement decisions. The current arms procurement policy aims at achieving a regional strategic balance in the Southern Cone and at finding reliable and diversified sources of procurement.

ts abstracts of the 63 working papers on which the chapters of this book are
d by SIPRI Research Assistants Eva Hagström and Oscar Schlyter.

PATTILLO, G., 'The decision-making process in the acquisition of arms systems: an approach', SIPRI Arms Procurement Decision Making Project, Working Paper no. 67 (1997), 18 pp.

Operating costs are covered in Chile's national budget law, which stipulates that the government's contribution must be at least equal in real terms to the funds received by defence agencies in the 1989 budget. Arms procurement is financed by a special tax that apportions 10 per cent of the net profit on exports of copper and copper by-products. This amount is divided equally between the three branches of the armed services. The system of funding has guaranteed a certain level of resources without political debate. However, it is inefficient since the three services do not all have the same needs.

PORRAS, L., 'The influence of equipment modernizations, building national arms industry, arms export intentions and capabilities on national arms procurement policies and procedures', SIPRI Arms Procurement Decision Making Project, Working Paper no. 68 (1997), 24 pp.

Chile's defence industry has been shaped by the needs of the armed forces. This gives the forces direct influence over strategic industrial developments and gives the industry easy access to defence contracts. The existence of government-owned defence firms and the small market have impeded the establishment of a private-sector defence industry.

ROBLEDO, M., 'Domestic considerations and actors involved in the decision-making process of arms acquisition in Chile, 1990–97', SIPRI Arms Procurement Decision Making Project, Working Paper no. 69 (1998), 19 pp.

During the 'agreed' transition to democracy in Chile the military was granted considerable autonomy—irremovable commanders-in-chief, a guaranteed minimum military budget and special funding for arms procurement. Vast powers were also given to the president, who has a veto over the military budget and arms procurement decisions. However, in order to safeguard civil–military consensus, the role of Congress has been limited to a right to be informed and to initiate new legislation.

THAUBY, F., 'Research and development policies in the Chilean defence industry', SIPRI Arms Procurement Decision Making Project, Working Paper no. 70 (1997), 15 pp.

Chilean arms procurement policies prioritize the international procurement of second-hand combat systems, which are adapted and modernized locally. There is no specific policy or substantial funding for R&D, which is focused on the transformation, modification and maintenance of equipment.

ALIFANTIS, S., 'National defence in the aftermath of the Imia crisis: the concept of "flexible retaliation"', SIPRI Arms Procurement Decision Making Project, Working Paper no. 71 (1998), 7 pp.

The 1996 Imia crisis between Greece and Turkey over control of islands and islets in the Aegean Sea brought forward the need for a restructuring of Greece's defence policy and force structure. Turkey's new strategy of provoking crises and low-intensity conflicts has been difficult to deal with through the traditional Greek military doctrine. The new situation calls for a change to a doctrine based on 'flexible retaliation', which entails responding to Turkish threats on a equal level of intensity, focusing on vulnerable political targets and raising the political costs for Turkey.

DOKOS, T. and TSAKONAS, P., 'Perspectives of different actors in the Greek procurement process', SIPRI Arms Procurement Decision Making Project, Working Paper no. 72 (1998), 10 pp.

The current political system is designed to sustain civilian control over the armed forces. Nevertheless, the military plays a significant role in the arms procurement process through its influence on threat assessment. The other significant actor in the process is the government, through its ultimate responsibility for the preparation of national defence policy. The Parliament, the media and the general public have the least influence, although the former aims to play a greater role.

ement and
s Procure-
rking Paper

ar the USA
rce of for-
pport was
ation of
a con-
hid-
ce's
cause of
ment of the
oday Greece needs
to s of supply, negotiate
mo agreements and support its
dome ence industry.

MELETOPOULOS, M., 'The sociology of national decision-making behaviour', SIPRI Arms Procurement Decision Making Project, Working Paper no. 74 (1998), 15 pp.

Despite many similarities with Western democracies, Greece's socio-political situation is significantly different. Its particular characteristics include a strong communal tradition, extensive patron–client relations between the elected and the electors, the influence of the Orthodox Church, Byzantine traditions and a tradition of rebellion. The result is that decisions are formed primarily by the personal preferences of politicians and their relations with the various actors in the decision-making process. Public opinion on foreign and defence policy is largely shaped by Greece's relations with Turkey.

NARLIS, E. O., 'Arms development and defence R&D growth in Hellenic Republic', SIPRI Arms Procurement Decision Making Project, Working Paper no. 75 [1998], 5 pp.

Hitherto Greece has allocated a limited share of its defence budget to local R&D. However, the Ministry of National Defence recognizes that indigenous R&D can have a important role in improving national defence and should be given higher priority. What is needed is not only a significant increase in the budget for defence R&D but also a well-balanced allocation of resources within the R&D budget. Increased defence industrial cooperation within NATO and the EU is equally important.

VALTADOROS, C., 'The influence of foreign and security policies on Greek arms procurement decision making', SIPRI Arms Procurement Decision Making Project, Working Paper no. 76 (1998), 12 pp.

Membership of NATO and close security cooperation with the USA have enabled Greece's defence industry to import technological know-how. This has not, however, led to defence industrial self-sufficiency, owing to the structure of the defence industry and unfavourable bilateral agreements with arms-supplying countries. The dependence on the USA in terms of equipment and technology limits Greece's freedom in international relations, in particular at the regional level.

SHARIFAH MUNIRAH ALATAS, 'Government–military relations and the role of civil society in arms procurement decision-making processes in Malaysia', SIPRI Arms Procurement Decision Making Project, Working Paper no. 84 (1998), 49 pp.

Pre-colonial indigenous culture in Malaysia centred around the concept of authority, which gave the political leadership final decision-making authority in arms procurement. The influence of civil society is limited because of secrecy regarding military issues and defence budgets, middle-class complacency and a general lack of interest in defence issues.

SITI AZIZAH ABOD, 'Decision making process of arms acquisition in the Ministry of Defence, Malaysia', SIPRI Arms Procurement Decision Making Project, Working Paper no. 85 (1998), 9 pp.

Arms procurement decision making begins with the armed forces defining their equipment requirements in a five-year perspective plan. The plan is then examined by the Economic Planning Unit of the Prime Minister's Department, which determines the allocations for defence. Thereafter the Ministry of Defence prioritizes the requirements and submits the plan to the Cabinet for approval. Arms procurement deals below 5 million ringgits in value are implemented by the Ministry of Defence, others by the Treasury.

BALAKRISHNAN, K. S., 'Examine the institutionalisation of decision-making processes based on the principles of good governance: problems, apprehensions and barriers in building public awareness, public interest, transparency and accountability', SIPRI Arms Procurement Decision Making Project, Working Paper no. 78 (1998), 8 pp.

Although Malaysia has a well-developed democratic system there is still room for improvement, in particular with regard to laws on national security such as the Internal Security Act and the Official Secrets Act. These laws have reduced public awareness of national security issues by curbing the flow of information to the media. The situation has improved for the media, but the lack of public interest in national security issues is still an obstacle for accountability and transparency in government.

MAK, J. N., 'Security perceptions, transparency and confidence-building: an analysis of the Malaysian arms acquisition process', SIPRI Arms Procurement Decision Making Project, Working Paper no. 82 (1997), 22 pp.

The battle against communist insurgents and the colonial legacy have created a political and military culture of secrecy in Malaysia. Transparency in arms procurement is limited by the Internal Security Act and the Official Secrets Act, and public accountability is practically non-existent owing to a lack of interest in defence issues, a weak civil society and the overwhelming dominance of the ruling coalition in Parliament.

BALAKRISHNAN, K. S., 'Arms procurement budget planning process: influence of cost and supply source of alternative systems, procurement negotiations and methodologies, offset policies, contracting process and the issue of transparency', SIPRI Arms Procurement Decision Making Project, Working Paper no. 79 (1998), 12 pp.

Malaysia's financial planning for defence has been more influenced by economic than by other factors, although strategic factors such as threat perception and the changing security scenario play a vital role. In recent years the importance of offsets, in particular technology transfers, has increased in Malaysia's arms procurement plans.

ABDUL RAHMAN ADAM, 'Dynamics of force planning: the Malaysian experience', SIPRI Arms Procurement Decision Making Project, Working Paper no. 77 (1997), 18 pp.

Force restructuring in the 1990s began with a doctrinal shift from counter-insurgency warfare to conventional warfare in 1986. The aim is to strengthen conventional deterrence capability through reorganization of the armed forces, coordination of operations and substantial arms procurement. Priority areas are technology transfers and defence industrial cooperation.

FARIDAH JALIL and NOOR AZIAH Hj. MOHD AWAL, 'Control over decision-making process in arms procurement: Malaysia', SIPRI Arms Procurement Decision Making Project, Working Paper no. 81 (1997), 17 pp.

Arms procurement decisions are made by the Ministry of Defence as part of the formulation of defence policy and are accepted by the government in the Ministry of Defence annual budget. The Parliament also participates in the budget debate, but the dominance of the ruling coalition and the ignorance of senior ministers have reduced its importance. Furthermore, the notion of collective responsibility within the ruling party serves to reduce criticism.

ROBLESS, R., 'Harmonizing arms procurement with national socioeconomic imperatives', SIPRI Arms Procurement Decision Making Project, Working Paper no. 83 (1997), 19 pp.

Malaysia needs to strike a balance between defence spending and other socio-economic needs in order to avoid building up defence at the expense of development. At the same time, defence planning should be long-term and capability-driven since defence investments can contribute to national growth. Defence budgeting in Malaysia consists of five-year plans and an annual budget developed by the Ministry of Defence.

SUKUMARAN, K., 'Defence research and development (R&D) and arms procurement decision making', SIPRI Arms Procurement Decision Making Project, Working Paper no. 86 (1998), 30 pp.

The Malaysian Defence Science and Technology Centre was set up in 1968 to do defence R&D in Malaysia. It has not worked efficiently for a number of reasons. It has a human resources problem, having too few employees with not enough education; its funding has been inadequate; and there are bureaucratic obstacles in the way of its accessing the funds approved for it in the annual budget.

SUPIAN ALI, 'Harmonizing national security with economic and technology development in Malaysia', SIPRI Arms Procurement Decision Making Project, Working Paper no. 87 (1997), 40 pp.

Defence investments can provide employment, provide new technology and stimulate growth. However, in the case of Malaysia none of these effects has been demonstrated. Statistics show that social and economic spending have maintained their proportionate shares and that defence spending has not stimulated growth.

GOLEMBSKI, F., 'Threat perceptions, security concepts, operational doctrines and force planning', SIPRI Arms Procurement Decision Making Project, Working Paper no. 88 (1998), 15 pp.

The Polish defence doctrine recognizes that the main threat to the country stems from political instability in neighbouring countries, organized crime and uncontrolled population flows. However, Polish defence planning is perhaps more influenced by budgetary constraints and membership of NATO. NATO membership will be costly because it will require interoperability, but it may also limit the number of threats to Poland's security.

KOSCIUK, L., 'Process of arms procurement and exports: Polish experiences', SIPRI Arms Procurement Decision Making Project, Working Paper no. 89 (1998), 7 pp.

During the cold war Poland's defence industrial and disarmament policies were largely determined by the Soviet Union. With the end of the cold war Poland gained greater control over these issues but it is still dependent on Russia for spare parts and repairs. Its ambition to join the EU and NATO has limited its choices with regard to both procurement of equipment and foreign policy, and cuts in Polish defence have put the domestic defence industry in a difficult situation.

KROLIK, J. and SZYDLOWSKI, A., 'The alternative methodologies for military audit in terms of the performance, operability and serviceability of weapon systems', SIPRI Arms Procurement Decision Making Project, Working Paper no. 90 (1998), 15 pp.

The Ministry of National Defence has the main responsibility for audit of arms procurement in Poland. However, when a foreign company is involved, procurement is carried out by institutions outside the Ministry and the auditing is limited. Another problem is that auditors often lack the competence to perform their tasks efficiently. Responsibility for audit should be transferred completely to the Ministry of National Defence and the levels of competence required should be formalized.

MESJASZ, C., 'Restructuring of defence industrial, technological and economic bases in Poland, 1990–97', SIPRI Arms Procurement Decision Making Project, Working Paper no. 91 (1998), 23 pp.

The dramatic changes brought by end of the cold war directly affected the Polish defence industry, including the transition to a market economy, NATO enlargement and defence industrial restructuring. In order to deal with these developments Poland plans to increase private ownership in the defence industrial sector, to increase international and civil–military cooperation, both in the arms industry and in the R&D sector, and to focus more on offsets. So far, restructuring has not been as extensive as planned because of economic difficulties, political instability and the lack of a clear strategy.

MISZALSKI, W., 'Alternative procedures for technology assessment and the selection of equipment', SIPRI Arms Procurement Decision Making Project, Working Paper no. 92 (1997), 21 pp.

During the Warsaw Pact era technology assessment of defence equipment was limited to small-scale analyses mainly carried out to justify decisions already taken. The emergence of a more democratic society has improved the situation but there are still no standardized guidelines for technology assessment as part of the arms procurement process. Poland needs to increase levels of expertise among both civilian and military actors in the process and to improve and clarify legislation related to the arms procurement process.

MISZALSKI, W., 'Characteristics of acquisition procedures in terms of the organizational structures involved', SIPRI Arms Procurement Decision Making Project, Working Paper no. 93 (1997), 15 pp.

Since the Public Procurement Act of 1994 was enacted, the Polish public procurement system has been in a state of development. The Ministry of Defence has drafted several documents regarding arms procurement and inter-ministerial coordination. However, the process is too complicated and lacks transparency. The main areas in the Public Procurement Act which need improvement are inter-ministerial coordination and the rules regarding procurement from foreign suppliers.

STACHURA, J., 'Arms procurement decision making in Poland', SIPRI Arms Procurement Decision Making Project, Working Paper no. 94 (1998), 21 pp.

Civilian control over national defence has not been completely implemented despite being generally accepted. After several years of dispute a bill providing for civilian control over defence was passed, and since then the General Staff has gradually lost its authority. Parliamentary oversight of the military is mainly carried out by the Parliamentary Defence Commissions through their role in the budget process. Another important actor is the Highest Chamber of Control, or National Audit Office, which can investigate the activities of government-related institutions.

TARKOWSKI, M., 'Arms procurement decision making: process, pressure groups, inter-elite controversies and choices', SIPRI Arms Procurement Decision Making Project, Working Paper no. 95 (1998), 14 pp.

Poland has no tradition of organized lobbying or pressure groups. However, these forms of influence have gradually emerged since the fall of communism. The two most influential groups are the Polish Industrial Lobby and the trade unions. The period since 1989 has seen a significant growth in the mass media, which also shape decisions regarding defence issues. The most salient topic in the Polish defence debate is that of cooperation between domestic and foreign defence producers, in particular with regard to offset deals.

TARKOWSKI, M., 'Balancing arms procurement with national socio-economic imperatives', SIPRI Arms Procurement Decision Making Project, Working Paper no. 96 (1997), 16 pp.

Since 1989 Poland has tried to improve the position of its defence industry. Defence industry restructuring has entailed changes in ownership with privatization, financial support to key companies, and efforts to concentrate the defence industry in fewer, larger companies. However, development has been slow, mainly because of resistance to privatization on the part of the labour unions and the political elite and a lack of financial resources.

TREJNIS, Z., 'The sociology of national decision-making behaviour', SIPRI Arms Procurement Decision Making Project, Working Paper no. 97 (1998), 19 pp.

During the cold war Polish national defence was governed by the Communist Party and the Supreme Command of the Warsaw Pact. At the end of communist rule the old elite managed to transfer its political influence to economic power. It thus remains a powerful force. Another factor influencing defence decision making in Poland is the lack of laws governing the control of the armed forces, which has led to strains and conflicts between civilian and military decision makers.

WIECZOREK, P. and ZUKROWSKA, K., 'The influence of equipment modernization, building a national arms industry, arms export intentions and capabilities on national arms procurement policies and procedures', SIPRI Arms Procurement Decision Making Project, Working Paper no. 98 (1998), 31 pp.

The current structure of the Polish defence industry is based on traditions from the 1930s. Since the end of the cold war, however, it has undergone significant restructuring because of changes in the political situation in Europe and the subsequent drop in demand for armaments. It is undergoing privatization, conversion and downsizing. International cooperation through joint ventures is also increasing.

WIECZOREK, P. and ZUKROWSKA, K., 'Determinants of recipient dependence on a single source or predominant arms supplier, exemplified by the Polish experience', SIPRI Arms Procurement Decision Making Project, Working Paper no. 99 (1998), 21 pp.

During the cold war Poland was almost exclusively dependent on arms supplies from the Soviet Union. This had negative effects on the defence industry, in particular with regard to R&D and competitiveness. Currently, the main countermeasure against import dependence is interdependence, rather than diversification of supply. In order to acquire defence technology and facilitate arms exports Poland must cooperate with foreign partners, which will improve the economic situation of the defence industry and keep down the costs of modernization.

BATCHELOR, P., 'Balancing arms procurement with national socio-economic imperatives', SIPRI Arms Procurement Decision Making Project, Working Paper no. 100 (1997), 33 pp.

Since the ending of apartheid and the lifting of the UN embargo in May 1994, procurement decision making in South Africa has been transformed. Decision-making processes have been restructured and there is a much greater degree of transparency and public accountability. With budget constraints and changing priorities, procurement decision making is now largely determined by socio-economic and financial considerations rather than by the imperatives of national security. The choice of sources for procurement has also widened to include foreign sources.

BUYS, A., 'The influence of equipment modernization, building a national arms industry, arms export intentions and capabilities on South Africa's arms procurement policies and procedures', SIPRI Arms Procurement Decision Making Project, Working Paper no. 101 (1997), 26 pp.

According to the 1996 White Paper on Science and Technology, the essence of the new strategy of the South African armed forces is to convert the current force into a smaller but technologically more capable one. The Ministry of Defence policy is to support and contract the national defence industry and to run a limited number of long-term core programmes to enable it to maintain its engineering and production skills. Local suppliers receive a price preference, but the decline in defence expenditure and especially in the budget for arms procurement has been a major constraint in building a national defence industry.

CALLAND, R., 'An examination of the institutionalization of decision-making processes based on principles of good governance', SIPRI Arms Procurement Decision Making Project, Working Paper no. 102 (1997), 41 pp.

On paper South Africa's arms procurement decision-making process is the most transparent and controlled in the world. However, the ability to hold the government accountable for its actions is limited by the institutional weakness of the Joint Standing Committee on Defence, which lacks research support, comprehensive records of meetings and a system of dissent. The fusion of the executive and the legislature which is the result of the strength and party discipline of the African National Congress (ANC) also weakens the Parliament in general.

CILLIERS, J., 'Defence research and development in South Africa', SIPRI Arms Procurement Decision Making Project, Working Paper no. 103 (1997), 16 pp.

Recent reductions in the defence budget and the decline of the defence industry make defence R&D a strategic resource in South Africa. A defence technology base is needed to enable informed decisions on prospective defence procurement. Although Armscor, the Ministry of Defence and the Department of Arts, Science and Technology recognize the crucial importance of defence R&D, they are unable to sustain adequate levels of funding.

CRAWFORD-BROWNE, T., 'Arms procurement decision making during the transition from authoritarian to democratic modes of government', SIPRI Arms Procurement Decision Making Project, Working Paper no. 104 (1997), 24 pp.

During the Defence Review and at the Cameron Commission, the Anglican Church called for a total prohibition of arms exports from South Africa and the disbanding of Armscor and Denel. It has argued that the arms industry is heavily subsidized and thus diverts public resources from socio-economic priorities such as housing, education and health services.

GRIFFITHS, B., 'Arms procurement decision making', SIPRI Arms Procurement Decision Making Project, Working Paper no. 105 (1997), 20 pp.

The process of arms procurement budgeting allocates resources to programmes which meet the broad strategic goals of the South African National Defence Forces rather than directly to the armed forces as in the past. It begins with a top-down environmental analysis, which forms the basis for the bottom-up establishment of requirements by the armed forces. The process of contracting has changed to include aspects of life-cycle costing, multi-sourcing, accreditation of suppliers, affirmative procurement and industrial participation (offset).

HATTY, P., 'The South African defence industry', SIPRI Arms Procurement Decision Making Project, Working Paper no. 106 (1997), 33 pp.

Its defence industry is a strategic and operational asset as well as a significant technological and industrial asset for South Africa. It has developed advanced technological capabilities which enable it to compete effectively in the international market. Partnership and joint ventures with industries in other countries are providing valuable transfer of technology and expanded marketing opportunities.

LIEBENBERG, I., 'A socio-historical analysis of national decision-making behaviour', SIPRI Arms Procurement Decision Making Project, Working Paper no. 107 (1997), 28 pp.

The centralized decision-making process and secrecy of the apartheid regime, which the liberation movements were also forced to adopt as a result of severe repression, remain a problem in the transition to democracy. Increased civilian oversight of the military, education of political leaders and the development of a vociferous civil society and independent media are important mechanisms to improve the decision-making process.

MILLS, G. and EDMONDS, M., 'New purchases for the South African military: the case of corvettes and aircraft', SIPRI Arms Procurement Decision Making Project, Working Paper no. 108 (1997), 30 pp.

With the transition to democracy, civil and parliamentary control over South Africa's arms procurement decision making has increased. The case of procurement of corvettes for the navy marks a watershed: failure to take political considerations into account and involve the Parliament eventually halted the process. Meanwhile, the decision to procure Rooivalk helicopters for the air force—a decision which largely reflected the interests of the military—shows the limits of civilian control in the Department of Defence.

OMAR, Y. A., 'Different perspectives on the relationship between national security, military security and military capability objectives', SIPRI Arms Procurement Decision Making Project, Working Paper no. 109 (1997), 23 pp.

South Africa's defence policy is guided by a foreign policy which emphasizes human rights, regional development and security. Facing the Southern African region, which is still wracked by internal instability and economic underdevelopment, the South African National Defence Forces need several capabilities. The great transparency of and outside involvement in the process of developing the defence policy will enhance the prospects for regional confidence-building measures, which the Inter-State Defence and Security Committee of the Southern African Development Community (SADC) has begun to examine.

SPARRIUS, A., 'Quality in arms procurement', SIPRI Arms Procurement Decision Making Project, Working Paper no. 110 (1997), 24 pp.

Audit of arms procurement should focus on the quality of the procurement process, in which bid evaluation and source selection are the critical elements, rather than on the weapon systems delivered. In exercising oversight, the Joint Standing Committee on Defence first needs to review the military budget and thereafter, with the help of technical experts, audit selected procurement decisions.

TRUSCOTT, E., VAN DER MERWE, W. and WESSELS, G., 'Alternative procedures for technology assessment and equipment selection', SIPRI Arms Procurement Decision Making Project, Working Paper no. 111 (1997), 52 pp.

Focused technology assessment can only be performed if the future long-term equipment requirements are known, since technology development must be directed towards a future equipment requirement goal. The force structure plan of the SANDF and a strategic long-term vision of defence policy, which has been discussed in the White Paper and in the process of the Defence Review, must be the basis for scientific technology assessment and equipment selection.

VAN DYK, J. J., 'The influence of foreign and security policies on arms procurement and decision making', SIPRI Arms Procurement Decision Making Project, Working Paper no. 112 (1997), 66 pp.

Through the process of the Defence Review, public opinion has influenced South African arms procurement. It is performed by the Ministry of Defence, the Defence Secretariat, the SANDF, Armscor and the Parliamentary Joint Standing Committee on Defence. However, foreign policy does not influence the arms procurement process because there is no officially accepted foreign policy concept.

WILLIAMS, R., 'South African force planning', SIPRI Arms Procurement Decision Making Project, Working Paper no. 113 (1997), 30 pp.

South African force planning has progressed from the apartheid era threat analyses, which were subject to large political, personal and institutional influences. The White Paper on Defence and the Defence Review process have instead adopted a threat-independent approach in which the justification for the retention of the armed forces is the need for a core defence capability and the secondary functions the armed forces can perform. The inclusive processes of force planning have enabled hitherto excluded groups to influence policy formulation and implementation.

CHEN, E. I-HSIN, 'Security, transparency and accountability: an analysis of ROC's arms acquisition process', SIPRI Arms Procurement Decision Making Project, Working Paper no. 114 (1998), 19 pp.

Arms procurement in Taiwan has been shaped and driven by the security interests of the USA and also to a great extent by the government's political priorities rather than the objective requirements of national defence. An attempt was made to reform the procurement process by setting up a Procurement Bureau in 1995, but the continuing need for greater accountability and transparency was highlighted in 1998 by several scandals connected to arms contracts.

CHENG-YI LIN, 'Taiwan's threat perceptions and security strategies', SIPRI Arms Procurement Decision Making Project, Working Paper no. 115 [1998], 29 pp.

The locus of Taiwanese security and defence policy making is the National Security Council, a constitutional advisory body under the presidency. Security policy is almost completely aimed at repelling an armed assault from the People's Republic of China (PRC). Since the end of the cold war Taiwan has changed to a defensive rather than offensive posture. Despite striving for self-sufficiency, it remains dependent on the USA if the PRC should start an offensive.

CHIH-CHENG LO, 'Secrecy versus accountability: arms procurement decision making in Taiwan', SIPRI Arms Procurement Decision Making Project, Working Paper no. 116 [1998], 14 pp.

Before the late 1980s there was no public debate regarding defence policy in Taiwan. The situation has improved with increased democratization but is not satisfactory in terms of accountability and transparency. The Ministry of National Defense enjoys much greater authority in deciding which information should be kept secret than other state agencies. Attempts to increase the transparency and accountability of defence policy making have been resisted by the military.

CHIN-CHEN YEH, 'Arms acquisition decision making in Taiwan', SIPRI Arms Procurement Decision Making Project, Working Paper no. 117 (1998), 21 pp.

Taiwan's arms procurement process is based on a model known as the integrated strategy analysis framework, which seeks to combine relevant variables such as security and defence industrial considerations, budget aspects, resources and technology. The budget is governed by the planning, programming and budgeting system, which tries to combine military strategies, force structures and financial allocations in order to make efficient use of national resources.

LIN CHI-LANG, 'Policy analysis of land force arms procurement: the case of the Republic of China's army', SIPRI Arms Procurement Decision Making Project, Working Paper no. 118 (1998), 11 pp.

Since the early 1970s Taiwan has striven to create a credible and independent defence capable of resisting attacks from the PRC. It has focused on quality rather than quantity. To ensure power and independence, Taiwan has promoted R&D and the domestic arms industry. In order to achieve a more efficient arms procurement process, reforms such as increased coordination of the three branches of the armed forces and improved oversight and review mechanisms have been introduced.

LUNG KWANG PAN, 'Weapon acquisition and development under foreigner influence: trajectory of Taiwan's highest military research institute', SIPRI Arms Procurement Decision Making Project, Working Paper no. 119 (1998), 13 pp.

Taiwan has received a large part of its defence equipment from the USA. To counter the threat from the PRC it has emphasized the need to develop an indigenous defence industry, setting up a R&D institute, the Chung Shan Institute of Science of Technology (CSIST), in 1969. When the USA changed its arms export policy towards Taiwan in the early 1980s the CSIST was given an enhanced role. Although the USA has begun to export arms to Taiwan again, the CSIST remains an essential element in the national defence.

WEN-CHENG LIN, 'Taiwan's arms acquisition dependence and its effects', SIPRI Arms Procurement Decision Making Project, Working Paper no. 120 [1998], 13 pp.

Taiwan strives for self-sufficiency in defence R&D but also seeks to diversify its sources of weapon supply. The domestic defence industry still lags behind and pressure from the PRC has made Taiwan heavily dependent on the USA for arms supplies. This has led not only to Taiwan being very vulnerable to changes in the relations between the USA and China but also to the USA having major influence on Taiwanese domestic and foreign policy.

WONG MING-HSIEN, 'Influence of the ROC's foreign and security policy on its arms procurement decision making', SIPRI Arms Procurement Decision Making Project, Working Paper no. 121 [1998], 20 pp.

Since the end of the cold war, Taiwan's arms procurement has focused on economic survival and development. Other important considerations in arms procurement decision making are relations with the USA and Taiwan's geographical position. Arms procurement policy seeks to establish independence by prioritizing domestic procurement and diversifying sources of arms supplies.

WU, S. SHIOUH GUANG, 'Problems in Taiwan's arms procurement procedure', SIPRI Arms Procurement Decision Making Project, Working Paper no. 122 (1998), 10 pp.

Problems in arms procurement procedures in Taiwan have led to corruption in connection with arms deals. The main reasons are a serious lack of effective supervision mechanisms, an excessively hierarchical military organizational structure and the involvement of organized crime. Since 1995 a number of actions have been taken by the Taiwanese Government to increase transparency and efficiency in the arms procurement procedures by limiting the power of the military.

YANG, A. NIEN-DZU, 'Arms procurement decision-making: the case of Taiwan', SIPRI Arms Procurement Decision Making Project, Working Paper no. 123 [1998], 20 pp.

Lack of adequate domestic R&D and arms production capabilities has made Taiwan dependent on imported weapon systems, in particular from the USA. To counter this, attempts have been made to strengthen domestic R&D and production through technology transfer and offset agreements. The problem of US influence on threat assessment and the structure of the Taiwanese defence have led to a government decision to diversify the sources of supply of weapon systems.

YANN-HUEI SONG, 'Domestic considerations and conflicting pressures in Taiwan's arms procurement decision-making process', SIPRI Arms Procurement Decision Making Project, Working Paper no. 124 (1998), 31 pp.

The two main external factors which influence the arms procurement process in Taiwan are the perceived threat from the PRC and dependence on the USA for arms supplies. The process is also influenced by domestic factors such as a lack of cooperation between the different branches of service and institutions, the balance between domestic production and imports of arms, and the conflict between the need for secrecy and the demand for transparency.

Annexe C. About the contributors

Chile

Carlos Castro (Chile) is Professor at the National Academy of Political and Strategic Studies and at the Diplomatic Academy of Chile. He is a retired Air Force Colonel and previously served as a High Command Official and as a Professor of Personnel and High Command Studies at the Air Force War College. He has published several reports and articles on regional security issues.

Mireya Davila (Chile) is Assistant in Public Politics to the Minister-Secretary General of the Government of Chile. She holds a degree in history from the Catholic University of Chile and has previously been a Researcher at the International Relations and Military Area of the Facultad Latinoamericana de Ciencias Sociales (FLACSO)–Chile.

Claudio Fuentes (Chile) is a researcher in the International Relations and Military Area at FLACSO–Chile and a PhD candidate in political science at the University of North Carolina, Chapel Hill. He was formerly editor of *Fuerzas Armadas y Sociedad* [The armed forces and society] and is the author of several publications on regional security and civil–military relations.

Gabriel Gaspar (Chile) is an Undersecretary for War in the Chilean Ministry of Defense. He was formerly Associate Research Fellow in the International Relations and Military Area of FLACSO–Chile and an Assistant in the Department of Planning of the Ministry for Foreign Affairs. He is the author of several publications on the political process in Latin America and regional security issues.

Emilio Meneses (Chile) is Professor of Political Science at the Catholic University of Chile, Santiago. He was formerly a Consultant and Senior Adviser to the Chilean Air Force and has been a senior member of the *Libro de la Defensa Nacional* project in the Ministry of Defense.

Miguel Navarro (Chile) is a Professor at the National Academy of Political and Strategic Studies of the Chilean Ministry of Defence. He also teaches at the Chilean Air Force Staff College, is an adviser to the Defence Committee of the Chamber of Deputies, and is a Visiting Professor at the Diplomatic Academy of the Ministry of Foreign Affairs.

Guillermo Pattillo (Chile) is Associate Professor at the Department of Economics of the University of Santiago and the Institute of Political Science of the Catholic University of Chile. He was formerly an economist at the Department of Economic Analysis of the General Staff of National Defense.

Leopoldo A. Porras (Chile), a retired General of the Chilean Air Force, was formerly Director of the Aeronautics and Space Studies Center of the Chilean Air Force and is a private researcher in defence and security issues.

Marcos Robledo (Chile) is a journalist and editor of the international section of *La Epoca*. He publishes articles on political issues, civil–military relations, Chilean defence policy, foreign policy and international relations. He holds a Master's degree in Military Sciences from the Army War College.

Francisco Rojas Aravena (Chile) is Director of FLACSO–Chile and Professor at the Institute of International Studies of the University of Chile and at Stanford University, Santiago Campus. He is Director of *Fuerzas Armadas y Sociedad* and has published extensively on strategic issues and confidence-building measures in the region.

Fernando G. Thauby (Chile), a retired naval captain, is a Professor at the Naval War College of Chile and Director of *Revista de la Marina*, in which he writes on regional security, military strategy and force determination.

Greece

Stelios Alifantis (Greece) is Assistant Professor of International Relations at Panteion University of Social and Political Science in Athens. He is currently a Special Adviser on security policy issues to the Minister of National Defence. He has published and co-edited several books and articles in the field of security and foreign policy with a focus on the Balkans and Greece.

Thanos P. Dokos (Greece) is Director of Research at the Strategic Studies Division of the Greek Ministry of National Defence. His recent publications include *Negotiations for a CTBT: 1958–94* (1995, in English) and [Security problems in the Mediterranean] (1996, in Greek).

Haralambos C. Giannias (Greece) is a specialist researcher to the Advisory Board of the Ministry of National Defence and at the Foundation of Mediterranean Studies. He has published [Cost contributing factors in defence] (1997) and [Low-intensity conflicts] (1997) (both in Greek).

Christos G. Kollias (Greece) is Assistant Professor of Economics in the Department of Business Administration, Technological Education Institute, Larissa in Greece, and also lectures at the School of National Defence. He has published over 40 papers on defence economics, arms races, public-sector economics, labour economics and foreign-exchange markets and is the author of [The political economy of defence] (1998) and [Greece and Turkey: defence, economy and national strategy] (2000, both in Greek).

Meletis Meletopoulos (Greece) has a PhD in sociology from the University of Geneva and has lectured at the Greek National School of Public Administration since 1994. He previously taught at the University of Crete (1997–98) and Geneva University (1998) and is the editor of the journal *Nea Koinoniologia* [New sociology]. He has published widely on Greek politics, political history, political theory and political sociology. In 1993–94 he served as Adviser to the Minister of Public Order.

Ethon Narlis (Greece) is an electronics engineer. He worked in the Hellenic Aerospace Industry (EAB) in 1988–96, and in 1996–99 was Director of the Technology, Research and Development Directorate and Manager of the Scientific and Technological Cooperation Department of the General Directorate of Armaments in the Greek Ministry of National Defence. A founding member of the NATO Research and

Technology Organisation, he has represented Greece on the NATO Research and Technology Board.

Panayotis J. Tsakonas (Greece) is an adviser on defence and strategic issues in the Greek Ministry of National Defence and was previously a Research Fellow at NATO and at the Institute of International Relations in Athens. He has published several articles on regional security in South-Eastern Europe, Greek foreign policy and the theory of alliances.

Christos Valtadoros (Greece) holds a Master's degree in political science from the University of Lancaster and currently works as a political consultant. He has worked as Head Coordinator of Industrial Security Planning in the General Directorate of Armaments. He specializes in Russian defence policy and Greek–Albanian relations.

Malaysia

Sharifah Munirah Alatas (Malaysia) is an analyst at the Institute for Strategic and International Studies (ISIS), Kuala Lumpur, where her research focuses on the socio-political aspects of economic development. She is co-author with Zakaria Haji Ahmad of 'Malaysia: in an uncertain mode', ed. J. W. Morley, *Driven by Growth: Political Change in the Asia–Pacific Region* (1998).

Noor Aziah Hj. Mohd Awal (Malaysia) is an Associate Professor at the Law Faculty, Universiti Kebangsaan Malaysia. Her area of expertise is comparative family law, child law, and women and the law. She is the author of [Women: law and leadership] (1996, in Malay); 'Annotated Statutes of Malaysia: Adoption Act 1952', *Malayan Law Journal*, 1997; and 'Annotated Statutes of Malaysia: Child Protection Act 1991', *Malayan Law Journal*, 1998; and she was co-editor with Hjh Siti Faridah of *An Appraisal of the Environmental Law in Malaysia* (1996).

Siti Azizah Abod (Malaysia) is Under-Secretary in the Malaysian Ministry of Defence. She was previously Assistant Director at the Ministry of Trade and Industry and has also been Programme Coordinator at the National Institute of Public Administration.

K. S. Balakrishnan (Malaysia) is a lecturer at the Department of International and Strategic Studies, Universiti Malaya. A former Fellow at the Malaysian International Affairs Forum (1998) and a Senior Researcher at ISIS in Kuala Lumpur (1993–97), he is also an occasional guest lecturer at the Malaysia Armed Forces Defence College and the Malayian Defence Staff College in the Ministry of Defence. He is a member of the editorial board of the *Journal of Asian Defence and Diplomacy*.

Dagmar Hellmann-Rajanayagam (Germany) is a Senior Lecturer at the Strategic and Security Studies Unit at the Universiti Kebangsaan Malaysia. She was previously Associate Professor at the Department of History at the University of Kiel and Guest Lecturer at the University of Madras. She has published extensively on the civil war in Sri Lanka.

Faridah Jalil (Malaysia) is a Lecturer at the Faculty of Law at the Universiti Kebangsaan Malaysia, specializing in constitutional law, jurisprudence and private international law, and has served as Advocate at the High Court Malaya. Her recent publications include *Annotation on the Federal Constitution* (1998) and *Annotation on the Local Government Act, 1979* (1996).

Joon Num Mak (Malaysia) is Director of Research at the Malaysian Institute of Maritime Affairs (MIMA) and Head of its Maritime Security and Diplomacy Programme. His research focuses on regional security matters with a special emphasis on defence and naval affairs. He recently co-edited the SIPRI Research Report *Arms, Transparency and Security in South-East Asia* (1997) and is the author of 'The Asia–Pacific security order' in *Pacific Studies: Book 1* (forthcoming).

Abdul Rahman Adam (Malaysia) is a Lecturer at the Strategic and Security Studies Unit at Universiti Kebangsaan Malaysia and was previously Senior Military Intelligence Analyst at the Malaysian Ministry of Defence. He co-edited *United Nations Peacekeeping: Panacea or Pandora's Box?* (1994) with Zakaria Haji Ahmad.

Richard Robless (Malaysia) is a Consultant at the Aerospace Technology Systems Corporation, Kuala Lumpur. He has served as Assistant Chief of Air Force material and Commander, Third Air Division.

K. Sukumaran (Malaysia) was Assistant Secretary in the Malaysian Ministry of Defence from 1985 to 1997 and is currently Director of Drug Rehabilitation in government service.

Madya Supian Hj. Ali (Malaysia) is Associate Professor at the Faculty of Economics, Universiti Kebangsaan Malaysia, and was formerly Deputy Dean at the same university. He co-edited the Sixth Malaysian Plan (1994) and [Human resource development in Malaysia] (1996, both in Malay).

Poland

Franciszek Golembski (Poland) is a Professor at the Institute of Political Sciences, Warsaw University, and a former Research Fellow of the Polish Institute of International Affairs and Minister's Counsellor in the Polish Ministry of Foreign Affairs. He is the author of several books and articles in the fields of international relations, security and the cultural dimension.

Lech Kosciuk (Poland) is a political scientist, formerly at the Polish Institute of International Affairs, where he worked on military security, disarmament and arms control issues, and currently Senior Adviser at the Department of International Security of the Polish Ministry of National Defence. His 'Postzimnowojenny europejski uklad sil' [Post-cold war configuration of forces] is forthcoming in the journal *Sprawy Miedzynarodowe*.

The late **Jan Krolik** (Poland) was a Researcher at the Logistics Institute of the Military University of Technology, Warsaw.

Czeslaw Mesjasz (Poland) is working at the Cracow University of Economics and at the Jagiellonian Business School. He is the author of several publications on economic aspects of security since the end of the cold war.

Wlodzimierz Miszalski (Poland) is Commandant at the Logistics Institute and Professor of the Military University of Technology, Warsaw. A computer systems designer by training with a PhD in management and command, he is the author of over 200 publications on logistics management, maintenance organization, decision system engineering, disaster monitoring and relief systems command. He has represented Poland on the Systems Analysis and Simulation Panel of the NATO Research and Technology Organization.

Jadwiga Stachura (Poland), whose PhD is in political science, is a senior counsellor at the Department of Strategy and Policy Planning of the Polish Ministry of Foreign Affairs. Formerly a Researcher at the Polish Institute of International Affairs, she has published extensively on international relations and Polish foreign policy.

Adam Szydlowski (Poland) has been an Adjunct at the Logistics Institute of the Military University of Technology, Warsaw, since 1994. An engineer with an MSc and DSc in military science (command and control, specializing in camouflage), he is the author or co-author of more than 40 publications on command and logistics management, camouflage organization, decision-making systems and standardization.

Marian Tarkowski (Poland) works in the Press and Information Bureau at the Ministry of National Defence.

Zenon Trejnis (Poland) serves at the Institute of Humanities of the Military University of Technology, Warsaw.

Pawel Wieczorek (Poland) is a specialist in problems of European security. He is the author or co-author of 10 monographs and 50 papers, mainly on economic aspects of armaments and disarmament, including, most recently, *Europejskie Struktury Współpracy* [European structures of cooperation] (2000) and *Polska w NATO: Aspekty Ekonomiczno-Finansowe* [Poland in NATO: economic and financial aspects] (1999).

Katarzyna Zukrowska (Poland) is a Professor at the Institute of Development and Strategic Studies, Warsaw. She is also Director of the European Program at the Institute of International Studies, Warsaw School of Economics. She has published widely on defence economics, systemic transformation in East–Central Europe and EU integration.

South Africa

Peter Batchelor (South Africa) is the director of the Small Arms Survey project at the Graduate Institute of International Studies in Geneva. Formerly a Senior Researcher at the Centre for Conflict Resolution at the University of Cape Town and Coordinator of the Centre's Project on Peace and Security, he has published widely on military expenditure, the arms industry and arms trade issues in South and Southern Africa. He is co-author of *Small Arms Management and Peacekeeping in Southern Africa* (for UNIDIR, 1996) and the SIPRI volume *Disarmament and Defence Industrial Adjustment in South Africa* (with Susan Willett, 1998).

André Buys (South Africa) is Head of the Planning Division of Armscor in Pretoria. He has held various engineering and management positions in the South African defence industry, has served on various working groups to review the South African defence industry and the Ministry of Defence arms acquisitions policy, and was a member of the 1996–97 Defence Review.

Richard Calland (South Africa) is Project Manager of the Political Information and Monitoring Service at the Institute for Democracy in South Africa (IDASA), Cape Town. He previously practised at the London Bar, specializing in public law, was founder editor of *Parliamentary Whip* newspaper, and is the author of *The First Five Years: A Review of South Africa's Democratic Parliament* (1999) and *All Dressed Up and with Nowhere to Go? The Rapid Transformation of the South African Parliamentary Committee System* (1997).

Gavin Cawthra (South Africa) directs the Defence Management Programme at the Graduate School of Public and Development Management at the University of Witwatersrand, which offers academic courses to military officers and officials from Southern Africa. He also coordinates the research of IDASA's Defence and Security Programme. He is the author of *Sub-regional Security Co-operation: the Southern African Development Community in Comparative Perspective* (1997) and *Securing South Africa's Democracy: Defence, Development and Security in Transition* (1997).

Jakkie Cilliers (South Africa) is Executive Director of the Institute for Security Studies in Pretoria. He was previously an artillery officer and a systems analyst and divisional manager of operations research. He has published widely on security and defence issues and submitted several proposals on the restructuring of different defence institutions. His most recent publications are *Peace, Profit or Plunder? The Privatisation of Security in War-Torn African Societies* (ed. with P. Mason, 1999); *From Peacekeeping to Complex Emergencies: Peace Support Missions in Africa* (with G. Mills, 1999); and *Angola's War Economy: The Role of Oil and Diamonds* (ed. with C. Dietrich, 2000).

Terry Crawford-Browne (South Africa) is a former international banker who during 1985–89 advised Archbishop Desmond Tutu on the banking sanctions campaign against apartheid. He represented the Anglican Church at the Cameron Commission of Enquiry and at the South African Defence Review consultations, and is currently chair of the South African affiliate of Economists Allied for Arms Reduction.

Martin Edmonds (United Kingdom) is Director of the Centre for Defence and International Security Studies at Lancaster University. He has published widely on naval defence issues and is the author of *Beyond the Horizon: Defence, Diplomacy and South Africa's Naval Opportunities* (with Greg Mills, 1998) and *Defence Diplomacy and Out of Area Operations* (forthcoming).

Bryan N. Griffiths (South Africa) is Senior Manager of Aeronautical Logistics at Armscor. He has been associated with the South African defence industry since 1979 in engineering and management and is the author of several reports on out-sourcing logistics and decision making for contractor selection.

Paul Hatty (South Africa) is an industrial consultant who has worked with production and management in manufacturing industry, including the defence industry. He has been Chairman of the Industrial Policy Committee of the South African Chamber of Business, and was recently involved in various aspects of policy development for the defence industry.

Ian Liebenberg (South Africa) holds the MA degree in both Political Science and Development Studies from South African universities and is currently a Research Fellow of the Unit for African Studies, University of Pretoria. He was a pro-democracy activist in the 1980s in South Africa, working *inter alia* on democracy advocacy in IDASA and the Center for Intergroup Studies in the University of Cape Town from 1986 to 1990, and from 1990 to 2000 was a senior researcher/political analyst in the Human Sciences Research Council of South Africa. His current research interests include civil–military relations, public sector transformation, the truth and reconciliation processes, and public participation in young democracies, and he has published widely.

Greg Mills (South Africa) is National Director of the South African Institute of International Affairs, Braamfontein, and a Senior Research Associate at the Centre for Defence and International Security Studies at Lancaster University. He has published widely on South African foreign relations, *inter alia* as author of *The Wired Model: South Africa, Foreign Policy and Globalisation* (2000).

Y. Abba Omar (South Africa) is General Manager of Corporate Communications at Armscor and Secretary of the Defence Industry Interest Group of South Africa.

Ad Sparrius (South Africa) is a consultant in systems engineering and project management. He previously worked at the Council for Scientific and Industrial Research in South Africa and the University of Pretoria.

Erica Truscott (South Africa) is a Senior Consultant at Deloitte & Touche Consulting Group, Hatfield, South Africa. She is an engineer dealing with projects on optimal allocation of funds, transformation and restructuring.

Willie van der Merwe (South Africa) is Chief Executive Officer at SACAD Consultants, Pretoria, where he participates in management decision support projects using computer-aided decision support (CADS) and operational research techniques. He specializes in strategic management consulting, project management, electronically-based research, user requirement specifications, optimal allocation of funds and strategic management process development. He was previously Head of Industry Decision Support at Armscor/Denel and is co-author of the INFORMS (Institute for Operations Research and the Management Sciences) award-winning paper 'Guns and butter: decision support for determining the size and shape of the South African National Defense Force', *Interfaces* (1997).

Johan J. van Dyk (South Africa) is Senior Manager of the Countertrade Division at Armscor, and has worked previously on arms control issues at Armscor and at the South African Defence Secretariat.

Gys Wessels (South Africa) is a Partner at Deloitte & Touche Consulting Group, Hatfield, South Africa, where he works on projects using operational research and management science techniques. He is the co-author of 'Guns or butter: decision support for determining the size and shape of the South African National Defence Force', *Interfaces* (1997).

Rocklyn Williams (South Africa) is Director of Operational Policy at the Defence Secretariat. He is a former Umkhonto we Sizwe officer who integrated into the new South African National Defence Forces and has been involved in the annual force planning exercises since the Joint Military Coordinating Council period.

Taiwan

Edward i-hsin Chen (Taiwan) is Professor and Director of the Graduate Institute of American Studies at Tamkang University and has a PhD in Political Science from Columbia University. He was a member of the Legislative Yuan (Parliament) during 1996–99, specializing in relations between the USA, China and Taiwan.

Cheng-yi Lin (Taiwan) is Research Fellow and Director of the Institute of European and American Studies, Academia Sinica in Taipei. He received his PhD in foreign affairs from the University of Virginia. His major fields of interest are international relations in East Asia and the national security policy of the USA.

Chih-cheng Lo (Taiwan) is Associate Professor at the Department of Political Science, Soochow University, as well as Special Assistant to the President of the Institute for National Policy Research and Secretary-General of the Taiwanese Political Association.

Chin-chen Yeh (Taiwan) is Associate Professor at the Graduate Institute of International Affairs and Strategic Studies, Tamkang University, and has a PhD from Chiao Tung University. His most recent publications are 'The relationships among defense environment, strategies and resource allocation: the case of Taiwan ROC', *Journal of the National Defense Management College* (1995); and, with Chao-Jan Chang, 'Weapon systems acquisitions strategy model for developing country: case of ROC', presented at the Sixth National Conference on Defense Management, 1998.

Lin Chi-lang (Taiwan), a former artillery commander and associate professor at the War College of the Armed Forces University, is Dean of Student Affairs and associate professor at the Institute of Non-profit Organization Management, Nan Hau University, Taiwan. Among his most recent publications are 'Theoretical thinking and policy choice of the Revolution in Military Affairs', *Military Sociology Quarterly*, 1999; and 'Taiwan's strategy for participating in international NGO networks: strategic thinking of security issues, Track II diplomacy and connection with NGOs', presented at the conference on Taiwan NGOs: Marching Towards the 21st Century, July 2000.

Lung Kwang Pan (Taiwan) is Associate Professor of Mechanical Engineering at the Chung Cheng Institute of Technology. His research focuses on nuclear instrumentation and the constitution of military systems.

Wen-cheng Lin is Associate Professor teaching negotiation, international security and the foreign policy of the People's Republic of China at the Institute of Mainland China Studies at the National Sun Yat-sen University. He is the author of *The PRC's Theories and Practices in Negotiation* (2000) and of many articles and contributions to books on cross-Strait relations, Asia–Pacific security, US foreign policy, and the foreign policy and military strategy of the PRC.

Wong Ming-hsien (Taiwan) is Director of the Graduate Institute of International Affairs and Strategic Studies, Tamkang University. He has a PhD in Political Science from the University of Cologne and specializes in Sino-Taiwanese military relations.

Samuel Shiouh Guang Wu (Taiwan) is Associate Professor at the Department of Public Administration in Chengchi University, Taiwan. She has co-edited *Inherited*

Rivalry: Conflict across the Taiwan Straits (1995) with Tun-Jen Cheng and Chi Huang.

Andrew Nien-dzu Yang (Taiwan) is Secretary General of the Chinese Council of Advanced Policy Studies in Taiwan and Research Associate at the Sun Yat-sen Center for Policy Studies at Sun Yat-sen University. He contributed a chapter to *In China's Shadow: Regional Perspectives on Chinese Foreign Policy and Military Development*, ed. by Jonathan D. Pollack and Richard H. Yang (1998), and co-edited *Chinese Regionalism: The Security Dimension* (1994).

Yann-huei Song (Taiwan) is a Research Fellow and head of the Division of Legal and Political Studies at the Institute of European and American Studies, Academia Sinica in Taipei. His research interests are in the fields of international law of the sea, international fisheries law, environmental law, naval arms control and maritime security. He has published extensively, including 'China's missile tests in the Taiwan Straits: relevant international law questions', *Marine Policy* (1999, in English).

Index